T0326350

Erectile Dysfunction as a Cardiovascular Impairment

"At the heart of science is an essential balance between two seemingly contradictory attitudes—an openness to new ideas, no matter how bizarre or counterintuitive they may be, and the most ruthless skeptical scrutiny of all ideas, old and new. This is how deep truths are winnowed from deep nonsense."

—Carl Sagan

Erectile Dysfunction as a Cardiovascular Impairment

Robert Fried, Ph.D.
Emeritus Professor
Doctoral Faculty in Behavioral Neuroscience
City University of New York

Emeritus
American Physiological Society
Cardiovascular and Respiratory Divisions

AMSTERDAM • BOSTON • HEIDELBERG • LONDON
NEW YORK • OXFORD • PARIS • SAN DIEGO
SAN FRANCISCO • SINGAPORE • SYDNEY • TOKYO
Academic Press is an imprint of Elsevier

Academic Press is an imprint of Elsevier
525 B Street, Suite 1800, San Diego, CA 92101-4495, USA
32 Jamestown Road, London NW1 7BY, UK
225 Wyman Street, Waltham, MA 02451, USA

British Library Cataloguing-in-Publication Data
A catalogue record for this book is available from the British Library

Library of Congress Cataloging-in-Publication Data
A catalog record for this book is available from the Library of Congress

ISBN: 978-0-12-420046-3

For information on all Academic Press publications
visit our website at elsevierdirect.com

Printed and bound in the United States of America

14 15 16 17 18   10 9 8 7 6 5 4 3 2 1

Dedication

This book is dedicated with sincere appreciation to my wife, Robin Bushman, for her unwavering patience and support, accompanied by an uninterrupted supply of cups of hot coffee while I was at work on this book.

In one book, [Dr. Fried] provides us the fundamental strengths, as well as the dramatic changes, that have occurred in both physiological and pathological aspects of ACh/NO/cGMP pathway and endothelial function, which are found in hundreds of different subspecialty journals.

Osama Tayeh, M.D., FESC

[This] work by Dr. Fried is an excellent and timely work on the basic biochemical and physiological mechanisms underlying erectile dysfunction (ED). This well-referenced tome very nicely describes how impairments in nitric oxide (NO) production by the vascular endothelial cells contribute to ED but, more importantly, how this early impairment may be one of the first signs of the onset and progression of cardiovascular disease (the number 1 killer of men and women in the US).

Nathan S. Bryan, Ph.D.

Erectile Dysfunction as a Cardiovascular Impairment is a treasure. Dr. Fried has thoroughly explored the cutting-edge literature on preclinical and clinical research on endothelial dysfunction, and he masterfully explains that this dysfunction is the underlying pathophysiological mechanism behind numerous inflammatory and/or degenerative diseases.

Richard M. Carlton, M.D.

Contents

Three American scientists, Drs. Robert F. Furchgott, Louis J. Ignarro, and Ferid Murad were awarded the 1998 Nobel Prize in Medicine for the discovery of the biological role of nitric oxide (NO). NO is derived in the body mostly from the amino acid L-arginine by a process involving one or another nitric oxide synthase enzyme (NOS), depending on the target organ(s). This discovery of a "novel signaling molecule" opened a window into an entirely new world of events in cardiovascular function as it unraveled a parallel and hitherto unimagined world of endothelium-dependent blood vessel physiology, the nitric oxide/cyclic guanosine monophosphate (NO/cGMP) pathway.

Dr. L. J. Ignarro detailed the role of nitric oxide in erectile function and dysfunction, thereby realizing a new concept, "vasculogenic erectile dysfunction." It was not long before it became clear that the prevalence of this malady dominated all other known etiologies, thus also sealing the fate of traditional psychogenic theories of impotence.

Spurred by these discoveries, I initially set out to describe the research and its beneficial health implications for the general public in a book titled *The Arginine Solution*.[1] Research on NO and clinical therapeutics in health and diseases has since multiplied exponentially, yet without any substantial application outside the laboratory and the clinic despite it becoming increasingly obvious that raising dietary L-arginine intake promotes vascular and heart health.

One of the earliest clinical trials to support that contention was conducted by Rector *et al.* in 1996 in a study titled "Randomized, double-blind, placebo-controlled study of supplemental oral L-arginine in patients with heart failure."[2] In 2003, the National Institutes of Health (NIH), of the National Heart, Lung, and Blood Institute (NHLBI), conducted a study titled "Dietary Nitrate and Nitrite to Increase Nitric Oxide in Patients with Coronary Artery Disease."[3]

The race was on! Well... no, actually it wasn't. Despite continued efforts to show the salutary effects of increasing dietary L-arginine on cardiovascular and heart health, and erectile function, as well as countless other benefits including improving peripheral circulation, lipids profile, wound healing, insulin sensitivity, kidney function, immune system function, etc., precious little of this filtered out of the laboratory and clinic to benefit the public.

Given that it was now generally recognized that NO promotes cardiovascular and heart health and strengthens erectile function, and that dietary L-arginine is pretty much the major source of NO formation in the body, it was time to update "*The Arginine Solution*." So, in 2006, I co-authored

"*Great Food, Great Sex*" with Dr. Lynn Edlen-Nezin.[4] Here we supplied recipes that promote NO formation and also antioxidants to keep the vascular plumbing free of sludge. The diet promotes vascular health, but it is not haute cuisine. No healthy diet can be that. For one thing, it is not a weight loss-level, low-calorie diet, and it is also low in sodium and sugar; what's more, it is high in fiber. However, this diet promotes cardiovascular health by raising NO bioavailability and by reducing oxidative stress, thus protecting the vascular *endothelium* from free radical damage and function impairment.

After carefully pointing out that statistics on drugs like VIAGRA® show that there is little evidence for lack of sexual desire in American men, and that foods that support erectile strength are not aphrodisiacs but work by enhancing blood flow to the genitals, I was still asked predictably by media figures what foods enhance sexual desire in women. The answer that I gave, and that they did not want to hear, was, "There are no common aphrodisiac foods. But desire is driven by sex hormones that are derived in the body from dietary fat. So if you want to raise libido, eat fat." They thought that I was kidding.

If it can be said that, "all roads lead to Rome," it can likewise be said that all cardiovascular and heart pathways lead to the *endothelium*. . . better yet, from the *endothelium*. . . add erectile function as well. The explosion of findings about the inter-linkage of *endothelium* physiology to all cardiovascular and heart ills is truly staggering, and what's more, this has led me to some startling revelations, far from anything I could have imagined at the outset—most are counterintuitive.

Keep in mind that I invented nothing in this book. What I did was to arrange information from many sub-disciplines, and they began to appear to be pieces from a jigsaw puzzle that seemed to fit a pattern; and indeed, a pattern emerged. I was at somewhat of a disadvantage because I did not know beforehand that there was a pattern, much less what the pattern would look like. So, just as it became clear that for most with erectile failure, psychological therapies held no resolution, so did this emerging science also cause collateral damage to cherished medical dogma(s). To wit:

The evidence that, on average, cardiovascular and heart disease, and erectile dysfunction, are primarily the result of age-related decline with a timeline dictated by individual genetic inheritance tells us first that a good diet and healthy lifestyle can do little to slow it down. However, a poor diet and unhealthy lifestyle can dramatically accelerate it.

Second, the expression of the function-decline timeline appears to be due to damage to the *endothelium* caused by oxidative stress because as time goes by, the body also loses the ability to produce adequate antioxidants to combat that blight. Parenthetically, insulin resistance accelerates the timeline as elevated serum glucose has been shown to potentiate oxidative stress damage to the *endothelium*.

Third, due to elucidation of the NO/cGMP pathway, many authorities are now beginning to consider hypertension and atherosclerosis as symptoms of *endothelium* impairment rather than as disease entities. If that turns out to be the case, and that is the theme of this monograph, then erectile dysfunction is likewise a symptom of *endothelium* impairment, and also not a distinct disease entity, *per se*.

If a picture is worth a thousand words, then a word is worth one one-thousandth of a picture: What I tried to do here, metaphorically, is to take the thousand words and recreate the picture by paraphrasing and summarizing the thousands of publications that now speak to this theme.

Robert Fried
Carmel, NY

REFERENCES

1. Fried R, Merrell WC. *The Arginine Solution*. New York: Time/Warner Books; 1999.
2. Rector TS, Bank AJ, Mullen KA, Tschumperlin LK, Sih R, Pillai K, et al. Randomized, double-blind, placebo-controlled study of supplemental oral L-arginine in patients with heart failure. *Circulation* 1996;**93**(12):2135−41.
3. National Institutes of Health (NIH). National Heart, Lung, and Blood Institute (NHLBI). "Dietary Nitrate and Nitrite to Increase Nitric Oxide in Patients with Coronary Artery Disease": <http://clinicaltrials.gov/show/NCT00069654>; 2013 [accessed 4.11.13].
4. Fried R, Edlen-Nezin L. *Great Food, Great Sex*. New York: Ballantine Books; 2006.

Acknowledgments

I owe a debt of gratitude to Stacy Masucci, Senior Acquisitions Editor, Biomedical Research, Elsevier Publishing, for her unwavering support for this monograph.

In the early 1980s, I was director of the Rehabilitation Research Institute (RRI) of the ICD-International Center for the Disabled, New York City. Richard M. Carlton, MD, was the physician assigned to my unit. Over the course of four years as we developed some successful methods for helping patients to cope with idiopathic seizures intractable to conventional medications, Richard constantly irritated me with the obvious fact that his grasp of the basics of physiology and biochemistry, and his knowledge of neuroscience, are far superior to mine. In connection with the present monograph, he is at it again! Don't ever stop, Richard. Thank you for your help.

I wish to express my sincere appreciation to Dr. Nathan S. Bryan, Asst. Professor of Molecular Medicine at the University of Texas Health Science Center at Houston, and Founder/Chief Science Officer of Neogenis Labs, Inc., for reviewing this monograph. As this book will show the reader, I have assiduously and comprehensively reviewed the research and clinical "literature" on the basics and applications of NO science. I emphasize this because it gives weight to my assertion that I hold that there is no one in the field today who is more knowledgeable about that science than Dr. Bryan. I was very fortunate that he agreed to review this book.

Thanks also to Dr. Osama Tayeh, Assistant Professor of Critical Care Medicine, Cairo University, Cairo, Egypt. His unique expertise on clinical applications of the physiology of ADMA opened a new *window* for me into the workings of the endothelial nitric oxide synthase enzymes, and provided the foundation for that chapter.

Many thanks are due to Dr. Jacqueline Perle for her help with preparing different aspects of the manuscript and for her exquisite patience in securing permissions to reproduce and quote things in the book. Her dogged determination and dedication won the day in the face of countless mishaps that are heir to the shifting sands that are publications sources.

Last, and by no means least, I wish to thank my wife, Robin Bushman, who spent many hours proofreading the manuscript and improving it in many ways: Thank you, Robin.

DISCLAIMER

The materials provided in this book are intended solely to provide information to the reader. The author does not intend that information in this book be used, nor should it be used, to diagnose or treat any specific medical condition. The author cannot certify the accuracy, validity and safety of the medical diagnostic techniques described, nor can he endorse the value, safety, dosages and use of treatment substances reported in clinical trials. Commercially available products are cited for information purposes only and citation does not constitute endorsement of their value, safety, effective dosages or use.

Chapter 1

Introduction

"If we do not change our direction, we are likely to end up where we are headed."
—Ancient Chinese proverb

1.1 ERECTILE DYSFUNCTION—AS PRESENTLY DEFINED

The common term "erectile dysfunction" (ED) is "Male Erectile Disorder"
in the current *Diagnostic and Statistical Manual of Mental Disorders of the
American Psychiatric Association*. It is said to be a persistent or recurrent
inability to attain or to maintain an adequate erection to completion of the
sexual activity. It is a disturbance causing marked distress or interpersonal
difficulty, an ED not better accounted for by any other diagnostic categories
except mental retardation and personality disorder, and one that is not due
exclusively to the direct physiological effects of a substance, e.g. a drug
abused, a medication, or a general medical condition (APA—DSM-IV-TR,
2000).[1] The definition assigns etiology to other than medical causes thus
seemingly excluding vascular/*vasculogenic* ED, now considered to be the
most common form found among American men.

By no means is each and every case of chronic ED vasculogenic, but the
overwhelming majority of cases that appear to develop in adults—estimated
at about 90 percent in American men—and to follow an age-related timeline,
fit that definition. There are, of course, other causes of ED, some due to hormonal imbalance, and others with an emotional basis, but they are not the
concern of this monograph.

The recommendation to include diagnostic criteria for a medical basis of
ED in subsequent editions of the DSM is made in the *Journal of Sexual
Medicine*,[2] and it is consistent with the theme of this monograph. In fact it is
essential to make both the medical and the mental health services community
aware of that medical basis so as to improve health care delivery beyond
simply treatment of ED *per se*.

Robert Fried: Erectile Dysfunction as a Cardiovascular Impairment.
DOI: http://dx.doi.org/10.1016/B978-0-12-420046-3.00001-9

1.2 THE AIM AND SCOPE OF THIS MONOGRAPH

This monograph will make the general case and document the research and clinical evidence that ED is indeed caused by a medical condition: impaired vascular endothelium resulting in reduced bioavailability of nitric oxide (NO). NO is a biologically active gas, the link between sexual arousal and erection. It is derived in the body principally from the amino acid L-arginine and from dietary nitrates. NO mediates between the arousal-triggered release of acetylcholine (ACh), and the formation of the vasodilator, cyclic guanosine monophosphate (cGMP), in the ACh/NO/cGMP pathway to penis cavernosal relaxation, engorgement, and erection.

The advent of phosphodiesterase type-5 (PDE5) inhibitors including VIAGRA®, Cialis®, and Levitra®, have made the medical and psychiatric/psychological community aware that there are medical treatments for the most common form of ED, but it is not clear to most practicing clinicians how and why they work except in the most vague terms that they somehow influence blood flow and thus erection. Not being familiar with the ACh/NO/cGMP pathway, many suspect that these meds might just treat the psychological basis for ED.

There are precedents in medicine to parallel the disjunction between emotional disorder and medication treatment. For instance, bipolar disorder is a mood disorder that is treatable with medications. Thus, there is an understanding that medications can affect mood without most practitioners understanding the brain physiology basis for the mood, or the disorder, except in vague terms such as "neurotransmitter biosynthesis" that only specialists really know. Thus, it is not unreasonable to figure that meds can treat ED even if one holds that ED is basically an emotional disorder.

The reason for this information gap is that the clinical and basic research leading to our understanding of the ACh/NO/cGMP pathway to cavernosal relaxation and penis engorgement and erection, and related physiology, is found in thousands of scientific reports in hundreds of different journals straddling many dozens of different subspecialties, published over the past three decades. It would be an impossible task for any practicing clinician to keep up with this avalanche of data.

For example, there is an estimated 850,000 licensed physicians in the US. Less than half that number (about 300,000) subscribe to the *Journal of the American Medical Association* (JAMA). The *New England Journal of Medicine* (NEJM) has about 200,000 subscribers, and *Annals of Internal Medicine* has about 110,250 subscribers. These are major journals. Even so, basic research and clinical trials relevant to the present concerns appear in specialty journals such as *Archives of Biochemistry and Biophysics*, *Ultrasound Medicine and Biology*, and *Vascular Physiology*, where researchers basically talk to each other; and also in better known journals with a narrow-focus regular readership such as *Diabetes Care*, *Hypertension*, and the *Journal of the American college of Cardiology*, to name just a few. The information often doesn't filter down to the rank and file practitioners who really most need it to serve patients well.

This monograph synthesizes information from a wide range of subspecialties, and in most cases indicates the source by naming the journal where cited reports were published. Naming the source is especially important here because it establishes the credibility of information drawn from a wide range of research subspecialties, each providing what amounts to a handful of tiles to form the overall mosaic: ED as a cardiovascular impairment. Furthermore, naming the journal sources also gives the reader insight into the complexity of contemporary research: The simple days are pretty much a thing of the past when, for instance, penicillin was discovered when an *in vitro* bacterial culture was inadvertently destroyed by a contaminating spore (*penicillium*) in Dr. Alexander Fleming's lab, or the treatment of diabetes when Dr. Frederick Banting injected pancreatic extract into a patient, or when Dr. Barry Marshall established the cause of most ulcers by downing an "infectious broth" holding *Helicobacter pylori*.

1.3 THE PREVALENCE OF ERECTILE DYSFUNCTION

According to a 2007 report from the *Johns Hopkins Bloomberg School of Public Health*, more than 18 million American men 20 years old, or older, experience some degree of ED, from mild-episodic to severe-sustained ED.[3] The report pointed to causes ranging from cardiovascular and heart disease to diabetes. Advances in medical science are now shifting the attention of many clinicians from treating the mind to treating the body.

There is now overwhelming evidence that most men with ED can attribute this calamity to cardiovascular and metabolic disorders including hypertension, atherosclerosis, CHD, type-2 diabetes, and metabolic syndrome. It is also now known that the progress of this assortment of medical disorders is further enhanced by oxidative stress, by the progressive age-related decline of androgenic hormones, and by changes in energy storage and marshaling control-hormones.

For better or for worse, the timeline of these progressive age-related calamities for each one of us is pretty strictly subject to the dictates of genetics. It is pretty well established, and scientific proof abounds to show that we have little power to retard it. However, with the help of poor nutrition and an unhealthy life style we can readily accelerate it.

We are frequently cautioned by health authorities via media and other sources that unhealthy nutrition and sedentary lifestyle are at the very core of the cardiovascular afflictions that thwart vigorous and enduring erectile function. These reports bolster the idea that when it comes to lovemaking, our diet can either make us or unmake us because our diet is principally responsible for these medical disorders that result in ED. In fact, our diet by itself is not at fault. The most recent scientific findings teach us that:

- High blood pressure, atherosclerosis, insulin resistance, visceral adiposity (metabolic syndrome), and ED are natural, progressive, age-related cardiovascular conditions with a timeline determined principally by our genes.
- The progress timeline of these conditions can be accelerated by a diet that is poor in certain nutrients and micronutrients, and either too poor or, paradoxically, too rich in antioxidants.
- High blood pressure, atherosclerosis, and insulin resistance, leading to cardiovascular and heart diseases, diabetes, metabolic syndrome, and the ensuing ED, may not be "diseases" as we understand that term, but the "presenting clinical symptoms" of an underlying natural process of systematic degradation and ultimate dysfunction of the *endothelium*, the cell-layer lining the blood vessels throughout the body.
- Endothelium dysfunction adversely impacting the ACh/NO/cGMP pathway is causally linked to high blood pressure, atherosclerosis, insulin resistance, oxidative stress, metabolic syndrome, declining testosterone, changes in sensitivity/resistance of appetitive control hormones, and ultimately, ED, by interference with the formation of endothelium-derived NO.
- The endothelium serves a vital physiological control function that is undermined in part by oxidative stress and is in part linked to a diet that fails to supply certain essential nutrients and micronutrients that support the proper balance of endogenous and exogenous antioxidants.
- Endothelium impairment is now recognized as the primary cause of most cases of ED, and in fact, prescription medications such as VIAGRA are used to bolster its flagging function.
- While even the "healthiest" diet and lifestyle may not stop progressive aging, they can significantly keep it to its natural individual dictates, whereas poor diet and sedentary lifestyle can speed up endothelial aging as noted by rapidly progressing hypertension, atherosclerosis, and insulin resistance.

These findings may seem outside the mainstream of current intuition regarding what leads to what, that leads to ED. Nevertheless, while perhaps not widely known yet, these assertions constitute the consensus of contemporary scientists spanning those subspecialties examining the clinical and the laboratory links between cardiovascular risk factors and ED.

The epidemiological derivative of the links of cardiovascular disorders and ED to endothelium impairment gives rise to a formulation of their interrelationship that strongly suggests that they are all, in fact, symptoms of underlying blood vessel disease. Many investigators worldwide now hold that view. Thus, this book proposes no unconventional theory of the cause of ED, and no unconventional theory of its treatment outside the mainstream of up-to-date medical research. The ideas and assertions are those of the rank and file in medical laboratory and clinical research whose journal publications are abstracted, paraphrased, and cited here. Their individual research forms fractions of a whole, and the synthesis of these disparate findings into that coherent "whole" is, however, my main contribution to understanding what leads to what, that leads to ED, and how best to treat it.

1.4 IT IS MORE COMPLICATED THAN THE STANDARD AMERICAN DIET

The Standard American Diet (SAD) is often accorded the lion's share of blame for the common cardiovascular disorders that befall Americans. The prevention and treatment of ED would be greatly simplified and facilitated were it just a matter of nutritional intervention. The fact is that this is not the case. Unhealthy diet alone does not cause ED. ED results from the complex interaction of diet, heredity, endocrine function, natural progressive aging, and cardiovascular hazards.

This is not to say that nutrition is tangential. To the contrary, it plays a significant role in accelerating the progression of the factors that lead to ED, and it can play a significant role in strategies that correct the factors that speed up that time line. For example: If left to its natural dictates, and in the absence of certain nutrient supplements, the time course of hypertension, atherosclerosis, and insulin resistance will result in a predictable age-related prevalence of ED. This time course is detailed later in the book. However, nutritional intervention can stretch and extend that time line for any given individual, mostly by preventing it from accelerating.

For that reason, elements of nutrition, both what we consume and what we fail to consume, are examined in some detail later in different sections of this monograph. The selection of these nutrients is based on what their active constituents teach us about endothelial function and dysfunction. No attempt at a comprehensive survey is made here: samples serve as examples.

1.5 THE STANDARD AMERICAN DIET

The SAD is high in animal fats, high in saturated, hydrogenated fats, low in fiber, high in processed foods, low in complex carbohydrates, low in fruits and greens. The daily intake of important micronutrients (vitamins, minerals, etc.) is impossible to gauge with any degree of certainty. Were one to wish to meet the Required Daily Allowance (RDA) of vitamins and minerals as set by the US Food and Drug Administration (FDA), minimal levels at best, certainly the SAD, is not likely to provide it.

As an alternative to the SAD, and in order to forestall these health hazards that impair response to sexual arousal in both men and women, my co-author and I proposed "Three Food Factors" in our earlier book *Great Food, Great Sex*:[4] two factors to supply the body with adequate amounts of the amino acid L-arginine, and dietary nitrates, from which we derive the NO, proven by 1998 Nobel Prize in Medicine research to be the key to penis erection in men and clitoral engorgement and vaginal lubrication in women. The other factor is intended to reduce *oxidative stress* and inflammation by providing sufficient NO and antioxidants to keep our heart and blood vessels in tip-top condition as recommended by the National Heart Lung and Blood Institute (NHLBI) of the National Institutes of Health (NIH). Many of the recipes for success in that book were provided by the NHBLI studies promoting cardiovascular and heart health.[5]

This book will also explain that what is not consumed will also undermine sexual vitality: The effects of a lifetime of SAD that most can see beginning to downsize sexual performance were shown scientifically in many cases to be reversible by the dietary micronutrients and other supplements detailed in later chapters. Conventional medical research now focuses on what was once thought to be unconventional approaches to treatment of ED by zeroing in on the molecular basis of endothelial function support. For instance, a clinical study appearing in the *New England Journal of Medicine* reported on the salutary effect of coupling statin therapy with niacin (vitamin B3) in patients with low high-density lipoprotein cholesterol (HDL-C) levels.[6]

1.6 AN IMPORTANT NOTE TO THE READER

In the following sections, and indeed in subsequent chapters, there will be numerous medical science studies that are paraphrased and summarized. In some instances, there are direct quotations. These summaries will always include:

- AIM of the study—what the researcher intended to accomplish with the treatment(s); a description of
- PARTICIPANTS in the treatment plan, including age, gender, health status, etc;
- METHOD, i.e. procedure and protocols used in carrying out the treatment, including diagnostic laboratory and other test procedures used to evaluate pre-trial to post-treatment changes; and the researcher's
- RESULTS & CONCLUSION(S) about statistically significant findings.

Background information providing context for the studies is conventionally found in the INTRODUCTION in journal articles. It has been omitted here as the book itself provides that. In addition, things such as number of participants, the types of statistical analysis used, and the "significance level" have been omitted from the summaries. Their absence is neither an oversight nor an effort to obfuscate. Here, I exercise my editorial skills to determine what does, and what does not, explain or enlighten. The information omitted is usually science or statistics jargon intended to convince the editors of a given medical journal to which a study is submitted for publication that the research methodology is valid and sound, and that the data and conclusions are reliable and valid. Having passed their accepted-for-publication test, there is in most cases no further need for that sort of information. That the margin of error (significance level) is 1%, 2%, or 5% is trivial. The reader may rest assured that no study in this monograph is cited that did not meet criteria for statistical significance at a conventionally acceptable level.

The aim of this editing is to help to focus on the issues without the distraction of minutiae that contribute nothing to understanding what the study was about and what we can learn from it. For most readers, the aim of each study cited and the outcome and conclusion will be clear and comprehensible in context. Furthermore, each assertion and each study is carefully referenced in the text directly to the source on the Internet or in a library to support its credibility. The actual references can be found in the reference list at the end of each chapter.

1.7 WHAT IMPAIRS SEXUAL VITALITY?

This book does not address issues concerning dysfunctional sexual desire because there is no need to do that. Contrary to long held belief, ED is very rarely due to lack of desire, but rather to blood vessel dysfunction. The PDE5 inhibitor meds do not correct low or absent desire. There is no need for that: If men with ED had no desire, it would be difficult to explain how just VIAGRA saw sales of $2 billion in 2011. Lacking desire, who would buy it?

Most men experience occasional temporary loss of sexual desire and/or fluctuation in the strength of desire. They may also experience temporary inability to erect, and fluctuations in latency and strength of erection. Infrequent such events are not abnormal, although they cause alarm. The desire to explain consistent inability to attain erection has led to psychological hypotheses that assign these events to inhibiting subconscious motives.

The advent of the PDE5 inhibitors has shown that desire and performance are two separate, albeit overlapping, functions; desire is affected by the available level of testosterone in men, and in women, whereas performance is determined by the capacity of the endothelium to form NO when prodded to do so by sexual arousal.

It should be noted that the ontological development of sexual activity may depend on maturation-related testosterone levels, but its continuation in adulthood may not. Castration after puberty has, in most cases, relatively little impact on sexual activity in various animal species, as well as in men.[7] This suggests that in consistently sexually active men, declining available testosterone levels do not go very far to explain ED in such a vast proportion of the adult population.

1.8 SEXUAL AROUSAL AND "PERFORMANCE"

There have been, over the years, many theories to explain to the patients whose arousal fails to result in erection that, barring hormone insufficiency, the problem is either psychological or stress. There is now rarely mention of stress as a cause of ED as we have learned that stress does not cause erectile failure. In fact, erectile failure causes stress. Stress and anxiety actually heighten sexual arousal.[8]

In bygone days, when one mentioned to his physician that he was having an erectile problem, he might have been referred to a urologist who, if he couldn't find any obvious physical evidence of a problem, would either refer him for a test for available testosterone levels or make a further referral to a mental health professional. Today, he might also be examined and tested for cardiovascular and heart disease (CVD), and his blood tested for elevated cholesterol levels and fasting glucose level. That's what modern medicine now recommends because it has learned the hard way that ED is most often the first warning sign of impending CHD or diabetes, or both:

In a CNN (TV) program aired August 17, 2011, titled "Erectile dysfunction: the leading indicator of heart disease in men," Dr. Sanjay Gupta reported that ED that, according to the National Institutes of Health, affects at least 30 million men in the United States, is the leading indicator of heart disease in men. He quotes Dr. Terry Mason, urologist and chief medical officer at Cook County Hospitals in Chicago, who avers that ED, difficulty maintaining an erection sufficient for sex, is "the canary in the coal mine."

Medical authorities now recognize that ED is a sign that there's more widespread disease, and not just for the heart, because the endothelial cells in the penis spongy cavernosae chambers are a window to the health of endothelial cells in blood vessels throughout the body. The report further warns that the connection between ED and heart disease is especially important because it may help in early diagnosis and treatment before developing heart problems become serious. What's more, according to a 2009 study reported in the *Mayo Clinic Proceedings*, ED is even a strong predictor of death in men with cardiovascular disease, and those men with ED were found to be 1.6 times more likely to suffer from a serious cardiovascular problem such as a heart attack or stroke.[9] The importance of this announcement is that it may help focus attention on the new research into the cause of the most prevalent form of ED.

1.9 ERECTILE DYSFUNCTION AND CARDIOVASCULAR DISEASE

A clinical study titled "Erectile dysfunction and subsequent cardiovascular disease" appeared in The *Journal of the American Medical Association* (JAMA) in 2005. The incidence of ED was compared in men with varying degrees of cardiovascular disease (CVD). It concluded that, "Erectile dysfunction is a harbinger of cardiovascular clinical events in some men. Erectile dysfunction should prompt investigation and intervention for cardiovascular risk factors."[10]

1.10 ERECTILE DYSFUNCTION AND HEART DISEASE

The *Journal of the American College of Cardiology* featured a study titled "Heart disease risk factors predict erectile dysfunction 25 years later (the Rancho Bernardo Study)." In a prospective study of men 30 to 69 years old, seven classic coronary heart disease (CHD) risk factors (age, smoking, hypertension, diabetes, hypercholesterolemia, elevated triglycerides, and obesity) were assessed from 1972 to 1974.

In 1998, after an average follow-up of 25 years, surviving men participants were asked to complete certain components of the *International Index of Erectile Function* (IIEF-5), which allows stratification of ED into five groups. At the start, the average age of the ED study participants was 46 years. At follow-up, it was 72 years. Average age, body mass index, cholesterol, and triglycerides were each significantly associated with an increased risk of ED. Cigarette smoking was marginally more common in men with severe to complete ED, as compared with those without ED.

Blood pressure and fasting blood glucose were not significantly associated with ED, likely due to *selective mortality*. *Selective mortality* here means that those with the highest blood pressure and fasting glucose levels were not included in the final data because they died before the study ended. Selective mortality will crop up again in later statistics on average age-related cardiovascular risk prevalence.

The authors concluded that "Improving CHD risk factors in mid-life may decrease the risk of ED as well as CHD. ED should be included as an outcome in clinical trials of lipid-lowering agents and lifestyle modifications."[11] In a word, "statins."

1.11 "HYPERTENSION IS ASSOCIATED WITH SEVERE ERECTILE DYSFUNCTION"

That's the title of an article that appeared in the *Journal of Urology* in 2000. It was found that in addition to the observation that ED is more prevalent in patients with hypertension than in an age-matched general population, it is more severe in those with hypertension than in the general population.[12] In fact, one physician from the Heart Institute, Good Samaritan Hospital, USC, CA, stated that when examining patients, it should be kept in mind that:

- Patients with hypertension may also have ED. One should ask hypertensive patients about their sexual health.

- Patients with ED may have hypertension. Patients presenting with ED should be asked about their cardiovascular-risk factors, including hypertension and to obtain a blood pressure readings. These men should also be assessed for the presence of dyslipidemia, diabetes, smoking, obesity and lack of physical exertion.

1.12 ERECTILE DYSFUNCTION IS LINKED TO ELEVATED CHOLESTEROL AND CORONARY HEART DISEASE RISK

The journal *European Urology* reports the results of a prospective study comparing a group of patients with ED to matched non-ED patients with respect to known levels of risk factors for ED and CHD (serum lipids [total cholesterol], triglycerides, HDL-C, low-density lipoprotein cholesterol (LDL-C), and Total cholesterol/HDL-C ratio). The average age of the ED group was 57.6 years, and those with no ED, 59.7. The prevalence of clinically elevated cholesterol (TC greater than 200 mg/dL) was 70.6% in ED patients vs. 52% in the non-ED group. Increased 10-year CHD risk was 56.6% in the ED group compared to 32.6% in the non-ED group.

The authors contend that because hyperlipidemia is common in ED patients, the HDL-C and total cholesterol/HDL-C ratio (TC/HDL-C) are predictors of ED. These patients are at a high risk of later developing CHD. "Erectile dysfunction might therefore serve as a sentinel event for coronary heart disease."[13] Yet, no one really knows how nutrition can affect this progressive age-related cardio-sexual timeline.

Last, but certainly not least, ED is extremely common among type-2 diabetes patients: The journal *Diabetes Care* reported in 2002 that overall, 34% of patients had frequent erectile problems, 24% had occasional problems, and less than half (42%) reported having no erectile problems.

In that report, after adjusting for patient characteristics, ED was associated with higher levels of diabetes-specific health distress and worse psychological adaptation to diabetes, which were, in turn, related to worse diabetes metabolic control. Erectile problems were also associated with a dramatic increase in the prevalence of severe depressive symptoms, lower scores in the mental components of the SF-36®, and a less satisfactory sexual life. (The SF-36 is a multi-purpose, short-form health survey with only 36 questions. It yields an 8-scale profile of functional health and well-being scores.)

It is noteworthy that 63% of the patients reported that their physicians had never investigated their sexual problems.[14]

These are not isolated studies. There are now many thousands in online medical databases that address these four health factors and ED in medical terms. But in human terms, it is not about this number-over-that-number cholesterol, or blood pressure, or blood sugar numbers. It is also about quality and enjoyment of life.[14]

1.13 IT'S ALSO ABOUT QUALITY OF LIFE

Healthy sexual function is not only about cardiovascular health and longevity. Men think in terms of sexual "performance," and "vigor." For many of them the primary concern is not risk of future calamity, but it is also about satisfactory quality of life. Erectile failure, even once, makes many ask if they are losing an important aspect of their life, if they are giving an unintended "message" about loss of interest—a message they really don't mean, and certainly don't intend to convey. It results in frustration and dissatisfaction.

A 2011 article appearing in *The Journal of Gerontology* detailed the emotional consequences of the decline in sexual frequency among American men and women between the ages of 44 and 72. The authors concluded that regardless of how frequently men and women engage in sex to begin with, the rate at which sexual activity declines over the years is pretty much the same. A change in the association between happiness and frequency is parenthetically also a significant factor for both men and women.

In fact, this monograph will show that the timeline pattern of decline in sexual activity parallels the age-related progressive impairment of endothelial function, the progressive increase in resting blood pressure, the progressive accumulation of atheromas in blood vessel walls, the progressive severity of metabolic syndrome, and the progressive severity of insulin resistance and type-2 diabetes. Not only will it be shown that these disorders are a related cluster, but that this cluster may be causally linked to the adverse impact of normal-course-of-events oxidative stress on endothelial function. What's more, the SAD and sedentary lifestyle can accelerate the age-related timeline by increasing oxidative stress.

Following the findings of countless publications that support decline in sexual activity by men as they age, and the consequent loss of happiness, we now encounter only rare mention of loss of interest in sex, i.e. loss of sexual desire (libido). The principal culprit is typically poor cardiovascular health. That is a dramatic change, and in fact it is one that would not have been predicted just twenty-five years ago, certainly not before VIAGRA and the likes came to be generally known.

1.14 DECLINING SEXUAL DESIRE VS. DECLINING PERFORMANCE

Many American men who suffer varying degrees of ED rarely experience low or absent sexual desire. ED in adult men was most often, and erroneously, blamed on weak or absent libido due either to subconscious psychological issues, stress, or low serum levels of testosterone. The emotional basis of ED was steeped in Freudian psychobabble about subconscious *Oedipal* conflicts and similar myths. This mythology had enjoyed great popularity in the US despite the lack of any tangible evidence to support it.

Men who think the secret to addressing ED is rooted in emotional problems might consider the following: According to The National Institutes of Health (NIH), about 4% of men in their 50s, and nearly 17% of men in their 60s, experience a total inability to achieve an erection. The incidence jumps to 47% for men older than 75.[15]

Judging from these statistics, one would have to conclude that in a fair number of American men, subconscious emotional conflicts lie in wait, hidden out of sight until their mid-50s and 60s, when they suddenly surge as would an epidemic simultaneously now infecting almost half the population, and completing the infection by the time they reach the age of 75. There is neither logic nor scientific basis for concluding that subconscious psychological conflicts develop and grow in severity as we mature until, in our middle years, they finally worsen and bring us down.

It could of course be that sexual desire declines with age because testosterone, the "hormone of desire," declines with advancing age. While that is certainly true, it is not generally the case that serum-available testosterone levels decline precipitously in the middle years as shown in a later section of this monograph. In fact, age-related change causes an enzyme, 5-alpha-reductase, to convert testosterone to a more powerful form, dihydrotestosterone (DHT), that boosts libido and, by the way, causes benign prostate hypertrophy (BPH) as well as *pattern baldness*. We do, however, know now that there is an age-related decline in NO production that would directly affect performance.

Note: Conventional medical treatment for either BPH or pattern baldness, or both, consists mainly of meds that inhibit the action of the enzyme 5-alpha-reductase (for instance, Dutasteride®). These are, however, not without some risk of adverse impact on sexual desire (libido) and erectile function: In 2004, the *Journal of Urology* reported that DHT plays a central role in BPH, and that has led to the introduction of 5-alpha-reductase inhibitors that can slow the progress and even prevent the progression of BPH by suppressing DHT synthesis.

Although 5-alpha-reductase inhibitors decrease prostate volume, in so doing, they also improve symptoms and urinary flow, and decrease the risks of acute urinary retention and the need for BPH-related surgery. The commonly found adverse reactions of 5-alpha-reductase inhibitors are decreased libido, ED, and ejaculatory dysfunction. However, these are said to occur in a minority of men and tend to decrease with longer treatment duration.[16]

A more recent finding disagrees with the conclusion stated above: In 2012, the *Journal of Sexual Medicine* reported that, "Finasteride [Propecia®] has been associated with sexual side effects that may persist despite discontinuation of the medication. In a clinical series, 20% of subjects with male pattern hair loss reported persistent sexual dysfunction for [more than] 6 years, suggesting the possibility that the dysfunction may be permanent. ..."[17]

Testosterone does not fuel sexual performance. Average testosterone level of men between the ages of 59 and 69 is somewhat lower than that in men between 19 and 29. Concentration of plasma testosterone is quite variable,[18] with highest levels and least variability in men between the ages of 32 and 51. But even by the age of 70, while more variable and thus less predictable, testosterone levels are not all that low.[19]

While that holds for "free testosterone," the *proportion* of the *bound* form that is not effective may be higher. In other words, the serum concentration of the free form may be lower. The free form has indeed been shown to be lower, but not dramatically lower until about age 80.[20]

In 2009, Americans spent $807.7 million on erection-promoting prescription meds. Why would they do that if they lacked sexual desire?

1.15 DIURNAL VARIATION

On average, serum testosterone level in young men peaks at about 7:00 AM and is lowest around 8:00 PM. In older men, it peaks at about 8:00 AM, and then several more times during the day. Although the average serum testosterone level is somewhat lower in older men, the main difference from young men is the leveling of the diurnal peaks as rhythmicity is attenuated over the day. *Diurnal* (daily) variation in serum estradiol (estrogen) levels in women vary somewhat with the menstrual cycle, but on average, women peak at approximately 8:00 AM.[21]

Because estradiol indirectly drives desire in women, and heightened desire in women may be sensed by the man, this may tend to *favor* "older" men paired with women in their reproductive years because their daily testosterone peaks more nearly coincide. In women of reproductive age, estradiol exhibits a diurnal rhythm that tends to peaks in the early morning; the timing of the peaks is shifted later during the menstrual phase.[21]

Testosterone drives desire. Despite declining age, there is some degree of compensation with the onset of the dihydro form in the middle years. What thwarts "performance"? The answer now generally accepted is that when the endothelium is damaged, it fails to deliver an adequate supply of the crucial biologically active signaling molecule, NO, that triggers penile erection and sustains it throughout lovemaking.

Much of this monograph is dedicated to explaining the medical research that shows how damage occurs, and what can slow its progress and even possibly reverse it.

1.16 THE ACETYLCHOLINE/NITRIC OXIDE/CYCLIC GUANOSINE MONOPHOSPHATE PATHWAY

In the early 1900s, the Austrian pharmacologist Otto Loewi discovered a substance made by cells in the nervous system that he termed "vagus material." Some years later, Sir Henry Dale, with whom Loewi later collaborated at the University of London, and with whom he shared the 1936 Nobel Prize in

Medicine, named it "acetylcholine" (ACh). This was the beginning of modern neuropharmacology based on *neurotransmitters*, chemical signals by which cells in the nervous system and brain communicate. ACh was found to relax arterial smooth muscles, and this seemed a promising road to lower blood pressure to treat hypertension. However, it did so unreliably and unpredictably.

In 1980, Dr. Robert F. Furchgott, professor of pharmacology at Downstate Medical Center in Brooklyn, NY, published his findings in the journal *Nature*, that ACh relaxation of arterial smooth muscle depended on the simultaneous presence of a mysterious substance made by the endothelium. It was termed *"endothelium-derived relaxing factor"* (EDRF): ACh relaxed the vessels only when EDRF was also present.[22] The identity of EDRF was a mystery.

It did not take long to verify these findings, and in 1993, Dr. Salvador Moncada at Wellcome Research Labs, UK, who had previously first identified EDRF as the gas, NO, detailed its role in health and disease in the *New England Journal of Medicine* (NEJM).[23] This news was first met with disbelief—even derision. A gas, indeed! It was not known then that NO had a biological role, much less in blood vessels: NO is derived in the body principally from the food-borne amino acid L-arginine, and from nitrates as well.

In 1998, Dr. Furchgott and two colleagues were awarded the Nobel Prize in Medicine for the discovery of the biological role of NO in blood vessel, cardiovascular, and heart function, and the nervous system and brain. The immune system was soon added to the list of NO-dependent body systems.

One of the three recipients, Dr. Louis J. Ignaro, detailed how ACh in sexual arousal caused increased and sustained production of NO synthesized from the amino acid L-arginine by the endothelium lining the spongy chambers of the penis cavernosae. This causes them to relax (dilate), allowing increase blood inflow.

1.17 A FOOTNOTE TO HISTORY

Why doesn't ACh reliably cause vasodilation in human blood circulation or in blood vessel strips in the laboratory? What Dr. Furchgott discovered in strips of arterial blood vessels washed with ACh is that the strips did not relax when, on closer examination, it was found that laboratory preparation had damaged the lining, the endothelium, and damaged endothelium impaired NO production.

The observation that damaged endothelium cannot produce NO in amounts sufficient to cause dilation and maintain adequate blood circulation is now the explanation for how we come by cardiovascular disorders such as hypertension, CHD, and atherosclerosis; now we can add ED. All these are said to be due to, or aggravated by, oxidative stress. Each of these conditions is said to result from damage to the endothelium, and each in turn causes further damage to the endothelium. Oxidative stress and the cardiovascular disorders mentioned above impair endothelium function as surely as if it were mechanically damaged by mishandling in the laboratory. The SAD is also effective in increasing oxidative stress, and what's more, we seem to be doing it in ever growing numbers ... and in record time.

However, "carelessly" is not really to the point as no one had any idea that the endothelium had any biological function, so why spare it? It was thought then to be not much more than a sort of lining between the blood stream and the blood vessel walls—something like a bed-sheet—to keep undesirable stuff in blood out of the vessel walls.

1.18 THE ENDOTHELIUM—THE FIRST VASCULAR CONTROL SIGNAL SYSTEM

In ordinary circumstances, the degree of *tonus* of arterial blood vessels, and consequently, blood pressure, is determined by activity level in two control signal-systems:[1] the endothelium-dependent signal system based on the availability of L-arginine-derived NO, produced by the endothelium; and[2] the endothelium-independent system of action-hormones (adrenalin, noradrenalin, etc.) and diuresis.

Figure 1.1 is an electron micrograph of a cross section of a blood vessel: red blood cells are shown in the *lumen* through which blood flows. The fluted accordion-like structure is the endothelium cell lining, surrounded by the blood vessel wall. The folds, or fluting, facilitate relaxation and thus enlargement of the vessel increasing blood flow.

The ability of the endothelium to deliver NO in sexual arousal is the key to erectile function. However, availability of L-arginine is never the limiting factor in delivering NO. As will be shown in a subsequent chapter, it is typically due to conditions that unlink or otherwise disable the nitric oxide synthase (eNOS) enzyme that cleaves NO from L-arginine.

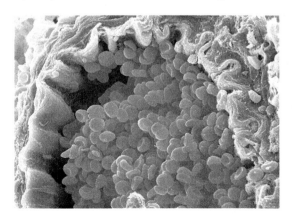

FIGURE 1.1 Electron micrograph of a section of arterial blood vessel showing the endothelium surrounding red blood cells flowing in the blood stream. *Reproduced with permission from Steve Gschmeissner/Science Source.*

Numerous health and lifestyle hazards can cause damage to the endothelium, thus impairing the delivery of endothelium-derived NO and hampering sexual function. The endothelium is the gateway to sexual arousal and sexual "performance" vitality.

1.19 VESSELS OF THE VESSELS—*VASA VASORUM*

In order to fully appreciate the function of the endothelium and its ability to form and deliver NO, it helps to know yet a little more about the structure and function of the blood vessels burdened with delivering increased blood flow to the penis in sexual arousal. Arterial blood vessel walls and the endothelium do not get oxygen and nutrients directly from the blood stream in the lumen of the vessel. Nor do they eliminate metabolic wastes that way. Instead, they are supplied by their own set of blood vessels that draw blood from the blood stream and deliver it to the vessel wall both inside and outside (Figure 1.2).

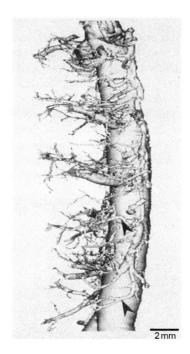

2 mm

FIGURE 1.2 Normal pig coronary artery showing origin and spatial distribution of *vasa vasorum*. Adventitial *vasa vasorum* in balloon-injured coronary arteries: visualization and quantitation by a microscopic three-dimensional computed tomography technique. Source: *Kwon, H. M., Sangiorgi, G., Ritman, E. L.,* et al. *(1998). Adventitial* vasa vasorum *in balloon-injured coronary arteries: visualization and quantitation by a microscopic three-dimensional computed tomography technique.* J Am Coll Cardiol, *32, 2072–2079. Reproduced with kind permission from* Journal of the American College of Cardiology.

There are two sets of these vessels, termed *vasa vasorum*, one set outside the walls and the other inside. It is well to keep this in mind as the last word has not been written on the cause and course of atherosclerosis thought to involve *vasa vasorum*.

1.20 *VASA VASORUM* DEPENDS EXCLUSIVELY ON ENDOTHELIUM-DERIVED VASORELAXATION

In a study titled "On the regulation of tone in *vasa vasorum*" published in *Cardiovascular Research*, it was shown that *vasa vasorum* relaxation was solely endothelium-dependent, that is, NO-dependent, whereas the arterial vessel responded to both endothelium-dependent and endothelium-independent relaxation signals.[24] In brief, the damage done by atherosclerosis to the endothelium intimately involves first changes in the *vasa vasorum*.

Writing in the journal *Medical Hypotheses*, in 2009, researchers contend that known risk factors for atherosclerosis, such as high blood pressure and nicotine, reduce blood flow in the end branches of the *vasa vasorum*. Local impaired blood flow affects the cells of the endothelium causing local inflammation. This makes the endothelium permeable to large particles such as various bacteria and LDLs and other fatty acids, which macrophages engulf, transforming them into foam cells.[25] Damage to the endothelium is caused by damage to the *vasa vasorum* that then deprives it of oxygen, thus allowing sludge to infiltrate the blood vessel walls.

NO is a vasodilator that controls the volume of blood flowing through blood vessels by controlling the diameter of the lumen. It signals the mechanism to relax blood vessels, allowing increased blood to flow though them. It also controls the blood flowing into the blood vessels by controlling the tonus of the *vasa vasorum*. But, what is the role of NO in sexual function?

1.21 THE PENIS

The information on how NO affects blood vessels and what this has to do with erection makes sense only if one understands that the penis is basically a bundle of blood vessels.

There are muscles at the base of the penis, attached to the pelvis, by means of which most men can slightly raise an erect penis. But, the penis is largely an encased pair of spongy cavernosa chambers, modified spongy arterial blood vessels, with blood-supplying arteries whose endothelium controls blood inflow.

A transverse section of the penis would show the two large central spongy *corpora cavernosae* and below those, a spongy chamber that encircles the urethra, the *corpora spongeosum*. The blood inflow arteries (*profunda*) are within the cavernosae that engorge with blood during sexual stimulation with increased NO signaling the endothelium to relax the vessel.

As the penis erects, the bulging chambers constrict the veins under the relatively inelastic outer *tunica*, thus causing less blood to flow out of the penis through the veins than comes into it through the arteries. This is basically a pressure regulated mechanical valve.

1.22 THE PATHWAY TO PENILE ERECTION

In sexual arousal, the brain sends signals via the nervous system and blood stream to the spongy expandable cavernosae of the penis:

1. Components of the central nervous system release the neurotransmitter ACh that signals the cell of the endothelium to form NO from L-arginine by eNOS facilitated by a co-factor, tetrahydrobiopterin (BH4).
2. In the vascular smooth muscle penis cavernosae, NO signals the release of another enzyme, *guanylate cyclase* (GC), that helps form *cGMP* from *guanosine-5'-triphosphate* (GTP).
3. cGMP is the vasodilator that actually relaxes the cavernosae smooth muscle, increasing blood inflow via relaxation that follows sequestration of myoplasmic Ca^{2+}.[26]
4. As blood inflow rises in the cavernosae, increased inflow and pressure cause shear stress on the endothelium, resulting in additional release of eNOS, and further maintenance of NO formation.
5. As blood fills the cavernosae, they expand, exerting pressure on the veins that lie between the *albuginea* and the *tunica*. This pressure constricts the veins, thus limiting blood outflow.
6. At the same time, there is also the formation and buildup of PDE5 that breaks down cGMP to end the erection cycle. PDE5 rapidly breaks down cGMP, causing reabsorption of the constituents.

1.23 PHOSPHODIESTERASE TYPE-5—FRIEND OR FOE?

Numerous body functions depend on NO regulation. Three specific (iso-) forms of nitric oxide synthase (NOS) are presently known: two are calcium-dependent constitutive forms (cNOS), one of which is the endothelium form (eNOS), and an inducible calcium-independent form (iNOS). These regulate NO formation from L-arginine in three specific body functions. In the endothelium NO regulates blood flow throughout the body by adjusting blood vessel caliber as required. There is a different form of NOS in the immune system, and yet another form in the nervous system and the brain.

By the same token, there are numerous forms of the enzyme PDE that control NO concentration in various body systems by controlling the concentration of NO-derived cGMP. In addition to *corpus cavernosum* smooth muscle, PDEs are also found in lower concentrations in other tissues, including blood platelets and skeletal muscle.

NO-derived cGMP leads to vasodilation. If it were produced uncontrolled in the body it could cause a potentially fatal medical emergency. In serious inflammation such as sepsis, when the immune system overproduces NO, the concentration is often fatal—it stops respiration when the respective PDEs are swamped. In extreme cases, they just can't invariably keep up with NO production. In erection, the cascade that follows the release of ACh resulting in endothelium-derived NO/cGMP relaxation of the cavernosae is modulated by PDE5.

PDE5 exists in various concentrations in many body tissues, but especially in the endothelium of the cavernosae. According to an article in the *Journal of Cell Biology*, it is said to regulate cGMP production mostly by acting directly on cGMP in what is said to be a "long lasting negative-feedback loop"—a form of homeostasis: As one goes up it causes the other to go down.[27,28]

If there is endothelium damage with atherosclerosis or diabetes, and one can generate NO/cGMP, but insufficient for satisfactory sexual "performance," then one can rely on disabling PDE5 with an agent such as VIAGRA (sildenafil citrate). However, in the face of a seriously impaired endothelium, even PDE5 inhibitors may not work.

1.24 CARDIOVASCULAR DYSFUNCTION AND SEXUAL DYSFUNCTION IN AGING

Americans are living longer now than they did just a few decades ago. According to the *CIA World Book*, in 2011, average life expectancy in the US was 78.37 years. Women live somewhat longer than men. There is little evidence of widespread loss of interest in sexuality in our "maturing" population. As we age, issues of sexuality for men and women diverge: While desire remains relatively high in older men, it can drop precipitously in most women in menopause.

Common myths make a compelling case for dismissing the interest, and lack of interest, of seniors in lovemaking as there is still a residual attitude that sexuality as we reach the "golden years" is really not altogether proper. Soaring sales of meds like VIAGRA tell a different story: It's not all in your head, and it's not all just your age, but, some of it is.

Seniors do face a challenge as the years pile up. There are anatomical changes in the cavernosal endothelium in men as they age, and in the comparable clitoral cavernosa in postmenopausal women, in whom decreased hormones of desire reduce both libido and the ability to maintain sexual intercourse comfortably.

ED is of great concern to many men, especially as they age. A number of misconceptions cloud our understanding of the extent of ED in American men. For instance, a press release dated August 5, 2003 from the Harvard School of Public Health, titled "Prevalence of erectile dysfunction increases with age,"

tells us that fewer than 2% of the men in the study experienced erection problems before age 40, and 4% between the ages of 40 and 49. From age 50
upwards, the percentage rose dramatically, with 26% between the ages of 50 to
59, 40% in those aged 60 to 69 years, and 61% for men older than 70.

The Harvard press release avers that, "... 61 percent [of] men older than
70 [have] experienced ED," but it does not define "experienced." In fact, in
a Massachusetts Male Aging Study (MMAS) (Figure 1.3) it was shown that
only about 15% of men in that age group were totally impotent, while less
than 30% of them experienced "moderate" ED. Clearly, the manner of presenting data can give a different impression of the facts.

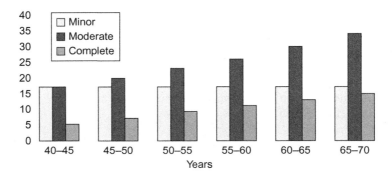

FIGURE 1.3   Age-related increase in incidence of ED by decade of life: MMAS Study. Source:
*Feldman, H. A., Goldstein, I., Hatzichristou, D. G., et al. (1994). Impotence and its medical and
psychosocial correlates: results of the Massachusetts Male Aging Study.* J Urol, *151, 54–61.
Reproduced with kind permission from the* Journal of Urolology.

The MMAS shows the likelihood that someone may experience either no
dysfunction, minimal dysfunction, moderate, or complete dysfunction. This
is a more meaningful way to describe sexuality in men as they age.

In women, testosterone also is pretty much minimal by age 45, whereas
in men, both free testosterone and free dihydrotestosterone remains relatively
high well past that age. In fact, free dihydrotestosterone levels in men remain
higher than free testosterone levels up to about age 70, as shown in
Figures 1.4(A) and 1.4(B).

Hormonal issues notwithstanding, the evidence clearly points to the fact
that the NO/cGMP mechanism affects the response to sexual arousal in both
men and women. Much has been published about that response in men, and
to some extent the research on the role of that mechanism in women has not
received the same degree of attention. However, it is known that there are
age-related changes in the clitoris cavernosae, just as they were found to
occur in the penis cavernosae in aging men.

The *Journal of Urology* reported a strong link between increasing age and decreased clitoral cavernosal smooth muscle fibers. In the post-mortem tissue samples studied, not only did that decrease significantly correlate with increase in age, but in the age group of 44 to 90 years, clitoral cavernosal fibrosis was found to be significantly greater in the presence of cardiovascular disease-related mortality compared with those without cardiovascular disease-related mortality.[29]

It's only a small jump from these data to the conclusion that women's physical arousal problems may also be linked to regularly resorting to SAD and sedentary lifestyle, accelerating endothelial impairment as happens in men. The authors did not pull their punches: "Vascular risk factors may adversely affect the structure of clitoral cavernosal tissue." In other words, it is impaired endothelium, and thus impaired NO/cGMP formation.

Aging has a significant adverse impact on sexual response in both men and women, and medical authorities now tell us that it is related to progressive impairment of the endothelium and its ability to serve us NO when and where needed. Parenthetically, sildenafil citrate (VIAGRA) increases clitoral and uterine blood flow in healthy postmenopausal women ... even without any erotic stimulation.[30]

The question remains: Does NO production *ipso facto* decrease with age in parallel with declining erectile function?

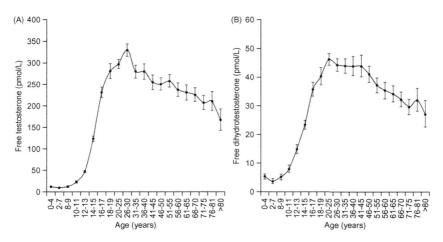

FIGURE 1.4 (A) The course of free testosterone serum concentrations over the life span. (B) The course of free dihydrotestosterone serum concentrations over the life span. Source: *Stárka, L., Pospíšilová, H., & Hill, M. (2009). Free testosterone and free dihydrotestosterone throughout the life span of men.* J Steroid Biochem Mol Biol, *116, 118–120. Reproduced with kind permission from the* Journal of Steroid Biochemistry and Molecular Biology.

1.25 NOTE ON TERMINOLOGY

Clinical experiments to determine the health of the endothelium and its ability to deliver the NO/cGMP stimulus to relax blood vessels usually refer to "endothelium-dependent vasodilation" because blood vessel relaxation depends on the endothelium forming NO and cGMP when signaled to do so by ACh. For the sake of brevity, it will here and henceforth mostly be called "NO/cGMP," and occasionally, "the ACh/NO/cGMP pathway." There are other means of dilating blood vessels, but as they do not involve NO/cGMP directly, they do not concern us presently.

According to an article in *Hypertension*, aging progressively impairs the ability of the endothelium to form NO to relax the blood vessels in the forearm. Studies in experimental (animal) models suggest that NO/cGMP relaxation is reduced with aging, and this circumstance may be relevant to the way that we develop atherosclerosis.

In people ranging in age from 19 to 90, measurement involved compressing the brachial artery with a blood pressure cuff so as to cut off circulation. A solution containing constituents that ordinarily trigger endothelial NO production was injected into the blood stream. Then, cuff pressure is released, and the temporal pattern of the restored blood flows in the forearm indicates how much NO/cGMP had been formed by the injected substances. This procedure is called Flow-Mediated Dilation (vessel relaxation; FMD) (see Chapter 2). In addition to age, blood lipids and resting blood pressure were also considered.

Figure 1.5 shows that NO/cGMP vasodilation of the brachial artery was progressively impaired with increasing age. NO formation declines

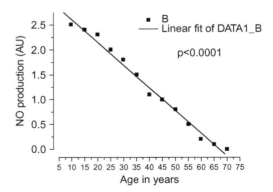

FIGURE 1.5 Endothelium dependent (NO/cGMP) blood vessel relaxation decreases with age. Source: *Gerhard, M., Roddy, M. A., Creager, S. J., et al. (1996). Aging progressively impairs endothelium-dependent vasodilation in forearm resistance vessels of humans.* Hypertension, *27, 849–853. Reproduced with kind permission from* Hypertension.

dramatically with age. The decline was already evident by the fourth decade (age 30 to 39 years). In parallel, FMD likewise declines with age.[31]

In addition to age, total cholesterol and LDL-C were also predictors of NO/cGMP vasodilation. Vasodilation triggered by other injected substances that do not result in NO/cGMP formation did not change with age.

Age remained the most significant predictor of endothelium-dependent (NO/cGMP) vasodilator decline. From these observations, it was concluded that NO/cGMP vasodilation declines steadily with increasing age *even in healthy persons* (my italics). The measures obtained in the forearm are of course, and for good reason, a window opened into the health of the endothelium throughout the body.

What accounts for that age related decline? For one thing, the prime suspect is thought to be oxidative stress damage to the endothelium caused by exogenous factors such as the SAD and inactivity. There is also evidence linking aging and lower testosterone to anatomical changes in corpus cavernosum. In addition, there is evidence that such changes relate to a reduction in NO availability.[32] Others implicate free radical damage that ages the endothelium and accelerates endothelial cell apoptosis.[33]

The next chapter details the main diagnostic tests and procedures commonly used to evaluate functionality before and after treatment:

- How did it function before treatment?
- What changed due to treatment?
- Is the change *statistically significant* or likely due to chance?

The observation/measurement techniques described are standard, and will appear in most of the clinical studies examined.

REFERENCES

1. American Psychiatric Association (Author). *Diagnostic and statistical manual of mental disorders (DSM-IV-TR)*. American Psychiatric Publishing, Inc.; 2000.
2. Segraves RT. Considerations for diagnostic criteria for erectile dysfunction in DSM V. *J Sex Med* 2010;**7**(2 Pt 1):654–60.
3. Selvin E, Burnett AL, Platz EA. Prevalence and risk factors for erectile dysfunction in the US. *Am J Med* 2007;**120**:151–7.
4. Fried R, Edlen-Nezin L. *Great food, great sex*. New York: Ballentine Books; 2006.
5. <http://clinicaltrials.gov/ct2/show/NCT00069654>; [accessed 13.11.13].
6. Boden WE, Probstfield JL, Anderson T, Chaitman BR, Desvignes-Nickens P, Koprowicz K, et al. Niacin in patients with low HDL cholesterol levels receiving intensive statin therapy. *N Engl J Med* 2011;**365**:2255–67.
7. Ford CS, Beach FA. *Patterns of sexual behavior*. New York: Harper; 1952.
8. Barlow DH, Sakheim DK, Beck JG. Stress heightens arousal. *J Abn Psychol* 1983;**92**:49–54.
9. Nehra A. Erectile dysfunction and cardiovascular disease: efficacy and safety of phosphodiesterase Type 5 inhibitors in men with both conditions. *Mayo Clin Proc* 2009;**84**:139–48.

10. Thompson IM, Tangen CM, Goodman PJ, Probstfield JL, Moinpour CM, Coltman CA. Erectile dysfunction and subsequent cardiovascular disease. *JAMA* 2005;**294**:2996−23002.

11. Fung MM, Bettencourt R, Barrett-Connor E. Heart disease risk factors predict erectile dysfunction 25 years later: the Rancho Bernardo Study. *J Am Coll Cardiol* 2004;**43**: 1405−11.

12. Burchardt M, Burchardt T, Baer L, Kiss AJ, Pawar RV, Shabsigh A, et al. Hypertension is associated with severe erectile dysfunction. *J Urol* 2000;**164**:1188−91.

13. Roumeguère T, Wespes E, Carpentier Y, Hoffmann P, Schulman CC. Erectile dysfunction is associated with a high prevalence of hyperlipidemia and coronary heart disease risk. *Eur Urol* 2003;**44**:355−9.

14. De Bernardis G, Belfiglio B, Di Nardo B, Greenfield S, Kaplan SH, Pellegrini F, et al. Erectile dysfunction and quality of life in type 2 diabetic patients: a serious problem too often overlooked. *Diabetes Care* 2002;**25**:284−91.

15. National Institutes of Health (NIH). Consensus Conference. Impotence. NIH Consensus Development Panel on Impotence. *JAMA* 1993;**270**(1):83−90.

16. Andriole G, Bruchovsky N, Chung W, Matsumoto AM, Rittmaster R, Roehrborn C, et al. Dihydrotestosterone and the prostate: the scientific rationale for 5alpha-reductase inhibitors in the treatment of benign prostatic hyperplasia. *J Urol* 2004;**172**(4 Pt 1):1399−403.

17. Irwig MS. Persistent sexual side effects of finasteride: could they be permanent? *J Sex Med* 2012. Available from: http://dx.doi.org/10.1111/j.1743-6109.2012.02846.x.

18. Wagner G. Erection. In: Wagner, Green R, editors. *Impotence.* New York: Plenum Press; 1981. p. 25−36.

19. Vermeulen A, Rubens R, Verdonc K. Testosterone secretion and metabolism in male senescence. *J Clin Endocrinol* 1972;**34**:730−5.

20. Vermeulen A. The male climacterium. *Ann Med* 1993;**25**:531−4.

21. Bao AM, Liu RY, van Someren EJ, Hofman MA, Cao YX, Zhou JN. Diurnal rhythm of free estradiol during the menstrual cycle. *Eur J Endocrinol* 2003;**148**:227−32.

22. Furchgott RF, Zawadzki JV. The obligatory role of endothelial cells in the relaxation of arterial smooth muscle by acetylcholine. *Nature* 1980;**288**:373.

23. Moncada S, Higgs A. The L-Arginine-nitric oxide pathway. *NEJM* 1993;**1993**(329):2002−12.

24. Scotland R, Vallance P, Ahluwalia A. On the regulation of tone in *vasa vasorum. Cardiovasc Res* 1999;**41**:237−45.

25. Järvilehto M, Tuohimaa P. *Vasa vasorum* hypoxia: initiation of atherosclerosis. *Med Hypothe* 2009;**73**:40−1.

26. Murphy RA, Walker JS. Inhibitory mechanisms for cross-bridge cycling: the nitric oxide-cGMP signal transduction pathway in smooth muscle relaxation. *Acta Physiol Scand* 1998;**164**:373−80.

27. Kass DA, Takimoto E, Nagayama T, Champion HC. Phosphodiesterase regulation of nitric oxide signaling. *Cardiovasc Res* 2007;**75**:303−14.

28. Mullershausen F, Friebe A, Feil R, Thompson WJ, Hofmann F, Koesling D. Direct activation of PDE5 by cGMP: long-term effects within NO/cGMP signaling. *J Cell Biol* 2003;**160**:719−27.

29. Tarcan T, Park K, Goldstein I, Maio G, Fassina A, Krane RJ, et al. Histomorphometric analysis of age-related structural changes in human clitoral cavernosal tissue. *J Urol* 1999;**161**:940−4.

30. Alatas E, Yagci AB. The effect of sildenafil citrate on uterine and clitoral arterial blood flow in postmenopausal women. *Med Gen Med* 2004;**13**(6):51.

31. <http://cardiovascres.oxfordjournals.org/content/46/1/28/F3.expansion.html>; [accessed 13.11.13].

32. Tomada I, Tomada N, Almeida H, Neves D. Androgen depletion in humans leads to cavernous tissue reorganization and upregulation of Sirt1-eNOS axis. Age (Dordr) 2011;**35**:35−47.

33. Brandes RP, Fleming I, Busse R. Endothelial aging. *Cardiovasc Res* 2005;**66**:286−94.

Chapter 2

Measuring and Evaluating Function, Impairment, and Change with Intervention

2.1 INTRODUCTION

The previous chapter detailed some of the factors contributing to accelerating the impairment of sexual function. Most authorities now agree that, bottom-line, they damage blood vessel endothelium, and that in turn jeopardizes the ability to make and keep on making nitric oxide/cyclic guanosine monophosphate (NO/cGMP) as sexual arousal demands. It has been pretty well established also that the problem facing most men with erectile dysfunction (ED), except those in their late seventies and older, is a *performance* issue, not a *desire* issue. Serum levels of testosterone can therefore in many cases usually be easily ruled out as the main problem.

To conclude that damaged endothelium is at the core of the problem facing men with various degrees of ED, it is necessary to show that it is damaged endothelium, and only damaged endothelium, that causes ED. We typically determine it by inference from indirect measures of vessel wall thickness, or function impairment as indicated by blood flow response to standard "challenges" by acetylcholine (ACh), papaverine, or sodium nitroprusside, for instance. This chapter addresses some of the more conventional means at our disposal to observe, measure, or count factors that affect endothelium function or impairment and erectile function or impairment.

2.2 WHEN DO WE KNOW THAT WE "KNOW" IN BIOLOGICAL SCIENCE?

Dr. R. F. Furchgott shared the 1998 Nobel Prize in Medicine for discovering the biological role of nitric oxide (NO). He did not, strictly speaking, identify the role of "nitric oxide." He discovered a *mysterious something* made by the endothelium that relaxes blood vessels when signaled to do so by ACh. He termed it "*endothelium-derived relaxing factor*" (EDRF). Investigators

Robert Fried: Erectile Dysfunction as a Cardiovascular Impairment.
DOI: http://dx.doi.org/10.1016/B978-0-12-420046-3.00002-0

that followed didn't know what it was either, and they called it *"non-adrenergic−non-cholinergic"* (NANC) relaxing factor.

Dr. S. Moncada solved the puzzle: EDRF = NANC = NO. In his report titled "The L-Arginine-Nitric Oxide Pathway" published in the *New England Journal of* Medicine, Dr. Moncada stated that, "The discovery that mammalian cells generate nitric oxide, a gas previously considered to be merely an atmospheric pollutant, is providing important information about many biological processes. Nitric oxide is synthesized from the amino acid L-arginine by a family of enzymes, the nitric oxide synthases, through a hitherto unrecognized metabolic route—namely, the L-arginine−nitric oxide pathway."[1] The "mammalian cells" that generate nitric oxide (NO) are those that line the endothelium.

Medical science now avers that hypertension, atherosclerosis, coronary heart disease, diabetes, and metabolic syndrome are the major culprits in ED, and that what they have in common is endothelium impairment downgrading blood circulation by reducing NO biosynthesis. This chapter describes some of the means available to measure things in the body that help us to know how cardiovascular disorders are linked to endothelium impairment and ultimately to ED.

2.3 THE CLINICAL TRIAL (OR CLINICAL STUDY)

The studies cited in this book are reports of clinical trials, aka clinical studies published in conventional medical journals. To meet criteria for publication by a peer-reviewed medical journal editorial board, they must have a certain form and meet certain standards and specifications. Specific information about the structure and format of clinical trials can be obtained from *ClinicalTrials.gov*, a service of the US National Institutes of Health.[2]

To appreciate the synthesis of results from many clinical trials concerning apparently different cardiovascular disorders here said to form a cluster including ED, it is essential to establish the credibility of the data proper, and of the sources from which data were drawn. To that end, here follows a recap of some defining aspects of research reporting and data analysis that may perhaps be somewhat unfamiliar to practicing clinicians who do not engage in research.

2.4 NOTE ON FORM OF CITATION

All the information needed to find a research article online or in libraries is given for any citation: author(s), the date of publication, the title of the article, abbreviated name of the journal or book, the volume, and text pages. All authors of a given reference are listed in the reference section at the end of each chapter. In the text, only up to the first three authors are listed followed by *"et al."* and the date of publication.

2.5 NOTE ON VALIDITY OF RESEARCH REPORTS

Every reasonable effort went into making sure that each study cited in this book conforms to the rules and regulations that govern how clinical trials and studies are to be conducted and in what form the results are to be made public. Citation of a study appearing in a conventional medical journal usually assures those goals, as the studies must meet these criteria *a priori* for consideration for publication. Every study cited in this book was published in a peer-reviewed conventional med-science journal. There are no exceptions.

2.6 STATISTICS SIMPLIFIED

In summarizing studies in context, the details and jargon of statistical analyses including confidence criteria (*alpha*-level) were for the most part omitted. The *alpha*-level, or *level of significance*, tells us the odds that a given outcome may be due to chance. In citing the *alpha*-level criterion, let's say, P (outcome) < 0.01, the authors assert that the probability of the outcome of their study being due to chance is less than 1 percent. Because it is uncommon to choose an *alpha*-level greater than 5%, it is not necessary to state the *alpha*-level of each study. Suffice it to say that the comparisons attained statistical significance: in other words, the probability that chance factors account for the observed differences is less than *alpha*. Thus, the term "significant" has to do with whether the observations or change are significantly different from what one might expect based on random chance events alone and that has nothing to do with whether it is an important or valuable outcome.

2.7 HYPERTENSION

ED is linked to hypertension, chronically elevated resting blood pressure. In any given individual, blood pressure (BP) will be found to be different each time it is sampled because it is in fact variable within relatively narrow limits. Small variations in BP are usually missed when taken with a mercury-column *sphygmomanometer* such as the one typically used in a physician's office or hospital, whereas electronic units accurately sample BP for a few successive heartbeats and then average the result. The average will depend on where you start measuring in the breathing cycle because BP rises with inhale, and declines with exhale (Hering−Breuer reflex).

Very small variations in BP are normal, and systolic blood pressure varies to a greater degree than diastolic blood pressure.[3] However, wide variations may spell trouble. In fact, one authority contends that the goal of successful anti-hypertensive medication is lowering both systolic and diastolic blood pressure, and reducing variability.[4]

TABLE 2.1 Standard Values for Blood Pressure Ranging from Optimal to Stage II Hypertension

Blood Pressure	Range	Values
Systolic BP	Optimal	less than 120 mm Hg
	Normal	120 to 130 mm Hg
	High–Normal	130 to 140 mm Hg
	Stage I Hypertension	140 to 160 mm Hg
	Stage II Hypertension	greater than 160 mm Hg
Diastolic BP	Optimal	less than 80 mm Hg
	Normal	80 to 85 mm Hg
	High–Normal	85 to 90 mm Hg
	Stage I Hypertension	90 to 100 mm Hg
	Stage II Hypertension	greater than 100 mm Hg

Table 2.1 lists the commonly accepted values for blood pressure, ranging from optimal to stage II hypertension, that are standard fare in virtually all conventional clinical studies. What we cannot know from the nomenclature, "optimal" to "Stage 2," is whether the risk of ED rises in proportion to blood pressure increase, or is the risk of ED maximum at any value above "optimal" BP, and after what duration.

The incidence of ED in the population rises as the incidence of hypertension in the population rises. But, we can't predict for any given person when ED might first emerge in the future based on the severity and duration of hypertension. Some phenomena allow only retrospective analysis.

In a study titled "Erectile dysfunction in hypertensive subjects. Assessment of potential determinants," published in the journal *Hypertension*, the authors compared medical and sexual history, depression, hormonal profile, penile nocturnal tumescence, penile vascular supply, and nerve conduction in a group of hypertensive men with ED to the same factors in a group of normotensive men with ED.

The hypertensive men were older, had higher body mass index, and used more medications than the normotensive men, and they also had a marginally higher rate of ischemic heart disease. Circulating testosterone levels were lower in the ED group, but bioavailable testosterone levels were not different. The authors concluded that, excepting lower circulating testosterone levels, there was little difference between hypertensive and normotensive men with respect to a wide range of classic determinants of erectile function. "Direct study of the local vascular erectile apparatus appears necessary for further elucidation of the mechanisms underlying erectile dysfunction in hypertensive men."[5]

There is no non-invasive "direct" assessment of the *"local vascular erectile apparatus"* suitable to conventional clinical treatment. However, clinical assessment techniques described below speak to indirect, albeit useful, clinical assessment. Of course, in 1996, the link between "marginally higher rate of ischemic heart disease," a condition entailing blood flow impairment, and ED as related to endothelium function, was not yet obvious. That came somewhat later.

2.8 HYPERTENSION—PARADOXICALLY BENEFICIAL IN ERECTILE DYSFUNCTION?

A report in the *British Journal of Urology International* proposed a theory of hypertension and ED that seems paradoxical. The authors investigated the effect of blood pressure on (ED) in patients in the initial stages of peripheral arterial disease (PAD). Patients were assessed with the International Index of Erectile Function (IIEF), the Ankle-Brachial Index (ABI), and measurement of arterial blood pressure. It was reported that hypertensive patients in the initial stages of peripheral arterial disease had less severe grades of ED than normotensive patients. The investigators concluded that: The progression of ED parallels that of chronic arterial insufficiency. Systemic arterial hypertension in the initial stages of peripheral arterial disease might protect against ED, but peripheral arterial disease constitutes an aggravating factor for ED, and thus hypertension might exert a paradoxical effect in this stage of the disease.[6]

What is proposed here is that based on the results of assessments, blood vessels appear in some individuals to age and produce "chronic arterial insufficiency" leading to ED. The overall clinical picture suggested that blood pressure rises as a compensatory adjustment perhaps to maintain sexual function. This creates a quandary in the long run as rising blood pressure appears yet to cause further damage to the endothelium.

2.9 ATHEROSCLEROSIS

Atherosclerosis is manifest in several ways. First, anatomically as a collection of calcifying lipids, immune system and platelets cell detritus, and other sludge in the blood vessel wall that cause inflammation, loss of elasticity, as well as wall enlargement and "remodeling." When the vessel narrows, one observes impaired blood flow and hypertension.

Second, physiologically: Atherosclerosis is closely linked to impaired endothelium function. There are several ways to track the extent of this gradual but often inexorable process as it has been shown to (a) reduce NO bioavailability in ordinary blood vessel function, and also (b) as it causes blood vessels to respond abnormally to *acetylcholine* (ACH) or *nitroprusside* "challenge."

The endothelium is not ordinarily visible in a living man. Even if it were, it might provide few obvious clues to its function. When the endothelium was visible in the laboratory for many years in man and beast, it did not lead to any understanding of its function until Furchgott *et al.*, came along in the 1980s. Yet, knowing how well it functions is a crucial index to health, in general, and to erectile strength in particular.

It is not within ready reach of most medical facilities to determine exactly the degree of plaque formation in blood vessels, and clearly it is beyond that of most individuals practicing medicine. Nevertheless, knowing something about how much gunk is encroaching into the walls of blood vessels is of some consequence as this has also been closely linked to ED. The "*Methods*" section of clinical trials reports invariably detail the measurements procedures used to track aspects of atherosclerosis. Here follow some of the techniques employed. Unfortunately, all that is now known—and much is known—about methods that can help to determine the severity of atherosclerosis in routine patient care is rarely available to most of them.

2.10 FLOW-MEDIATED (VASO-)DILATION—A WINDOW INTO ENDOTHELIAL HEALTH

The control of blood circulation and blood pressure is attributed to two distinct major physiological control mechanisms: On the one hand is the *endothelium-independent* system that relies mostly on so-called action hormones such as noradrenaline, and kidney/diuresis hormones such as renin-angiotensin-aldosterone to control blood flow and blood pressure.

On the other hand, what R. F. Furchgott and colleagues discovered—that earned them the 1998 Nobel Prize in Medicine—is the *endothelium-dependent* ACh/NO/cGMP pathway. These two systems exist side-by-side and, to some extent, they interact.

There is yet a third distinct blood pressure control system, the "flow-mediated dilation" (FMD) system. Just as the term implies, blood pressure is also autoregulated to some degree by the flow-forces, or fluid dynamics, exerted in the vessel lumen on the vessel wall, as the blood proper courses through the vessels. These flow-forces are said to cause shear stress. The autoregulation of FMD is affected both by the flexibility of the vessel and its ability to distend, and that flexibility relies on its being clear of major plaque content as well as healthy endothelium function. Measuring vessel elasticity is detailed in a later chapter as a function of indices of reactive hyperemia (RH).

Many of the studies on the effects of the dietary supplements cited in a later chapter were selected because they showed that the regular consumption of the supplements improved endothelium function. In order to assess the effects of a given supplement, researchers have a number of conventional assay options including blood pressure, and blood tests for inflammation (nitrate/nitrite excretion) that will also be indicated in context, Doppler and ultrasound "imaging" of blood flow, and a marker of endothelium function, FMD, sometimes also called "flow-mediated (blood vessel) relaxation".

Blood flow rising in pressure in the vessel causes "shear stress," the fluid equivalent of *friction*. Shear stress also commonly occurs where blood flow changes velocity and course, such as at vessel branching. Shear stress is itself a factor in NO formation: NO bioavailability increase with shear stress is due to increased endothelial nitric oxide synthase (eNOS) formation. This is a homeostatic adjustment intended to maintain NO concentrations over variations in blood flow (see also[7]). The ability of blood vessels to self-regulate via flow-mediation is one of many factors affecting endothelial function that decline with age, and it potentiates impaired endothelium formation of NO/cGMP.

The FMD test depends on knowing how the endothelium responds to "challenge" to produce NO over the human lifespan. Any deviation from the age-related expected value is a marker of endothelium impairment. FMD data on how the endothelium responds to ACh challenge over the average lifespan can be used to evaluate the difference between expected response and observed response as a biomarker of endothelium health or impairment.[8]

FMD is now a common tool for inferring dysfunction thought to spring from coronary heart disease, for example. FMD correlates with coronary artery flow-mediated vasodilation, shown to predict long-term cardiovascular events. FMD has become a frequently employed measure of endothelial function. How is the FMD test conducted?

The FMD test may take several forms. In the most common form, ultrasound measures of blood flow are taken through the brachial artery. The brachial artery is then compressed for 5 minutes with a blood pressure cuff. After 5 minutes, the pressure is released and the diameter of the artery is assessed again by ultrasound. Endothelial dysfunction causes significantly reduced post-release vasodilation.[9]

In 2002, the *Journal of the American College of Cardiology* published a report titled "Guidelines for the ultrasound assessment of endothelial-dependent flow-mediated vasodilation of the brachial artery: A report of the International Brachial Artery Reactivity Task Force"[10,11]: high-frequency ultrasonographic imaging of the brachial artery to assess FMD was first developed in the 1990s to look at NO/cGMP formation and measure resulting blood vessel dilation as an index of endothelium function. The technique is non-invasive and can be repeated over time to study the outcome of different interventions that may affect blood vessel function (see also[12]). The specifics of measurement are not relevant here, and are therefore omitted. However, one may access them online.[13]

In a healthy person, approximately 30 seconds after the cuff is depressurized and blood begins to flow through the artery again, there is a surge, and then the vessel gradually returns to the previous state of tonus. In disease states involving the endothelium, both the time and the degree of return to previous state may be seen to change dramatically.

Figure 2.1 shows the time-course of blood flow and pressure recovery after cuff release in a healthy participant. The time course is characteristically different in persons with impaired endothelium function and/or in those with inadequate NO/cGMP formation (Sidebar 2.1).

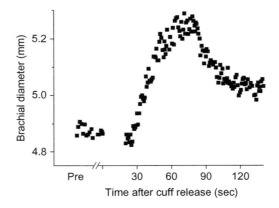

FIGURE 2.1   Time course of brachial artery FMD in a healthy individual. Source: *Corretti, M. C., Anderson, T. J., Benjamin, E. J., et al. (2002). Guidelines for the ultrasound assessment of endothelial-dependent flow-mediated vasodilation of the brachial artery: A report of the International Brachial Artery Reactivity Task Force.* J Am Coll Cardiol, *39, 257–265. Reproduced with kind permission from* Journal of the American College of Cardiology.

Sidebar 2.1

Caveat: First, for any given absolute change in the diameter, an initially larger baseline diameter results in a smaller percent change. Reporting absolute change in diameter will minimize this problem. Second, smaller arteries appear to dilate relatively more than do larger arteries. For studies where comparisons are made before and after intervention in the same individuals, percent change might be preferable if baseline diameter remains stable over time. However, the best policy may be to measure and report baseline diameter, absolute change, and percent change in diameter.

2.11 ARTERIES AND AGING

With age, the diameter of most arteries increases, whereas their elasticity declines. A study appearing in the *Journal of the American College of Cardiology* in 1996 titled "Aging-associated endothelial dysfunction in humans is reversed by L-arginine" describes several different "challenges" of the endothelium to form NO.[8] It consisted of exposing the endothelium to a substance known to result in measurable increased NO formation. The investigators chose two *endothelium-independent* vasodilators (papaverine and glyceryl trinitrate), and the *endothelium-dependent* vasodilator, ACh. Figure 2.2 shows the response to ACh.

The most obvious, and perhaps unsettling, information that we get from Figure 2.2 is that, other factors equal, the dilation response to ACh challenge decreases quite dramatically as we age. This is shown by the reduction in percent diameter-change from pre- to post-ACh challenge. This finding independently corroborates evidence presented in Chapter 3 that the endothelium actually forms less NO/cGMP as we age.

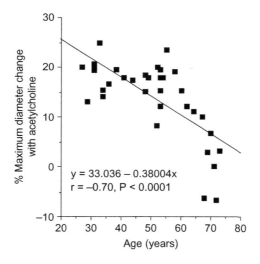

FIGURE 2.2   Scatterplot of correlation of age and maximal (percent of control values) diameter response to acetylcholine. Source: *Chauhan, A., More, R. S., Mullins, P. A., et al. (1996). Aging-associated endothelial dysfunction in humans is reversed by L-Arginine.* J American Coll Cardiol, *28, 1796–1804. Reproduced with the kind permission of* Journal of the American College of Cardiology.

Figure 2.3 shows two trend lines: The steep trend line is similar to the trend line is Figure 2.2. It is the pre-intervention response of coronary artery to ACh challenge as we age. After intervention with infusion of L-arginine, the age-related trend line is now much less steep: There is lesser decline in blood vessel response with aging, with the administration of L-arginine.

Following infusion of L-arginine, the slope of the regression line was less steep, resembling that in younger individuals, causing the authors to conclude that they had witnessed reversal of aging. In conclusion, they aver that there is a selective impairment of endothelium-dependent vasodilation with aging. This selective endothelial dysfunction may be reduced by intracoronary administration of L-arginine, the precursor of NO.[8]

Another study appearing in the *Journal of the American College of Cardiology* cautions that the effect of L-arginine administration on

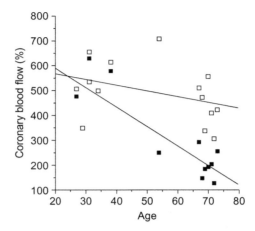

FIGURE 2.3 Scatterplot of correlation of age and peak (percent of control values) coronary blood flow response to ACh before (solid squares) and after infusion of L-arginine (open squares). *Chauhan, A., More, R. S., Mullins, P. A.*, et al. *(1996). Aging-associated endothelial dysfunction in humans is reversed by L-Arginine.* J American Coll Cardiol, *28, 1796–1804. Reproduced with kind permission from the* Journal of the American College of Cardiology.

ACh-induced vasodilation may be different in the coronary and forearm vasculatures in patients with coronary artery disease and hypertension. The participants in the Chauhan study cited above[8] were patients with "atypical chest pain" but normal coronary arteries without coronary spasm. Hypotheses for that difference are given by Hirooka, Egashira, Iamizumi, *et al.*[14] This issue of the difference in coronary and forearm vasculatures response will come up again: ED studies show that the penis cavernosae endothelium may exhibit impairment before endothelial impairment is observed elsewhere in the vascular system. Thus, differences in FMD in different regions of the vascular system beg the question: Is the endothelium seamless throughout the vascular system? If what one finds depends on where one looks, it becomes essential to know where to look.

FMD is a window into endothelium health because it is itself a physiological mechanism to control blood flow and blood pressure.

2.12 BLOOD PRESSURE/HYPERTENSION

In the journal *Circulation* article titled "Loss of flow-dependent coronary artery dilatation in patients with hypertension," there is this conclusion: Flow-mediated coronary artery dilatation is lost in hypertensive patients, and this may be the reason for impairment of normal blood flow observed in response to an increase in metabolic demand of the heart.[15]

2.13 CHOLESTEROL/ATHEROSCLEROSIS

In the *American Journal of Cardiology* article "Effect of a single high-fat meal on endothelial function in healthy subjects": Change in post-prandial FMD at 2, 3, and 4 hours correlated with elevation in 2-hour serum triglycerides. A single high-fat meal temporarily impairs endothelial function: a high-fat diet may be atherogenic independent of induced changes in cholesterol.[16]

In the *Journal of Clinical Investigation* article "Oral L-arginine improves endothelium-dependent dilation in hypercholesterolemic young adults": dietary supplementation with L-arginine significantly improves endothelium-dependent blood vessel dilation in young adults with elevated serum cholesterol, and this may help avert atherosclerosis.[17]

In the journal *Circulation* article "Atherosclerosis impairs flow-mediated dilation of coronary arteries in humans": The dose-dependent dilation in response to increasing blood flow is grossly impaired in the coronary arteries of patients with angiographic evidence of atherosclerosis. However, dilation in response to the endothelium-independent vasodilator, nitroglycerin, remains intact, therefore this failure of dilation is likely due to endothelial dysfunction.[18]

In the *Journal of the American College of Cardiology* article "Hyperglycemia rapidly suppresses flow-mediated endothelium-dependent vasodilation of brachial artery": Hyperglycemia in response to oral glucose loading rapidly suppresses endothelium-dependent vasodilation, probably through increased production of oxygen-derived free radicals. These findings strongly suggest that prolonged and repeated post-prandial hyperglycemia may play an important role in the development and progression of atherosclerosis.[19]

2.14 HEART DISEASE

In the journal *Circulation* article "Oral magnesium therapy improves endothelial function in patients with coronary artery disease": Oral magnesium therapy in coronary artery disease (CAD) patients is associated with significant improvement in brachial artery FMD and exercise tolerance.[20]

2.15 DIABETES

In the journal *Circulation* article "Acute hyperglycemia attenuates endothelium-dependent vasodilation in humans *in vivo*": Elevated blood sugar may contribute to the endothelial dysfunction in patients with diabetes mellitus (DM), as shown in Figure 2.4.[21] Ten minutes after baseline, patients with type-2 diabetes show lower brachial artery blood flow than do controls.

Many patients with type-2 diabetes, and those with impaired glucose tolerance (IGT), have fasting blood glucose within normal limits, and

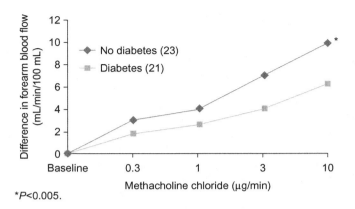

*P<0.005.

FIGURE 2.4 NO-mediated vasodilation is impaired in type-2 diabetes. Source: *Williams, S. B., Cusco, J. A., Roddy, M-A., et al. (1996). Impaired nitric oxide-mediated vasodilation in patients with non-insulin-dependent diabetes mellitus.* J Am Coll Cardiol, *27, 567–574. Reproduced with kind permission from* Journal of the American College of Cardiology.

hyperglycemia occurs only after meals. Because endothelial dysfunction has been shown to occur in patients with diabetes, and chronic hyperglycemia is implicated as a cause of endothelial dysfunction, would endothelial dysfunction occur when acute hyperglycemia is induced by oral glucose loading? This question was addressed in a study published in the *Journal of the American College of Cardiology.*

The study participants were admitted to the hospital for coronary artery disease. All were under 70 years old, and none of them had been previously diagnosed with diabetes. Fasting glucose levels of all the patients were less than 140 mg/dL, and a 75 g oral glucose tolerance test (OGTT) was performed to examine their risk factors for coronary artery disease. Both IGT and DM were diagnosed according to World Health Organization criteria.

Brachial artery FMD was measured in a fasting state and one hour and two hours after the administration of glucose (Figure 2.5). For procedure details, see Kawano, Motoyama, Hirashima, *et al.*, 1999).[19]

It is a complex and multipart study, but here are the findings that are most important to our understanding ED as a cardio-sexual impairment: First, Figure 2.5 tells us that the three participant groups, the normal glucose tolerance (NGT) group, the IGT group, and the DM group, had quite different FMD responses in the fasting stage even before the intervention trial even began.

The IGT group had the most profound change after glucose intervention, but the *caveat* must be kept in mind that change is affected by the starting point: the higher the starting point, the further it can drop. The DM group began lower, so it did not have as far to drop but, most importantly, two hours later, it was still at the lowest point. Figure 2.6 shows clearly that there is a near-linear relationship between FMD response and plasma glucose level.

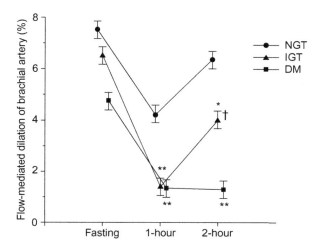

FIGURE 2.5   Effects of oral glucose loading on the FMD of the brachial artery. There was a significant difference in FMD among the three groups: DM, IGT, and NGT. Source: *Kawano, H., Motoyama, T., Hirashima, O.,* et al. *(1999). Hyperglycemia rapidly suppresses flow-mediated endothelium-dependent vasodilation of brachial artery.* J Am Coll Cardiol, *34, 146–154. Reproduced with kind permission from* Journal of the American College of Cardiology.

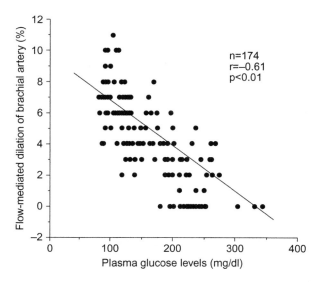

FIGURE 2.6   Correlation between the FMD of the brachial artery and the plasma glucose levels. Source: *Kawano, H., Motoyama, T., Hirashima, O.,* et al. *(1999). Hyperglycemia rapidly suppresses flow-mediated endothelium-dependent vasodilation of brachial artery.* J Am Coll Cardiol, *34, 146–154. Reproduced with kind permission from the* Journal of the American College of Cardiology.

The lower the plasma glucose level, the higher the FMD response. Conversely, the higher the level of plasma glucose, the lower was the FMD response. What's more, it is pretty clear that the decline is dose-dependent. However, the scatter-plot data suggest that a least-squares linear regression line may, in fact, obscure a curvilinear relationship. That being the case, the predictive reliability at the high and low ends is different from that in the middle. A test for significance of the difference between the obtained correlation coefficient (r_{xy}) and a hypothetical population coefficient of $rho_{XY} = 0$ prior to regression analysis could not be found.

2.16 EFFECT OF FLOW-MEDIATED DILATION VALIDATION OF THE CARDIOVASCULAR BENEFIT OF A MICRONUTRIENT

Archives of Biochemistry and Biophysics recently reported a clinical trial titled "Effect of cocoa/chocolate ingestion on brachial artery flow-mediated dilation and its relevance to cardiovascular health and disease in humans."[22] Numerous studies have found that high intake of dietary flavanols, such as those in cocoa/chocolate, is associated with lower rates of cardiovascular event and mortality. These effects include lower blood pressure, inhibition of platelet aggregation, lower risk of thrombosis, and lower incidence of inflammation. Figure 2.7 shows the effects of several dosages of cocoa on brachial artery FMD. Note that a high dose of flavanols (963 mg; squares) significantly increased FMD.[22]

The authors of the study conclude that increases in FMD after cocoa/chocolate ingestion appear to be dose-dependent, and increases in FMD are observed after consumption of larger quantities. The mechanisms underlying these responses are likely increased NO bioavailability, and vascular health benefits of cocoa/chocolate on the endothelium may underlie reductions in cardiovascular risks.

One may reasonably wonder why glucose has such a profound effect on FMD and endothelium function. It seems that the blood vessels are balking at increased free radicals, or so say the authors: "Hyperglycemia in response to oral glucose loading rapidly suppresses endothelium-dependent vasodilation probably through an increased production of oxygen-derived free radicals in humans. These findings strongly suggest that prolonged and repeated post-prandial hyperglycemia may play an important role in the development and progression of atherosclerosis."[19] In fact, the journal *Circulation Research* reported that high glucose concentrations increase the permeability of the endothelium and causes holes to form in it, thus impairing its ability to form NO.[23,24]

The *Journal of Nutrition* reported that post-prandial elevated glucose impairment in blood vessel endothelial function in healthy men was reduced by oxidizing cholesterol and decreasing NO bioavailability through an oxidative stress-dependent mechanism.[25]

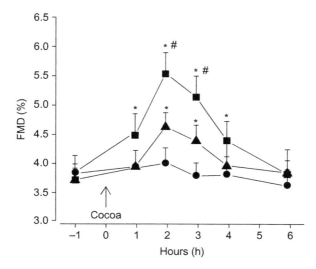

FIGURE 2.7 Time course of acute changes in brachial artery FMD on ingestion of flavanol-containing cocoa in medicated diabetic patients. Pre-ingestion FMD (Hours −1) was similar in all groups. After ingestion of the cocoa drinks containing either a medium (371 mg; triangles) or a high (963 mg; squares) dose of flavanols, FMD increased significantly, while ingestions of the control drink (75 mg flavanols; circles) had no effect. Effects observed after ingestion of a medium and high dose of flavanols appeared dose-dependent. Source: *Balzer, J., Rassaf, T., Heiss, C., et al. (2008), Sustained benefits in vascular function through flavanol-containing cocoa in medicated diabetic patients: A double-masked, randomized, controlled trial.* J Am Coll Cardiol, *51, 2141−2149. Reproduced with kind permission from* Journal of the American College of Cardiology.

What FMD tells us is that there are average age-related changes in blood vessels, as predicted from FMD observations of the response to "challenge" of the brachial artery. The assumption underlying all these observations is that FMD reflects endothelial function. There is no reason to doubt that. What's more, endothelial function is not simply a function of age, it is a function also of other factors that impact it as we age.

FMD has gained widespread use, and the examples above illustrate how this technique is used to clarify the role of factors that impact on endothelial function and impairment.

2.17 PENILE NITRIC OXIDE RELEASE TEST—FLOW-MEDIATED DILATION IN CAVERNOSA CONDUIT ARTERIES

A clinical trial reported in the *Journal des Maladies Vasculaires* (*Journal of Vascular Diseases*), aimed to assess the penile NO content of the conduit arteries supplying blood to the cavernosae, by measuring FMD: Patients with normal erections and those with various degrees of ED were administered the International Index of Erectile Function (IIEF) and the erectile activity index (IAE).

Two sub-groups emerged: Patients with mainly organic cause, and those with mainly psychogenic cause of ED. The diameter of one of the two cavernous arteries was measured by duplex-scan before and after a five-minute complete occlusion of penile vascular flow. The criterion retained was the maximum percentage of dilatation achieved within 45 to 90 sec after the release of the occlusion.

The mean percent increase in the control group was 65.2% vs. 34.9% in the ED group. The psychogenic subgroup with 68.2% was similar to the control group (65.2%), whereas the organic group had a highly significant lower average increase at 16.4%. Those in the control group and those in the psychogenic subgroup with additional arterial risk factors exhibited a less pronounced response, 53.7%, compared to those free of any risk factors, 80.3%.

Age, pre-occlusion arterial diameter, and DHEA-S, as well as IIEF and IAE scores, correlated with the magnitude of the reaction (DHEAS, testosterone, and several other androgens are used to evaluate adrenal function and to distinguish androgen-secreting conditions that are caused by the adrenal glands from those that originate in the ovaries or testes.)

The author of the study concludes that flow-dependent vasodilatation of the cavernous artery is extremely strong, stronger in control subjects and in those patients with predominantly psychological ED. An increase of 30% seems to be a critical level because only patients in the organic group were below that level. Thus, organic ED seems to be linked to the decrease of NO production by the cavernous bodies. It was further proposed that post-occlusive vasodilatation of the cavernous artery might be adopted as a non-invasive diagnostic and prognostic test in the study of ED. The test was called PNORT, for "penile NO release test."[26]

What such studies strongly imply is that the predictable progress of aging noted in the FMD studies may only be aging absent intervention to prevent accelerating it. These studies represent only a very small sample of clinical trials using FMD and variant forms as measurement of the factors that affect ED. There are other techniques that examine endothelium function, including Echo Tracking Sonography. However, in interpreting FMD observations in clinical trials, there are certain factors that need to be kept in mind.

For instance, the *Journal of the American College of Cardiology* reported a finding begging the question: Are there consistent subpopulation measures in FMD observations? The title of the study is "Reduced endothelium-dependent and -independent dilation of conductance arteries in African Americans." This is particularly noteworthy as it has been established that there is a higher incidence of hypertension in African Americans.

Figure 2.8 shows that there is indeed a higher incidence of hypertension in African Americans than in the comparison groups. Is that incidence related to differences in NO/cGMP formation? According to a study appearing in the *Journal of the American College of Cardiology*, African Americans show reduced responsiveness of arterial blood vessels to both endogenous and exogenous NO, compared to Caucasian Americans.

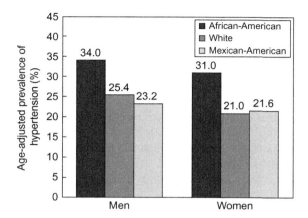

FIGURE 2.8   Age-adjusted prevalence of hypertension in African-American, white, and Mexican-American men and women. *Burt, V. L., Whelton, P., Roccella, E. J., et al. (1995). Prevalence of hypertension in the US adult population. Results from the Third National Health and Nutrition Examination Survey, 1988−1991.* Hypertension, *25, 305–313; and He, J., & Whelton, P. K. (1997). Epidemiology and prevention of hypertension.* Med Clin North Am, *81, 1077−1097. Reproduced with kind permission from* Medical Clinics of North America.

Based on brachial artery response to endothelium-dependent and -independent challenge, baseline diameter was found to be similar in blacks and in whites. FMD was significantly lower in black individuals compared with white individuals. Nitroglycerin-mediated (-dependent) dilation was also significantly lower in black individuals compared with white individuals. These findings point to racial differences in vascular function, and suggest a tentative physiological explanation for the increased incidence and severity of cardiovascular disease observed in African Americans.[27]

If one makes the case that African Americans have a somewhat lower response to NO/cGMP challenge and, not surprisingly, a higher prevalence of hypertension, then might it follow that one would expect to find a higher prevalence of ED in that subpopulation?

The *Journal of Sexual Medicine* reported a cross-sectional, population-based, nationally representative probability survey in the general community setting facilitated equivalent representation among US non-Hispanic white, non-Hispanic black, and Hispanic men aged 40 and older by using targeted phone lists to oversample the minority populations. The survey estimated prevalence of moderate or severe ED, defined as a response of "sometimes" or "never" to the question: "How would you describe your ability to get and keep an erection adequate for satisfactory intercourse?"

The estimated prevalence of ED was found to be 22.0% overall and 24.4% in African Americans. The odds ratio increased with increasing age and with co-morbid diabetes and hypertension.[28]

2.18 ECHO TRACKING SONOGRAPHY

The journal *Echosound in Medical Biology* reported a study where Echo Tracking Sonography was used to study the gender difference and change with age of the mechanical properties of the brachial artery. *Distensibility coefficients* (DC) relating to stretch, and *stiffness* and *compliance coefficients* (CC), were calculated in participants ranging in age between 9 and 82 years.

It was found that *stiffness* and *compliance* declined with age in both genders, but CC was higher in men. *Stiffness* increased and *distensibility* decreased with age in an exponential manner, without any gender differences (Figures 2.9(A) and 2.9(B)).[29]

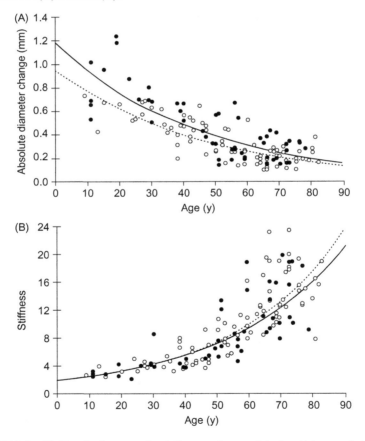

FIGURE 2.9 (A) The absolute age-related diameter change of the brachial artery in healthy men (•) and women (o); and (B) The DC of the brachial artery related to age in healthy men (•) and women (o). *Bjarnegård, N., Ahlgren, A. S., Sandgren, T., et al. (2003). Age affects proximal brachial artery stiffness; differential behavior within the length of the brachial artery?* Ultrasound in Med Biol, 29, 1115–1121. *Reproduced with kind permission from* Ultrasound Medicine and Biology.

Not surprisingly, as arteries narrow with age, they become less flexible and lose some ability to dilate. This phenomenon must be factored in with the other variables that account for progressive age-related rising blood pressure.

Figure 2.10 shows the relationship between decreased absolute diameter change of the brachial artery and aging. Absolute diameter change is larger in men than in women. Decrease in *distensibility* (DC) is a function of age. There is no gender difference. As we age, presumably other factors equal, arterial blood vessel walls thicken and the lumen narrows; their ability to distend with pressure changes decreases as they stiffen.

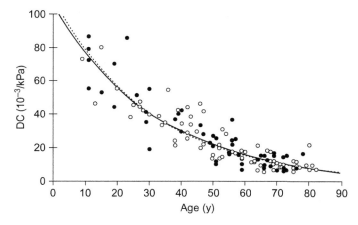

FIGURE 2.10   Age-related stiffness of the brachial artery in healthy men (•) and women (o). Stiffness increases with age. There are no gender differences. *Bjarnegård, N., Ahlgren, A. S., Sandgren, T., et al. (2003). Age affects proximal brachial artery stiffness; differential behavior within the length of the brachial artery?* Ultrasound in Med Biol, *29, 1115–1121. Reproduced with kind permission from* Ultrasound Med Biol.

2.19 ANKLE-BRACHIAL PRESSURE INDEX

While it is by no means definitive, there is a simple test procedure that offers a clue to the viability of the endothelium as it relates to ED. It is the test for the Ankle-Brachial Index (ABI), alternatively called the Ankle-Brachial Pressure Index (ABPI). It is doubtful that it is routinely performed in general-practice medicine, and yet it is highly recommended as the following excerpts from selected studies show.

A report in *Circulation* informs us that peripheral arterial disease (PAD), as determined by ABI testing, is an excellent measure of potential adverse cardiovascular prognosis. However, PAD frequently remains asymptomatic or undiagnosed. ED has been associated with subclinical atherosclerotic vascular disease, but whether ED is a marker of asymptomatic PAD is unknown. The authors contend that ED is a marker for previously undiagnosed PAD, and thus ED may identify men who would benefit from screening ABI testing.[30]

The journal *Atherosclerosis* reported a study titled "The association between erectile dysfunction and peripheral arterial disease as determined by screening ankle-brachial index testing." Men who had been referred for stress testing, and were without known peripheral artery disease (PAD), were prospectively screened for ED and PAD using the IIEF questionnaire and ABI, respectively. ED was defined by a score less than or equal to 25 on the ED domain of the IIEF, and PAD was defined as an ABI less than or equal to 0.9.

It was concluded that in men referred for stress testing, ED is an independent predictor of PAD as determined by screening ABI examination, and increasing severity of ED is associated with increasing prevalence of PAD. They proposed that men with ED might be targeted for routine screening ABI evaluation.[31] ED may be an early sign also of PAD, and should therefore be given medical attention.

2.20 PROCEDURE

The ABI, sometimes also called ABPI, is the ratio of the blood pressure in the lower legs to that in the arms, and it is calculated by dividing the systolic blood pressure at the ankles by the systolic blood pressures in the arms.[32] Using an electronic self-inflating cuff—in a supine position the blood pressure cuff is placed and inflated near the artery in the arm or in the leg as shown in Figure 2.11.

Inflation is allowed to cycle and when it ends, the BP reading is recorded. This procedure is followed for the right and left arm and leg. The index is calculated as follows (Eq. 2.1):

$$ABPI_{Leg} = \frac{P_{Leg}}{P_{Arm}} \tag{2.1}$$

where P_{Leg} is the systolic blood pressure of *dorsalis pedis* or posterior tibial arteries and P_{Arm} is the highest of the left and right arm brachial systolic blood pressure.

Interpretation: The higher systolic reading of the left and right arm brachial artery is generally used in the assessment. The pressures in each foot posterior *tibial* artery and *dorsalis pedis* artery are measured with the higher of the two values used as the ABI for that leg (Sidebar 2.2).[33]

Sidebar 2.2

ABI scores should be interpreted as follows:
- Greater than 0.90 = normal
- 0.71−0.90 = mild obstruction
- 0.41−0.70 = moderate obstruction
- Less than 0.40 = severe obstruction

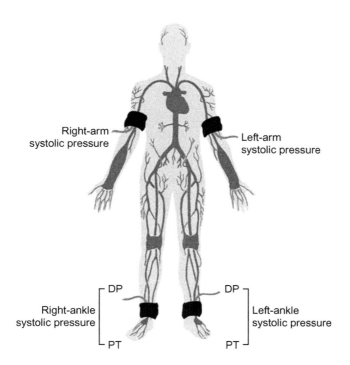

Right-arm
systolic pressure

Left-arm
systolic pressure

DP

DP

Right-ankle
systolic pressure

Left-ankle
systolic pressure

PT

PT

FIGURE 2.11   ABI. Blood pressure cuffs are placed on both arms and ankles. Systolic blood pressures are measured at the brachial artery and dorsalis pedis artery with the assistance of an ultrasound Doppler. Two measurements are taken from each arm and leg and then recorded as an average. The ABI is then calculated when the systolic blood pressure in the ankle is divided by the systolic blood pressure in the arm. *Reproduced with kind permission from Springerimages.com/Images/MedicineAndPublicHealth/5-10.1186_1750-4732-4-5-0.*

Alternatively, there is an ABI Chart that can be useful in interpreting ABI.[34] One can determine ABI by finding the appropriate obtained brachial blood pressure in the column on the left, and the corresponding leg blood pressure in the row at the top of the chart. The column/row intersection is the ABI.

Studies have shown the sensitivity of the ABPI to be about 90%, with a corresponding 98% specificity for detecting serious abnormal narrowing in major leg arteries, defined by angiogram.[35] The ABI typically reported in published clinical studies is not obtained with an automated blood pressure cuff, but rather with a sphygmomanometer or other medical-grade blood pressure device; observing systolic pressure is facilitated by a Doppler device.

Archives of Cardiovascular Diseases reports a study titled "Accuracy of ankle-brachial index using an automatic blood pressure device to detect peripheral artery disease in preventive medicine." The authors concluded that correlations between the automatic and Doppler methods were good in both the left and the right legs. In participants with abnormal indices, correlations with Doppler indices were good in both legs. The automatic method had good sensitivity (92%), specificity (98%), positive predictive value (86%), negative predictive value (99%), and accuracy (97%) compared to the Doppler method.[36]

There are, however, concerns about the trustworthiness of the index in certain clinical conditions where, for instance, there may be significant hardening of the arteries, and in some cases, in diabetes. They do not detract from the present message that ABI has shown to be a useful index to the severity of conditions that impair endothelial function and lead to ED. The following studies illustrate applications of ABI in conditions related to, or revealing, the endothelium impairment basis of ED.

A study titled "Ankle-arm index as a marker of atherosclerosis in the Cardiovascular Health Study" was conducted by the Cardiovascular Heart Study (CHS) Collaborative Research Group, and published in *Circulation*: An inverse response relationship was found between the ABI (here labeled "ankle-arm index") and the cardiovascular risk factors and subclinical and clinical CVD among older adult participants. The lower the ABI, the greater the increase in CVD risk. However, it was also the case that even those asymptomatic, with modest reductions in the ABI (0.8 to 1.0), appear to be at increased risk of CVD.[37]

Atherosclerosis reported, "The association between erectile dysfunction and peripheral arterial disease as determined by screening ankle-brachial index testing." It was thought that because ED is a marker for undiagnosed PAD, it might identify men who would benefit from screening ABI. They screened men referred for stress testing, and based on their findings concluded that ED is an independent predictor of PAD as determined by screening ABI examination, and what's more, increasing severity of ED is associated with increasing prevalence of PAD.[31]

The journal *Neuroendocrinological Letters* reported a study titled "Prevalence of risk factors of cardiovascular diseases in men with erectile dysfunction. Are they as frequent as we believe?" based on participants with no history of CVD, diabetes, or kidney disease, seeking medical treatment for "vascular ED." Based on medical, ABI, and CIMT evaluation (Sidebar 2.3), patients with vascular ED were found to have prevalence of CVD risk factors similar to that of the general population.[38]

ED has long been associated with heart disease risk factors and large-vessel lower extremity arterial disease (LEAD). A study published in *Atherosclerosis* is the first community-based study of the association between ED and small-vessel LEAD. Although ED has been associated with heart disease risk factors and large-vessel LEAD, no community-based studies have reported association between ED and small-vessel LEAD, despite the similar size of the arteries affected.

Sidebar 2.3

CIMT (Carotid Intima-Media Thickness) is a non-invasive ultrasound test recommended by the American Heart Association (AHA) and the American College of Cardiology (ACC) to screen for heart disease in apparently healthy individuals ages 45 or older.

A high-resolution B-mode ultrasound transducer records the combined thicknesses of the intimal and medial layers of the carotid artery walls: The relationship between degree of carotid artery atherosclerosis and that in a coronary artery is the same as in any two coronary arteries. Thus carotid atherosclerosis is an index of degree of coronary atherosclerosis in any individual. CIMT is an independent predictor of future cardiovascular events, including heart attacks, cardiac death, and stroke.

The *Journal of Hypertension* reports that hypertensive patients with ED have higher common CIMT and lower FMD of the brachial artery, a higher level of inflammatory markers associated with subclinical atherosclerosis, and impairment of blood vessel endothelium function.

Source: *Vlachopoulos, C., Aznaouridis, K., Ioakeimidis, N.,* et al. *(2008). Arterial function and intima-media thickness in hypertensive patients with erectile dysfunction.* J Hypertens, 26, 1829–36.

Is small-vessel LEAD associated with ED, and is this association independent of cardiovascular risk factors and medications? The investigators concluded that the severity of small-vessel LEAD is significantly and independently associated with the severity of ED. This may point to dysfunction in small-vessel arterial beds as well as in extremity large vessels.[39]

2.21 PERIPHERAL ARTERIAL TONE TECHNOLOGY

The Peripheral Arterial Tone (PAT) technique indirectly measures endothelial function after RH yielding measurable plethysmographic changes, usually in finger pulse volume. Plethysmography measures changes in volume. RH is the transient increase in blood flow that occurs following a brief period of blood vessel occlusion and ischemia. It can be produced by undoing a tourniquet, thus restoring blood flow. In a finger, for instance, it parallels blood vessel autoregulation (Sidebar 2.4).

A study reported in the *American Journal of Physiology* aimed to determine the contribution of endothelium-derived NO to RH in the forearm of normal subjects using the NOS synthase inhibitor, NG-monomethyl-L-arginine (L-NMMA). Forearm ischemia was induced by suprasystolic blood pressure cuff inflation for 5 min, and subsequent hyperemic flow was recorded for 5 min with venous occlusion strain-gauge plethysmography. The efficacy of NO blockade was tested by comparing the dose−response relationship to the endothelium-dependent agonist, ACh (3, 10, and 30 mg/min), before and after intra-arterial infusion of up to 2,000 mg/min of L-NMMA.

Sidebar 2.4

Autoregulation is a manifestation of local blood flow regulation. It is the intrinsic ability of an organ to maintain a constant blood flow despite changes in perfusion pressure. For example, if perfusion pressure declines in an organ (e.g. by partially occluding the arterial supply to the organ), blood flow initially falls, and then returns toward normal levels over the next few minutes. This autoregulatory response is intrinsic to the organ, and it occurs in the absence of neural and hormonal influences. When perfusion pressure (arterial minus venous pressure, PA−PV) initially decreases, blood flow (F) declines because of the following relationship between pressure, flow, and resistance:

$$F = (P_A - P_V)/R$$

When blood flow falls, arterial resistance (R) falls as the resistance vessels dilate. As resistance decreases, blood flow increases despite the presence of reduced perfusion pressure.

L-NMMA produced a significant downward and rightward shift in the dose−response relationship to ACh, and a 39% reduction in response to the maximum dose. With L-NMMA, peak hyperemic flow was reduced 16%, and the minimum forearm vascular resistance was increased by 22.8%. Total hyperemia, calculated from the area under the flow vs. time curve, at 1 min and 5 min after cuff release, was 17% and 23% lower following L-NMMA. It was concluded that endothelium-derived NO is involved in both RH and in maintenance of the hyperemic response following ischemia in the forearm.[40]

The non-invasive technique was also reported in the *American Journal of Cardiology* to show impaired endothelial function by decreased NO bioavailability in a way comparable to, and as an alternative to, intra-arterial infusions of vasoactive agents in the assessment of resistance vessel endothelial function in forearm circulation.[41]

2.22 THE AUGMENTATION INDEX

The AIx is based on blood pulse-wave reflection, and is an accepted measure of arterial stiffness and risk factor for cardiovascular disease.[42] The AIx is commonly accepted as a measure of the enhancement (augmentation) of central aortic pressure by a reflected pulse wave (shown in the second, smaller waveform in Figure 2.12).

The AIx reflects endothelium function: This study appearing in the *Indian Heart Journal* sought to establish the correlation of non-invasive estimation of arterial wall stiffness by pulse-wave velocity (PWV) and its association with endothelial dysfunction in subjects at higher risk for atherosclerosis. Men and women patients (mean age 51 years) included those with

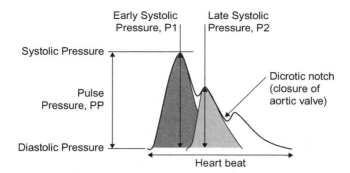

FIGURE 2.12   Peripheral augmentation index (pAI) is defined as the ratio of late systolic pressure (P2) to early systolic pressure (P1): pAI = P2/P1. *Reproduced with kind permission from Uscom (http://www.uscom.com.au/) For information on the Pulsecor Cardioscope, visit http://www.pulsecor.com/augmentation-index.html.*

hypertension, type-2 DM, concomitant type-2 DM and hypertension, and primary dyslipidemia without DM and hypertension.

PWV was measured by the Vascular Profiler 1000 (VP-1000) waveform analysis and vascular evaluation system, an automated, non-invasive, screening device. Endothelial function was assessed by brachial artery FMD. Brachial-artery diameter was measured on B-mode ultrasound images, with the use of a 7.0 MHz linear-array transducer.

Results: Mean brachial artery PWV on the right extremity was 1,699 cm/s, and on the left, 1,694 cm/s. Mean FMD was 3.6 ± 8.4%. Mean brachial artery PWV in the right and left extremities, and the higher value of brachial artery PWV of the two extremities, showed a significant negative correlation with brachial artery FMD. Mean heart-brachial PWV also showed a significant negative correlation with brachial artery FMD. Mean arterial stiffness was 36.2 ± 22%. Arterial stiffness in the right extremity, and the higher value of the two extremities, showed a significant negative correlation brachial artery FMD. The authors contend that the measures relating PWV and brachial artery FMD and arterial stiffness make the technique a valuable diagnostic tool.[43]

The AIx in hypertension: A study appearing in *Hypertension Research* aimed to evaluate the reproducibility of PWV and the AIx as indices of arterial stiffness in hypertension. Brachial blood pressure, brachial-ankle PWV (baPWV), and carotid Augmentation Index (cAIx) twice: First at the baseline, and then 4 weeks after the baseline.

The mean intraobserver−intersession difference was 29.0 cm/s with (SD = 201.6 cm/s for baPWV), and 0.5% (SD = 5.9% for cAIx). Bland-Altman plots demonstrated the good reproducibility of baPWV and cAIx. The average of the first and the second measurements of both baPWV and cAIx were significantly correlated with age, systolic BP (SBP), and pulse pressure; however, these factors were not correlated with each other.

The cAIx was correlated with height, heart rate (HR), total cholesterol, and low-density lipoprotein cholesterol (LDL-C).

Age, SBP, and HR were shown to be significant independent predictors of baPWV, while height, SBP, HR, and LDL-C were significant independent predictors of cAIx. Both PWV and AIx were reproducible, and their associated risk factors were different. The authors concluded that, "Automated simultaneous measurement of these arterial stiffness indices may be useful for risk stratification of hypertensives."[44]

The AIx in ED: A study appearing in *Andrologia* aimed to determine the diagnostic usefulness of RH and AIx as evaluated by PAT in men with ED of undifferentiated origin. Diagnosis included dynamic penile duplex ultrasound (PDU) and PAT device. Patients were assigned by vascular and nonvascular etiology. Control participants were assigned by presence or absence of vascular risk factors, and they were studied in a separate analysis.

It was concluded that an increased AIx, but not an impaired RH-PAT, is present in men with vascular ED independently of vascular risk factors (VRFs), and may represent an early detection of vascular impairment that may precede endothelial dysfunction in populations at low risk for developing vascular ED.[45]

AIx and hyperlipidemia: Patients with hypercholesterolemia have a higher central pulse pressure and stiffer blood vessels than matched controls, despite similar peripheral blood pressures. These hemodynamic changes may contribute to the increased risk of cardiovascular disease associated with hypercholesterolemia, and assessment may improve risk stratification.

In a study published in the *Journal of the American College of Cardiology*, radial artery pressure waveforms were recorded, and corresponding central waveforms were generated using pulse-wave analysis. Central pressure, AIx (a measure of systemic stiffness), and aortic PWV were determined.

Although no significant difference in peripheral blood pressure between the two groups was observed, central pulse pressure was significantly higher in the group with hypercholesterolemia (37 ± 11 mm Hg vs. 33 ± 10 mm Hg). AIx was also significantly higher in the patient group with hypercholesterolemia ($24.8 \pm 11.3\%$ vs. $15.6 \pm 12.1\%$), as was also the estimated aortic PWV. Age, short stature, peripheral mean arterial pressure, smoking, and LDL-C correlated positively with AIx, while there was a negative correlation with heart rate and male gender.

The authors concluded that patients with hypercholesterolemia have a higher central pulse pressure and stiffer blood vessels than matched controls despite similar peripheral blood pressures, and that these hemodynamic changes may contribute to the increased risk of cardiovascular disease associated with hypercholesterolemia.[46]

Both the finger-tip plethysmography (PAT) technique and the AIx that can be calculated from the same sampling technique have seen considerable application. Here are more sample references to applications of the methods.

2.23 PERIPHERAL ARTERIAL TONE METHODOLOGY

For details see:

- Assessing endothelial vasodilator function with the Endo-PAT 2000.[47]
- Assessment of endothelial function (nitric oxide) at the tip of a finger.[48]
- Variability of peripheral arterial tonometry in the measurement of endothelial function in healthy men.[49]
- Test–retest reliability of pulse amplitude tonometry measures of vascular endothelial function: Implications for clinical trial design.[50]
- Evaluation of the EndoPAT as a tool to assess endothelial function.[51]

. . . and endothelium—

- Assessing endothelial vasodilator function with the Endo-PAT 2000.[47]
- Nitric oxide synthase-dependent vasodilation of human subcutaneous arterioles correlates with non-invasive measurements of endothelial function.[52]
- Evaluation of endothelial function using finger plethysmography.[53]
- Assessment of endothelial function using digital pulse amplitude tonometry.[54]
- Evaluation of the EndoPAT as a tool to assess endothelial function.[51]
- Assessment of vascular endothelial function with peripheral arterial tonometry information at your fingertips?[55]

. . . and hypertension—

- AIx association with RH as assessed by Peripheral Arterial Tonometry in hypertension.[56]

. . . and Type-2 diabetes—

- Differential effects of low-carbohydrate and low-fat diets on inflammation and endothelial function in diabetes.[57]
- Endothelial dysfunction in type-2 diabetic patients with normal coronary arteries. A digital RH study.[58]

. . . and atherosclerosis—

- Non-invasive identification of patients with early coronary atherosclerosis by assessment of digital RH.[59]

- Small LDL-C concentration is a determinant of endothelial dysfunction by Peripheral Artery Tonometry in men.[60]

... and cardiovascular impairment—

- Non-invasive vascular function measurement in the community: cross-sectional relations and comparison of methods.[61]
- Endothelial function in a cardiovascular risk population with borderline ankle-brachial index.[62]
- Augmentation index derived from peripheral arterial tonometry correlates with cardiovascular risk factors.[55]

... and metabolic syndrome—

- Plasma leptin levels and digital pulse volume in obese patients without metabolic syndrome - A pilot study.[63]

... and erectile function—

- Application of digital pulse amplitude tonometry to the diagnostic investigation of endothelial dysfunction in men with erectile dysfunction.[45]
- A spontaneous, double-blind, double-dummy cross-over study on the effects of daily vardenafil on arterial stiffness in patients with vasculogenic erectile dysfunction.[64]
- Chronic administration of Sildenafil improves markers of endothelial function in men with type-2 diabetes.[65]
- The endothelial-erectile dysfunction connection: An essential update.[66]
- Assessment of endothelial function in the patient with erectile dysfunction: an opportunity for the urologist.[67]

2.24 CAN WE MEASURE NITRIC OXIDE PRODUCTION DIRECTLY?

To gain knowledge of endothelium health and function, especially in connection with erectile function, we look at derived measures such as FMD and various other techniques. These tell us by inference the capability of the endothelium to deliver NO/cGMP when and where needed. These techniques also help us to look at the mechanical structure-status of blood vessels to see whether they can respond and comply with messages from the endothelium and/or flow characteristics to dilate as needed.

The answer to an additional question would be very useful: Is the body actually forming sufficient NO (adequate volume)? There are a number of different ways that biosynthesis of NO could be inadequate:

- First, there might be an insufficiency of dietary intake of L-arginine and/or nitrates to trigger cellular NO release. This is unlikely, but see arginine paradox.

- Second, the enzyme eNOS has become unlinked from the process of forming NO. This is more likely the case.
- Third, is the endothelium too damaged to synthesize a continuous and adequate volume of NO? This is all too often the case.

There are several ways to measure NO formation directly. The earliest methods measured concentration in breath because it was shown that in ordinary circumstances L-arginine supplementation raised exhaled NO.[68] It is, parenthetically, also associated with an increase in concentration of nitrate in plasma. However, it did not prove readily feasible to disentangle the proportion of NO produced by immune system cells in the lungs in the case of various types of pulmonary infections from the proportion produced by the immune system and the endothelium elsewhere in the body.

There are several articles and reviews that address measurement of NO:

- Measurement of nitric oxide in biological models.[69]
- Different plasma levels of nitric oxide in arterial and venous blood.[70] This study demonstrated that endothelium derived NO comes primarily from arteries and not from veins.
- Measuring nitric oxide production in human clinical studies.[71]
- L-arginine increases exhaled nitric oxide in normal human subjects.[68]

There is, however, a method under development that seems to bear some promise for a reasonable estimate of endothelium-derived NO production. It relies on the fact that NO is rapidly metabolized to nitrite in the body, and further converted to nitrate in body fluids. The nitrate is then converted to nitrite because the conventional method of assay can only detect nitrite. Nitrate/nitrite appears as a marker of inflammation in fluid such as urine, saliva, etc.

Interpretation of results from such tests is not invariably clear. For instance, an article in *Circulation* cites a few examples: Diet is a potential source of nitrate and nitrite and patients need to consume a diet low in green leafy vegetables for several days before measurements. Intestinal bacteria can be a source of plasma nitrate; plasma NO in the blood can result from the activity of three different NOS enzymes; and there are a number of other caveats.[72] Nevertheless, many authorities recognize the value of assessing endothelial NO formation, and especially its impact and conveyance in blood.[73]

Dr. N. S. Bryan at The Institute of Molecular Medicine, the University of Texas School of Medicine in Houston, is pioneering the use of salivary nitrite as a marker of human NO status. Dr. Bryan tells us that up to now "... there have not been any new developments in the use of NO biomarkers in the clinical setting for diagnostic or prognostic utility. In fact, NO status is still not part of the standard blood chemistry routinely used for diagnostic purposes. This is simply unacceptable given the critical nature of NO in many disease processes and new technologies should be developed in humans." Dr. Bryan proposes sampling salivary nitrite as an accurate representation of total body NO production/availability with Nitric Oxide Diagnostics™ Test Strips (see Neogenis® Labs).[74]

In fact, in a review titled "Nitric oxide and its metabolites in the critical phase of illness: rapid biomarkers in the making," appearing in *Open Biochemistry Journal*, the authors suggest focusing on assessing NO and its oxidative metabolites, nitrite and nitrate, because these are readily detectable in body tissues and in fluids, and because they are linked to many pathophysiological processes. It is now relatively easy and inexpensive to analyze these with methods such as high-performance liquid chromatography (HPLC) and chemiluminescence.

The review emphasizes diagnostic rapidity for promoting quality of acute care. The authors present a strong case for developing these biomarkers more as point-of-care assays with potential of *color gradient test strips* for rapid screening of disease entities in acute care and beyond.[75]

For more information, visit the Neogenesis Labs website.[76] This site provides valuable information on dietary supplement products tailored to promote NO formation to support healthy cardiovascular and erectile strength. In addition, Neogenis Labs developed and markets the Nitric Oxide Diagnostics Test Strips that are simple to use: Saliva is applied to a strip and its coloration is then compared to a color code on the side of the strips container to yield a measure of NO concentration.[77]

Note: A high reading (NEO Optimal) may reflect a number of factors, including recently consuming high ʟ-arginine foods such as meats, fish, nuts or beans, and/or high nitrate foods such as green leafy vegetables, or beans, beets, and carrots. However, most importantly, it may also indicate that one is suffering from a lung, bowel, or other form of inflammatory disease. Immune system cells go into overdrive production of NO in serious infections and inflammation.

On the other hand, a low reading may be a clear indication that endothelial NO formation is inadequate. Because of the implication for the health risks, hypertension, atherosclerosis, CVHD, and diabetes, leading to ED, the value of discovering that one has a LOW NO reading cannot be underestimated. Such a reading may be interpreted to mean that there may be an underlying problem affecting endothelial function and health in general.

2.25 MEASURING PENILE RIGIDITY AND NOCTURNAL ERECTION

In an article titled "Evidence-based assessment of erectile dysfunction," in the *International Journal of Impotence*, Dr. G. A. Broderick contends that we need impotence testing because it is the clinician's obligation to establish the etiology of ED as either end-organ vascular failure, neurologic dysfunction, or psychosexual dysfunction, and to classify the severity of that dysfunction, and select a therapy that is not only acceptable to the patient, but also effective in treating his disorder.[78]

History records a very small, overlooked, mostly discarded item that should have heralded things to come in treatment of ED: In 1985, a brief case-history appeared in the journal *Annals of Internal Medicine* titled "Transdermal nitrate, penile erection, and spousal headache." The authors described the case of a man with a long history of cardiovascular and heart disease and, not surprising, a long history of ED, who suffered also from angina pectoris, for which he was prescribed a nitrate chest patch.

The patch gave him headaches, and so he thought logically enough that by distancing it from his head that might alleviate the headaches. And so one day, experimenting with different locations on his body, he applied the used patch to his penis and he rapidly attained erection.

The description of the case report in part reads as follows: "Within 5 minutes, he had a semi-rigid erection and became sexually aroused. Sexual intercourse with his wife followed. Several minutes later, she wondered why she had the worst headache she'd ever had in her life. The patient then told her of his experiment, and its apparent success. His wife was not impressed and strongly discouraged any more investigation in this area."[79]

The authors did not follow up on this potentially serendipitous event, and instead they concluded that it is doubtful "that further research in this area will be done." It was not for some years that chemically induced erection appeared on the scene, and it did so with a proverbial bang.

2.26 THE BRINDLEY "DEMONSTRATION"

The *nitrate-causes-erection* phenomenon did not immediately lead to treatment for ED. Today, it would be obvious to medical researchers that nitrates are NO-donors: In any case, VIAGRA® is not an NO-donor, it is an NO-preserver.

The clue that erection must be thought of as principally a vascular phenomenon was an accidental observation: Patients given papaverine to lower blood pressure were noted to get erections. Subsequent papaverine injection directly into the penis produced the same results.

In 1983, Sir Giles Brindley, a British physiologist, presented the results of his study of intra-cavernosal papaverine injections to the Urodynamics Society in Las Vegas. Laurence Klotz, a urologist and Editor-in-Chief of the *Canadian Journal of Urology*, attended the presentation and related his experience in the *British Journal of Urology International* in 2005 in an article titled "How (not) to communicate new scientific information: a memoir of the famous Brindley lecture," abstracted here.

Prof. Brindley had performed a series of experiments on himself wherein he had injected vasoactive substances such as papaverine and phentolamine into his penis. He now presented a series of slides, photos of his penis, in various states of tumescence after injection with a variety of doses of phentolamine and papaverine. He explained that to demonstrate the effects of these vasoactive substances, he had injected his penis with papaverine in his hotel room before coming down to the lecture hall, and that was why he was wearing a loose-fitting track suit.

Next, he pulled tight his track suit and stepped around the podium to display the results of the injection, but proclaimed dissatisfaction with the demonstration, so he dropped his trousers and displayed his erect penis. Albeit unusual, and by some standards shocking, the demonstration effectively convinced urologists to consider the vasculogenic basis of ED and vasoactive substances such as papaverine as treatment.

The chronicler reports that a papaverine-induced erection would typically last 1−2 hours and the treatment could be safely used once a day, about 3 times a week, the main complication being priapism.[80] Dr. Brimley demonstrated one way of evaluating erectile strength. There are others that are somewhat more in keeping with conventional scientific methods.

Defining such criteria is not a simple matter. ED was long held to mean the failure to attain an erection sufficient for penetration (of a vagina, presumably), and to maintain it for some time of satisfactory intercourse. Sufficient? Barely sufficient? Maintain how long?

There are now several subjective rating scales for "sexual satisfaction," and the erectile function−dysfunction continuum that evaluate these in more or less conventional terms. One example among many is one often encountered in sex research, The IIEF, said by the authors to be a "multidimensional scale for assessment of erectile dysfunction."

2.27 THE INTERNATIONAL INDEX OF ERECTILE FUNCTION

The IIEF is a multidimensional scale that can be used to evaluate ED. It addresses the most relevant aspects of male sexual function, such as erectile strength, orgasm, desire, satisfaction with intercourse, and overall satisfaction. This questionnaire can be readily self-administered either in research or in clinical settings, and it has the requisite sensitivity and specificity needed to detect treatment-related changes in patients with ED. The test was translated and validated for use in more than ten different languages.[81]

The IIEF classifies the severity of ED into five categories stratified by score:

- No ED.[26−30]
- Mild.[22−25]
- Mild to moderate.[17−21]
- Moderate.[11−16]
- Severe.[6−10]

The questionnaire is composed of 15 items spanning the domains of male sexual dysfunction (ED, orgasmic function, sexual desire, ejaculation, intercourse, and overall satisfaction), and there is an abbreviated format of five questions that can also be used: the Sexual Health Inventory for Men (SHIM). The IIEF is said to have met conventional goals for internal consistency, validity, and reliability. Significant changes between baseline and post-treatment scores were observed across all five domains in the treatment responder cohort, but not in the treatment of non-responder cohort.[81] The IIEF is widely used and, as shown below, in conjunction with objective measures, used to evaluate the adequacy of sexual functioning.

The *International Journal of Impotence Research* reported a study aimed at comparing measures of nocturnal penile tumescence and rigidity (NPTR) with the erectile function (EF) domain score of IIEF, to see how well it serves to diagnose the severity of ED. No correlation was found between IIEF-EF domain scores and NPTR measures in patients tested. However, if IIEF-EF domain scores were normal, NPTR measures were also normal. Thus, if the initial IIEF-EF domain scores are normal, then it is not necessary to perform NPTR testing as described below.[82]

The *International Journal of Impotence* reported a study titled "Evidence based assessment of erectile dysfunction." It proposes a number of different approaches to evaluating ED requiring the medical practitioner to differentiate between:

- "end organ vascular failure" (meaning blood vessels in the penis),
- "neurologic dysfunction" (meaning nerves part of the sexual stimulation scheme), and
- "psychosexual dysfunction" (meaning emotional issues, absent evidence of the other two possibilities),

in order to classify the severity of dysfunction and select a therapy that is acceptable not only to the patient, but also addresses the actual problem.[78]

The article names the most common diagnostic tests for ED, namely presence or absence of nocturnal erection, the *nocturnal penile tumescence* (NPT) test, because it upholds both nerve and penile blood vessel integrity. Conversely, absence of nocturnal erection does not help to identify the cause or severity of impotence. Other tests and criteria are also mentioned, but the NPT is the most relevant to our concerns. This is a major point: presence of nocturnal tumescence eliminates an organic cause of presenting ED.

2.28 NOCTURNAL PENILE TUMESCENCE

Clinical treatment of ED mandates objective clinical evaluation of subjective complaints. One approach to determining whether the plumbing can function in the face of complaints of ED is to ascertain whether the individual has nocturnal erections, aka nocturnal tumescence. The theory here is that if one experiences nocturnal tumescence, then the problem cannot be entirely due to physical incapacity.

2.29 NOCTURNAL PENILE TUMESCENCE AND RAPID EYE MOVEMENT SLEEP

It was not until the 1950s that we began to learn that sleep is not simply a passive cycle in our daily life. Our brain is actually quite active while we sleep: There are five consecutive distinct stages of sleep, the fifth being *rapid eye movement* (REM) sleep. These stages progress in a cycle from stage 1 to REM sleep, then the cycle starts over again with stage 1. We spend almost 50% of our total sleep time in stage 2 sleep, about 20% in REM sleep, and the remaining 30% in the other stages. The first REM sleep period usually occurs about 70 to 90 minutes after we fall asleep.

On entering the REM sleep stage, breathing becomes more rapid, irregular, and shallow; eyes jerk rapidly in various directions, and limb muscles become temporarily paralyzed. Heart rate increases, blood pressure rises, and men develop penile erections (in women, comparable clitoral engorgement). A complete sleep cycle takes 90 to 110 minutes on average (based on Brain Basics: Understanding Sleep, NIH Publication No. 06-3440-c).

The attribution of ED in many men to emotional issues may have been the reason why the seeming paradox of erection by night, and impotence by day, did not lead to the conclusions that these are incompatible phenomena and deserve further elucidation. It is the advent of VIAGRA that largely unlinked ED from "emotional issues" and allowed a renewed look at the significance of the phenomenon: What nocturnal erection tells us is that the plumbing is sound and should, with a little help, work as intended.

While nocturnal erection (NPT) is time-linked to the REM phase of sleep, the consensus is that we have no idea what triggers erection at that point. The non-REM (nREM) phase also coincides with dreaming, and that may be the last hope of the psychoanalytic tradition to attribute it to erotic dreams. There is no evidence to support that notion.

More likely than not, erection corresponds to a physiological factor that peaks during REM sleep. One possibility is the sleep rhythm of ACh, as noted in Figure 2.13.

Nocturnal erections occur during phases of sleep, therefore they most often go unnoticed. However, there are several ways of evaluating them, and they are known collectively as the "nocturnal (penile) tumescence test." The journal *Endocrinology Review* reported the following defining criteria:

- A total night erection time greater than 90 min.
- An increase in penis circumference in excess of 2 cm.

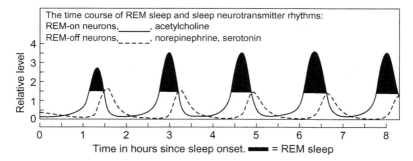

FIGURE 2.13   Schematic of the night's course of REM sleep in humans showing the occurrence and intensity of REM sleep as dependent upon the activity of populations of "REM-on" (= REM promoting neurons), indicated by the solid line. As the REM-promoting neuronal activity reaches a certain threshold, the full set of REM signs occurs (black areas under curve indicate REM sleep). The neurotransmitter acetylcholine is thought to be important in REM sleep production. Source: *McCarley, R. W. (2007). Neurobiology of REM and NREM sleep.* Sleep Med, *8, 302–330. Reproduced with kind permission from* Sleep Medicine.

- A change in circumference of 16 mm or 80 percent of a full erection is thought to reflect a sufficient degree of penile rigidity for vaginal intromission.
- Subsequently, a penile buckling pressure of 100 mm Hg using the manual tonometer, or 100 Penrig (unit used for the electronic dynamometer), was found to provide a more accurate assessment of the degree of penile rigidity required for vaginal penetration than the percentage change in circumference.
- A buckling pressure less than 60 mm Hg is thought to be inadequate for vaginal penetration.[83]

A number of means are available to determine that NPT actually occurs. Medical approaches include use (by prescription) of the RigiScan® (see below), and are ordinarily performed under medical supervision.

2.30 AGE-RELATED DECLINE IN NOCTURNAL PENILE TUMESCENCE

The aim of a study published in the *International Journal of Impotence Research* was to evaluate the quality of erectile episodes as a function of aging. Patients in the 20 to 71 years age range, initially reporting ED, underwent two nights of NPTR measurement with the RigiScan. Approximately three in four men were found to have normal NPTR.

The investigators evaluated the number of *normal* erectile episodes (penile tip rigidity greater than 60% for more than 10 min duration), RAU Tip, RAU Base, TAU Tip, TAU Base, average event rigidity of tip (percent), average event rigidity of base (percent), and duration of erectile episodes greater than or equal to 60% for 10 min in five age groups (group I, less than 30 years old; group II, 30 to 39 years old; group III, 40 to 49 years old; group IV, 50 to 59 years old; and group V, 60 years old or older). Statistically significant differences were observed for all of the NPTR parameters with respect to age.

Aging reduced the quality of nocturnal erections, especially in men older than 50 years. The number of erectile episodes declined from a steady average of 2.5 per night up to the age of 50, and thereafter it declined to an average of a little more than one per night by age 60 and beyond.[84]

The age-related decrease in frequency and other characteristics of NPT parallels age-related worsening of cardiovascular disorders such as hypertension, atherosclerosis, etc., that likewise parallel the age-related increase in prevalence of ED.

Men who experience NPT often also experience erection when napping during the day. According to the *Journal of Urology*, a study of men with ED found that those who did not have NPT also did not have erections while napping during the day, whereas those men who had NPT also had erections while napping during the day.[85]

2.31 RELIABILITY OF NOCTURNAL PENILE TUMESCENCE TESTING

How reliable is the NPT test? This question was addressed in a study reported in the *Journal of Urology* titled "Nocturnal penile tumescence and rigidity monitoring in young potent volunteers: reproducibility, evaluation criteria and the effect of sexual intercourse." The method of choice employed the RigiScan device. The aim of the study was mostly to ascertain the reproducibility of nocturnal penile tumescence rigidity evaluation measures. The investigators concluded that at least two nights of recording are needed to evaluate NPT and rigidity recordings.[86]

There are many other methods of recording both daytime erection and NPT. In general, what do they tell us about erectile function and dysfunction and... caveat—What don't they tell us? The journal *Sleep Medicine Review* reported a study titled "Sleep-related erections: Clinical perspectives and neural mechanisms." The investigators measured erectile strength—as rigidity, in this case—using a Digital Inflection Rigidometer (DRI; presumably UROAN®[87]) (Sidebar 2.5).

Another study employing the penile axial rigidity measures obtained with the DRI method, also published in *Sleep Medicine Review*, addresses "Sleep-related erections: Clinical perspectives and neural mechanisms." The legend to their first illustration (not shown) tells us about a normal sleep-related erection (SRE) pattern for a 56-year-old man with psychogenic ED. Rigidity measured during the first SRE episode was 880 g.

Sidebar 2.5

Rigidometer yields values for "axial rigidity" in grams (g) interpreted with the following criteria:

- Axial rigidity below 400 g; insufficient axial rigidity.
- Axial rigidity of 400–500 g; borderline rigidity.
- Axial rigidity of 500–1000 g; sufficient axial rigidity.
- Axial rigidity + 1000 g; optimum rigidity.

Axial rigidity in grams can be used as a common measure that indicates degree of ED. Validation of the DIR was carried out by Virag et al. (2001).[a]

Caveat—The *Journal of Andrological Sciences* cautions that penile axial rigidity as measured by Digital Inflection Rigidometry (DRI) varies significantly with change of a patient's position.[b]

Sources: *(a) Virag, R., Colpi, G., Dabees, K. L., et al. (2001). Multicentric international study to validate the digital inflection rigidometer in the diagnosis of erectile dysfunction. Geneva Foundation for Medical Education and Research (GFMER), WHO Collaborating Center in Education and Research in Human Reproduction, Geneva. American Urological Association Convention, Anaheim, California, June 1–6, 2001. (b) Dehó, F., Saccó, A., Fabbri, F., et al. (2009). Digital inflection rigidometry and penile dynamic colour doppler ultrasound: analysis and correlations. Journal of Andrological Sciences, 16, 108–111.*

A second panel shows the sleep and erection pattern for a 64-year-old man with organic ED and co-morbid diabetes. Although SREs were well coordinated with REM sleep, circumference increase and maximum rigidity were below normal. Furthermore, SREs did not persist throughout the REM sleep episode, and rigidity measured during the patient's best erection was 220 g."[88] The text makes the point and the figure is not relevant, and it is therefore not shown here:

The 56-year-old man with "psychogenic" ED had a sleep-related erection (SRE/NPT) rigidity measure of 880 g. Because this indicates *sufficient axial rigidity*, it may be assumed that the diagnosis of "psychogenic" ED is based on the SRE/NPT. The description of the 64-year-old patient whose "best erection" weighed in at 220 g is well within the range of *insufficient axial rigidity*, and that is well within our understanding of the impact of the "co-morbid" diabetes.

The concept of "co-morbid" underscores the notion that ED is considered a disease entity on its own right, rather than a symptom of diabetes in this case. This monograph disputes that contention, and underscored the need to see the connection between diabetes and endothelial impairment. What do we learn from this example of the use of the Rigidometer in connection with ED and diabetes "co-morbidity"?

Figure 2.14 shows that penile rigidity is already lower in men with co-morbid diabetes by the age of 45, as compared to those without that disorder.

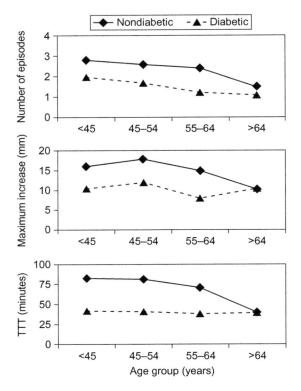

FIGURE 2.14 SREs in men with diabetes. Frequency, magnitude, and duration of SREs for men with ED, with and without co-morbid diabetes. Mean values across two nights for measures made at the penile base are illustrated for different age groups. The number of episodes, the maximum circumference increase, and the total tumescence time (TTT). Source: *Hirshkowitz, M., & Schmidt, M. H. (2005). Sleep-related erections: Clinical perspectives and neural mechanisms.* Sleep Med Rev, *9, 311–329. Reproduced with kind permission from* Sleep Medicine Review.

Rigidometer measures of axial rigidity of SRE/NPT in men with ED clearly show that diabetes co-morbidity significantly adds to the decrease in the total number of SREs, the maximum possible increase in axial rigidity, and the number of nightly episodes... up to the age of about 64. This suggests that whatever it is that causes ED in these patients, its cumulative effects gradually downgrade erectile strength to the age of about 60, and then precipitously to the age of 64.

The study also compared a group of men with diabetes to three other groups: normotensive vs. hypertensive, non-obese vs. obese, and non-smokers vs. smokers. Figure 2.15 teaches us that diabetes is the more damaging to erectile function.

The axial rigidity data for the men in the control group in all four quadrants is similar: about 600 g (sufficient nocturnal axial rigidity) below the

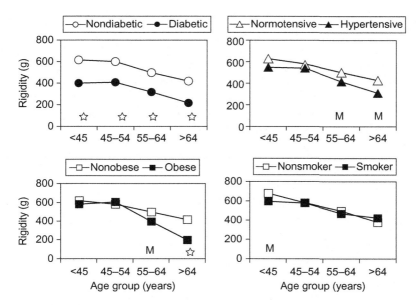

FIGURE 2.15   SREs in men with arterial risk factors as a function of age. Maximum rigidity during SREs in 911 men with ED. Rigidities are shown in different age groups for men with and without the arterial risk factors diabetes, hypertension, obesity, and cigarette smoking. Differences between patients with and without the co-morbid risk factor within each age group are shown with a star where statistically significant ($p < 0.05$), and with an M where marginally significant ($0.05 < p < 0.10$). *Hirshkowitz, M., & Schmidt, M. H. (2005). Sleep-related erections: Clinical perspectives and neural mechanisms.* Sleep Med Rev, *9, 311–329. Reproduced with kind permission from* Sleep Medicine Review.

age of 45, dropping to about 400 g (insufficient axial rigidity) by age 64. By comparison, diabetic patients are already at insufficiency by age 45.

It would seem from these data that diabetic patients, other factors equal, fare worse, and do so earlier when it comes to clinical markers of ED, than those who suffer from hypertension, those who are obese, or those who are smokers. These data may provide an interesting window into the relative depth and time course of endothelial damage of these conditions as well as other conditions that impact on cardiovascular—cardio-sexual—function.

A number of investigators have voiced concern about the use of auto-mated recording devices to assess nocturnal erections, as well as the value of the messages that the data send us about the patient's overall condition. For instance, a report in the *International Journal of Urology* questions whether RigiScan is a useful test in differentiating vascular from psychogenic ED.

Following a number of medical procedures, they found that 48% of patients with arterial failure, and 42% of the patients diagnosed with vascular impotence, actually had veno-occlusive dysfunction (venous channels that drain the corpora cavernosae). NPT recording revealed psychogenic ED in

39% and vascular ED in 61% of patients. NPT recording "has been regarded as the gold standard and, in our series, it showed 90.6% sensitivity and 88.2% specificity in differentiating the cause of erectile dysfunction." Rigidity measures and tumescence episodes were significantly higher in patients with psychogenic impotence, when compared to those with vascular impotence. Patients with arterial failure had fewer erections than those with veno-occlusive dysfunction. New means for recording NPT can precisely differentiate between organic and psychogenic ED. However, they cannot distinguish between subgroups with a vascular cause of ED.[89]

Another issue centers on whether RigiScan or a self-administered test such as the IEFF is a better predictor of impotence. The *International Journal of Impotence Research* reports that the IIEF is a useful tool and is helpful for follow-up of patients to evaluate efficacy of treatments for ED, but that it should not replace *objective* testing to diagnose the quality of ED.[90]

A complex multimodal study examined the predictive value of the patients' subjective assessment of early morning and nocturnal erections, the history of cigarette smoking, and presence/absence of vascular risk factors. The result of RigiScan NPT correlated with other measures of erectile function (peak systolic velocity, PSV), resistance index (RI) determined at color flow duplex Doppler ultrasonography, and the maintenance flow rate (Qm) determined at dynamic infusion cavernosometry and cavernosography (DICC)). The results were reported in the *International Journal of Impotence Research*:

A history of cigarette smoking and presence of vascular risk factors are good predictors of organic impotence, whereas the patient's subjective assessment of his own early morning erections is unreliable. Normal NPT predicts normal values of other objective measures (PSV, RI, and Qm), but it does not exclude organic impotence. Conversely, abnormal NPT predicts abnormal values of other objective measurements (abnormal PSV, RI, and Qm).[91]

2.32 CAVERNOSA PEAK SYSTOLIC BLOOD FLOW VELOCITY

A previously cited study proposed cavernosal blood velocity as a criterion measure of penile erectile function (PSV). The PSV measures correlating with NPT in that study are based on Duplex Ultrasound evaluation of cavernosal peak systolic blood flow velocity and waveform acceleration in the penis. The *International Journal of Andrology* reported that the combined duplex ultrasound assessment of PSV and waveform acceleration in the penile flaccid state can predict arterial dynamic inflow in the majority of patients with ED, with less time and expense and less discomfort for the patient.[92]

A report in *Acta Neurologica Belgica* reports that NPT predicts normal systolic blood velocity and normal measure of cavernosa blood flow.

However, this correlation is low when compared to diastolic blood velocity. On the other hand, a low correlation exists between abnormal NPT and abnormal systolic and diastolic blood flow or abnormal measures of cavernosa blood flow. Therefore, it is presumed that NPT, penile duplex, and infusion cavernosometry should be performed together in order to achieve a reasonably accurate diagnosis.[93]

On the other hand, the *Journal of Sexual Medicine* reports on "Use and abuse of RigiScan in the diagnosis of erectile dysfunction." The investigators corralled four clinicians with expertise in ED disorders and diagnosis and asked them for their opinion on the use of RigiScan in clinical practice. Here is their opinion: "The first held that NPT/RigiScan cannot be considered a useful diagnostic tool for differential diagnosis in ED. He is supported by the physiological considerations of the second expert, and by the experimental evidence produced and discussed by the fourth expert who questions the accuracy, reliability, and usefulness of these tools to measure adequate rigidity. The third expert suggested the use of the technique for at least the patient with no neurovascular risk factors who presents with a history suggesting psychogenic cause."[94]

The *International Journal of Impotence Research* reported a study aimed at comparing measures of nocturnal penile tumescence and rigidity (NPTR) with erectile function (EF) domain score of IIEF, to see how well it serves to diagnose the severity of ED.

No correlation was found between IIEF-EF domain scores and NPTR measures in patients tested. However, if IIEF-EF domain scores were normal, NPTR measures were also normal. Thus, if the initial IIEF-EF domain scores are normal, then it is thought unnecessary to perform NPTR testing.[82]

2.33 PENILE ARTERIAL WAVEFORM AMPLITUDE

The volume of blood flowing through our blood vessels is constant, but it is not uniform: Each systole sends a bolus of blood down the line. This bolus has a particular volume that distends the blood vessels as it passes through them. In diastole the vessel walls return to a state reflecting the lower blood volume flowing when the heart is relaxed.

The volume in a blood vessel can be observed at any readily accessible point in the body... in the penis, for instance. The instrument used to measure volume and volume changes is the plethysmograph. A normal PVR waveform is composed of a systolic upstroke with a sharp systolic peak followed by a downstroke that contains a prominent *dicrotic notch* (Figure 2.16). Mild to moderate disease is characterized by loss of the dicrotic notch and an outward "bowing" of the downstroke of the waveform.

With severe disease, the amplitude of the waveform is blunted, and according to a report in the *Journal of Urology*, with papaverine challenge, a

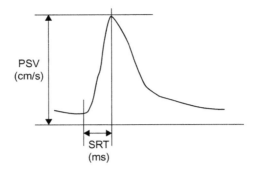

FIGURE 2.16 Typical arterial waveform shows PSV and SRT. Source: *Oates, C. P., Pickard, R. S., Powell, P. H.,* et al. *(1995). The use of Duplex Ultrasound in the assessment of arterial supply to the penis in vasculogenic impotence.* J Urol, *153, 354–357. Reproduced with kind permission from* Journal of Urology.

systolic rise time (SRT) greater than 110 milliseconds was found to be strongly indicative of disease in the arteries supplying the corpora caverno- sa.[95] SRT is also illustrated in Figure 2.16.[96,97]

　　Penile arterial waveform amplitude (PAWA) also detects testosterone insufficiency: Investigators reported in *Annals of Biomedical Engineering*, in a study to assess the validity of a Penile Arterial Waveform Analyzer (PAWA), that the Penile Brachial Index (PBI) correlated significantly with serum testosterone levels as well as key anthropometric and serum biochem- ical parameters, even in apparently healthy young adults, suggesting its potential as a sensitive tool in monitoring penile vascular function and risk for ED. Furthermore, the Penile Perfusion Index (PPI) was superior to the PBI in distinguishing the high- and low-risk ED participants.[98,99]

2.34 PENILE BRACHIAL INDEX

A report in the *Journal of Andrology* described an interesting analog of the ABI, the Dilatation Index (DI) and PAWA ratios for assessing systemic and penile endothelial function. This ratio is obtained by using an Air Pressure Sensing System (APSS) on the arm, and a Penile Arterial Waveform Analyzing System on the penis... a sort of PBI. "High-risk" participants in the study were selected on the basis of a criterion of total cholesterol/ high-density lipoprotein (TC/HDL) ratio greater than 4.1. Conventional blood tests were performed.

　　The DI correlated positively with serum testosterone and serum high- density lipoprotein (HDL) level, while it correlated negatively with total triglyceride and glycosylated hemoglobulin levels, body weight, waist cir- cumference, body mass index, and diastolic blood pressure. Likewise,

PAWA ratio correlated positively with serum testosterone and HDL levels, but negatively with body weight, waist circumference, and body mass index. Both the DI and PAWA ratio successfully identified subjects at high-risk for ED.

This procedure led the investigators to conclude that penile endothelial function can be indirectly assessed by evaluating systemic endothelial function in young healthy adults for early identification of risk for ED.[100]

Many of the assessment techniques described in this chapter come into play in the next chapters that assess the presence and severity of cardiovascular disorders that impinge on erectile function. However, it should be noted that the vast number of clinical trials that emerged in the past ten years that depend on measures of endothelial viability and blood flow parameters underscore the message of this monograph: ED is, for the most part, now generally accepted into the fraternity of cardiovascular disorders, and it is to be assessed, diagnosed, and treated as such.

The detailed description of the techniques supported here by only some representative research papers makes it clear that the methods used to evaluate both the potential risk and the actuality of hypertension and atherosclerosis, for instance, are indistinguishable from those used to assess ED.

REFERENCES

1. Moncada S, Higgs A. The L-arginine–nitric oxide pathway. *N Engl J Med* 1993;**329**: 2002–12.
2. <http://clinicaltrials.gov/ct2/info/understand>; 2013 [accessed 15.11.13].
3. Musini VM, Wright JM. Factors affecting blood pressure variability: lessons learned from two systematic reviews of randomized controlled trials. *PLoS ONE* 2009;**4**:e5673.
4. Parati G. Blood pressure variability: its measurement and significance in hypertension. *J Hypertens* 2005;**23**(*Suppl*):S19–25.
5. Jaffe A, Chen Y, Kisch ES, Fischel B, Alon M, Stern N. Erectile dysfunction in hypertensive subjects. Assessment of potential determinants. *Hypertension* 1996;**28**:859–62.
6. Spessoto LC, Cordeiro JA, de Godoy JM. Effect of systemic arterial pressure on erectile dysfunction in the initial stages of chronic arterial insufficiency. *BJU Int* 2010;**106**:1723–5.
7. Kelm M. Flow-mediated dilatation in human circulation: diagnostic and therapeutic aspects. *Am J Physiol − Heart Circulatory Physiol* 2002;**282**(1):H1–5.
8. Chauhan A, More RS, Mullins PA, Taylor G, Petch C, Schofield PM. Aging-associated endothelial dysfunction in humans is reversed by L-arginine. *J Am Coll Cardiol* 1996;**28**:1796–804.
9. Vogel RA. Measurement of endothelial function by brachial artery flow-mediated vasodilation. *Am J Cardiol* 2001;**88**:31E–4E.
10. Corretti MC, Anderson TJ, Benjamin EJ, Celermajer D, Charbonneau F, Creager MA, et al. Guidelines for the ultrasound assessment of endothelial-dependent flow-mediated vasodilation of the brachial artery: a report of the International Brachial Artery Reactivity Task Force. *J Am Coll Cardiol* 2002;**39**:257–65.
11. Tousoulis D, Antoniades C, Stefanadis C. Evaluating endothelial function in humans: a guide to invasive and non-invasive techniques. *Heart* 2005;**91**:553–8.

12. Stoner L, Sabatier MJ. Use of ultrasound for non-invasive assessment of flow-mediated dilation. *J Atheroscler Thromb* 2012;**19**:407–21.
13. <http://www.j-circ.or.jp/english/sessions/reports/64th-ss/shimokawa-l3.htm>; 2013 [accessed 15.11.13].
14. Hirooka Y, Egashira K, Iamizumi T, Tagawa T, Kai H, Sugimachi M, et al. Effect of L-arginine on acetylcholine-induced endothelium-dependent vasodilation differs between the coronary and forearm vasculatures in humans. *J Amer Coll Cardiol* 1994;**24**:948–55.
15. Antony I, Lerebours G, Nitenberg A. Loss of flow-dependent coronary artery dilatation in patients with hypertension. *Circulation* 1995;**91**:1624–8.
16. Vogel RA, Corretti MC, Plotnick GD. Effect of a single high-fat meal on endothelial function in healthy subjects. *Am J Cardiol* 1997;**79**:350–4.
17. Clarkson P, Adams MR, Powe AJ, Donald AE, McCredie R, Robinson J, et al. Oral L-arginine improves endothelium-dependent dilation in hypercholesterolemic young adults. *J Clin Invest* 1996;**97**:1989–94.
18. Cox DA, Vita JA, Treasure CB, Fish RD, Alexander RW, Ganz P, et al. Atherosclerosis impairs flow-mediated dilation of coronary arteries in humans. *Circulation* 1989;**80**:458–65.
19. Kawano H, Motoyama T, Hirashima O, Hirai N, Miyao Y, Sakamoto T, et al. Hyperglycemia rapidly suppresses flow-mediated endothelium-dependent vasodilation of brachial artery. *J Am Coll Cardiol* 1999;**34**:146–54.
20. Shechter M, Sharir M, Labrador MJ, Forrester J, Silver B, Noel Bairey Merz CN. Oral magnesium therapy improves endothelial function in patients with coronary artery disease. *Circulation* 2000;**102**:2353–8.
21. Williams SB, Goldfine AB, Timimi FK, Ting HH, Roddy MA, Simonson DC, et al. Acute hyperglycemia attenuates endothelium-dependent vasodilation in humans *in vivo*. *Circulation* 1998;**97**:1695–701.
22. Monahan KD. Effect of cocoa/chocolate ingestion on brachial artery flow-mediated dilation and its relevance to cardiovascular health and disease in humans. *Arch Biochem Biophys* 2012;**527**:90–4.
23. Hempel A, Maasch C, Heintze U, Lindschau C, Dietz R, Luft FC, et al. High glucose concentrations increase endothelial cell permeability via activation of protein kinase Cα. *Circul Res* 1997;**81**:363–71.
24. Mandal AK, Ping T, Caldwell SJ, Bagnell R, Hiebert LM. Glucose-induced endothelial damage in primary culture: possible mechanism and prevention. *Histol Histopathol* 2006;**21**:941–50.
25. Mah E, Noh SK, Ballard KD, Matos ME, Volek JS, Bruno RS. Postprandial hyperglycemia impairs vascular endothelial function in healthy men by inducing lipid peroxidation and increasing asymmetric dimethylarginine:arginine. *J Nutr* 2011;**141**:1961–8. [Epub 2011 Sep 21].
26. Virag R. Flow-dependent dilatation of the cavernous artery. A potential test of penile NO content. *J Mal Vasc* 2002;**27**:214–7.
27. Campia U, Choucair WK, Bryant MB, Waclawiw MA, Cardillo C, Panza JA. Reduced endothelium-dependent and -independent dilation of conductance arteries in African Americans. *J Am Coll Cardiol* 2002;**40**:754–60.
28. Laumann EO, West S, Glasser D, Carson C, Rosen R, Kang JH. Prevalence and correlates of erectile dysfunction by race and ethnicity among men aged 40 or older in the United States: from the male attitudes regarding sexual health survey. *J Sex Med* 2007;**4**:57–65.

29. Bjarnegård N, Ahlgren AS, Sandgren T, Sonesson B, Länne T. Age affects proximal brachial artery stiffness; differential behavior within the length of the brachial artery? *Ultrasound in Med Biol* 2003;**29**:1115−21.

30. Ward RP, Weiner JP, Ghani1 SN, Taillon LA, Min JK. Surgical treatment of valve disease. Abstract 3351: erectile dysfunction predicts peripheral arterial disease as measured by screening ankle brachial index testing. *Circulation* 2006;**114** II_712.

31. Polonsky TS, Taillon LA, Sheth H, Min JK, Archer SL, Ward RP. The association between erectile dysfunction and peripheral arterial disease as determined by screening ankle-brachial index testing. *Atherosclerosis* 2009;**207**:440−4. [Epub 2009 May 20].

32. Al-Qaisi M, Nott DM, King DH, Kaddoura S. Ankle brachial pressure index (ABPI): an update for practitioners. *Vasc Health Risk Managmt* 2009;**5**:833−41.

33. McDermott MM, Criqui MH, Liu K, Guralnik JM, Greenland P, Martin GJ, et al. Lower ankle/brachial index, as calculated by averaging the dorsalis pedis and posterior tibial arterial pressures, and association with leg functioning in peripheral arterial disease. *J Vasc Surg* 2000;**32**:1164−71.

34. <http://www.docstoc.com/docs/4875041/Ankle-Brachial-Index-Chart-Ankle-Pressure-mmHg-Brachial-Pressure>; 2013 [accessed 15.11.13].

35. Bernstein EF, Fronek A. Current status of non-invasive tests in the diagnosis of peripheral arterial disease. *Surg Clin North Am* 1982;**62**:473−87.

36. Benchimol D, Pillois X, Benchimol A, Houitte A, Sagardiluz P, Tortelier L, et al. Détection de l'artériopathie des membres inférieurs en médecine préventive par la détermination de l'index de pression systolique à l'aide d'un tensiomètre automatique [Accuracy of ankle-brachial index using an automatic blood pressure device to detect peripheral artery disease in preventive medicine]. *Arch Cardiovasc Dis* 2009;**102**:519−24.

37. Newman AB, Siscovick DS, Manolio TA, Polak J, Fried LP, Borhani NO, et al. Ankle-arm index as a marker of atherosclerosis in the Cardiovascular Health Study. Cardiovascular Heart Study (CHS) collaborative research group. *Circulation* 1993;**88**:837−45.

38. Prusikova M, Vrablik M, Zamecnik L, Horova E, Lanska V, Ceska R. Prevalence of risk factors of cardiovascular diseases in men with erectile dysfunction. Are they as frequent as we believe? *Neuro Endocrinol Lett* 2011;**32**(Suppl 2):60−3.

39. Chai SJ, Barrett-Connor E, Gamst A. Small-vessel lower extremity arterial disease and erectile dysfunction: the Rancho Bernardo Study. *Atherosclerosis* 2009;**203**:620−5. [Epub 2008 Aug 7].

40. Meredith IT, Currie KE, Anderson TJ, Roddy MA, Ganz P, Creager MA. Postischemic vasodilation in human forearm is dependent on endothelium-derived nitric oxide. *Am J Physiol* 1996;**270**:H1435−40.

41. Higashi Y, Sasaki S, Nakagawa K, Matsuura H, Kajiyama G, Oshima T. A noninvasive measurement of reactive hyperemia that can be used to assess resistance artery endothelial function in humans. *Am J Cardiol* 2001;**87**:121−5.

42. Janner JH, Godtfredsen NS, Ladelund S, Vestbo J, Prescott E. The association between aortic augmentation index and cardiovascular risk factors in a large unselected population. *J Hum Hyperten* 2012;**26**:476−84.

43. Jadhav UM, Kadam NN. Non-invasive assessment of arterial stiffness by pulse-wave velocity correlates with endothelial dysfunction. *Indian Heart J* 2005;**57**:226−32.

44. Matsui Y, Kario K, Ishikawa J, Eguchi K, Hoshide S, Shimada K. Reproducibility of arterial stiffness indices (pulse-wave velocity and augmentation index) simultaneously assessed by automated pulse-wave analysis and their associated risk factors in essential hypertensive patients. *Hyperten Res* 2004;**27**:851−7.

45. Aversa A, Francomano QD, Bruzziches R, Pili M, Natali M, Spera G. Application of digital pulse amplitude tonometry to the diagnostic investigation of endothelial dysfunction in men with erectile dysfunction. *Andrologia* 2009;**41**:1−7.

46. Wilkinson IB, Prasad K, Hall IR, Thomas A, MacCallum H, Webb DJ, et al. Clinical study: coronary artery disease increased central pulse pressure and augmentation index in subjects with hypercholesterolemia. *J Am Coll Cardiol* 2002;**39**:1005−11.

47. Axtell AL, Gomari FA, Cooke JP. Assessing endothelial vasodilator function with the Endo-PAT 2000. *J Vis Exp* 2010.10.3791/2167 44, pii: 2167.

48. Gerhard-Herman M, Creager MA, Hurley S. Assessment of endothelial function (nitric oxide) at the tip of a finger. *Circul* 2002;**102**(Suppl II). pii 851.

49. Liu J, Wang J, Jin Y, Roethig HJ, Unverdorben M. Variability of peripheral arterial tonometry in the measurement of endothelial function in healthy men. *Clin Cardiol* 2009;**32**:700−4.

50. McCrea CE, Skulas-Ray AC, Chow M, West SG. Test-retest reliability of pulse amplitude tonometry measures of vascular endothelial function: implications for clinical trial design. *Vasc Med* 2012;**17**:29−36.

51. Moerland M, Kales AJ, Schrier L van Dongen MG, Bradnock D, Burggraaf J. (2012). Evaluation of the EndoPAT as a tool to assess endothelial function. Int J Vasc Med, 2012. Available from: http://dx.doi.org/10.1155/2012/904141.

52. Dharmashankar K, Welsh A, Wang J, Kizhakekuttu TJ, Ying R, Gutterman DD, et al. Nitric oxide synthase-dependent vasodilation of human subcutaneous arterioles correlates with noninvasive measurements of endothelial function. *Am J Hypertens* 2012;**25**:528−34.

53. Faizi AK, Kornmo DW, Agewall S. Evaluation of endothelial function using finger plethysmography. *Clin Physiol Funct Imaging* 2009;**29**:372−5.

54. Hamburg NM, Benjamin EJ. Assessment of endothelial function using digital pulse amplitude tonometry. *Trends Cardiovasc Med* 2009;**19**:6−11.

55. Patvardhan EA, Heffernan KS, Ruan JM, Soffler MI, Karas RH, Kuvin JT. Assessment of vascular endothelial function with peripheral arterial tonometry information at your fingertips? *Cardiol in Rev* 2010;**18**:20−8.

56. Yang WI, Park S, Youn JC, Son NH, Lee SH, Kang SM, et al. Augmentation index association with reactive hyperemia as assessed by peripheral arterial tonometry in hypertension. *Am J Hypertens* 2011;**24**:1234−8.

57. Davis NJ, Crandall JP, Gajavelli S, Berman JW, Tomuta N, Wylie-Rosett J, et al. Differential effects of low-carbohydrate and low-fat diets on inflammation and endothelial function in diabetes. *J Diabetes Complications* 2011;**25**:371−6.

58. Gargiulo P, Marciano C, Savarese G, D'Amore C, Paolillo S, Esposito G, et al. Endothelial dysfunction in type-2 diabetic patients with normal coronary arteries. A digital reactive hyperemia study. *Int J Cardiol* 2013;**165**:67−71.

59. Bonetti PO, Pumper GM, Higano ST, Holmes Jr. DR, Kuvin JT, Lerman A. Noninvasive identification of patients with early coronary atherosclerosis by assessment of digital reactive hyperemia. *J Am Coll Cardiol* 2004;**44**:2137−41.

60. Okumura K, Takahashi R, Taguchi N, Suzuki M, Cheng XW, Numaguchi Y, et al. Small low-density lipoprotein cholesterol concentration is a determinant of endothelial zdysfunction by Peripheral Artery Tonometry in men. *J Atheroscler Thromb* 2012;**19**:897−903.

61. Schnabel RB, Schulz A, Wild PS, Sinning CR, Wilde S, Eleftheriadis M, et al. Noninvasive vascular function measurement in the community: cross-sectional relations and comparison of methods. *Circ Cardiovasc Imaging* 2011;**4**:371−80.

62. Syvänen K, Korhonen P, Partanen A, Aarnio P. Endothelial function in a cardiovascular risk population with borderline ankle−brachial index. *Vasc Health Risk Manag* 2011;**2011** (7):97−101.

63. Lin YH, Ho YL, Lee JK, Huang HL, Huang KC, Chen MF. Plasma leptin levels and digital pulse volume in obese patients without metabolic syndrome − A pilot study. *Clin Chim Acta* 2011;**412**:730−4.

64. Aversa A, Letizia C, Francomano D, Bruzziches R, Natali M, Lenzi AA. A spontaneous, double-blind, double-dummy cross-over study on the effects of daily vardenafil on arterial stiffness in patients with vasculogenic erectile dysfunction. *Int J Cardiol* 2011;**160**:187−91.

65. Aversa A, Vitale C, Volterrani M, Fabbri A, Spera G, Fini M, et al. Chronic administration of Sildenafil improves markers of endothelial function in men with type-2 diabetes. *Diab Med* 2008;**25**:37−44.

66. Costa C, Virag R. The endothelial−erectile dysfunction connection: an essential update. *J Sex Med* 2009;**6**(9):2390−404.

67. Tamler R, Bar-Chama N. Assessment of endothelial function in the patient with erectile dysfunction: an opportunity for the urologist. *Int J Impot Res* 2008;**20**:370−7.

68. Kharitonov SA, Lubec G, Lubec B, Hjelm M, Barnes PJ. L-arginine increases exhaled nitric oxide in normal human subjects. *Clin Sci (Lond)* 1995;**88**:135−9.

69. Archer S. Measurement of nitric oxide in biological models. *FASEB J* 1993;**2**:349−60.

70. Cicinelli E, Ignarro LJ, Schonauer LM, Matteo MG, Galantino P, Falco N. Different plasma levels of nitric oxide in arterial and venous blood. *Clin Physiol* 1999;**19**:440−2.

71. Granger DL, Anstey NM, Miller WC, Weinberg JB. Measuring nitric oxide production in human clinical studies. *Method Enzymol* 1999;**301**:49−61.

72. Dzau VJ. Markers of malign across the cardiovascular continuum: interpretation and application: established and emerging plasma biomarkers in the prediction of first athero-thrombotic events. *Circulation* 2004;**109**10.1161/01.CIR.0000133444.17867.56 IV-6−IV-19.

73. Lauer T, Kleinbongard P, Kelm M. Indexes of NO bioavailability in human blood. *News Physiol Sci* 2002;**17**:251−5.

74. <http://www.neogenis.com/Neogenis-Patent-Pending-Nitric-Oxide-Indicator-Strips-10ct_p_9.html>; 2013 [accessed 15.11.13].

75. Mian AI, Aranke M, Bryan NS. Nitric oxide and its metabolites in the critical phase of illness: rapid biomarkers in the making. *Open Biochem J* 2013;**13**:24−32.

76. <http://www.neogenis.com>; 2013 [accessed 15.11.13].

77. <https://secure.neogenis.com/test-strips.htmll>; 2013 [accessed 15.11.13].

78. Broderick GA. Evidence based assessment of erectile dysfunction. *Int J Impot Res* 1998;**10** (Suppl 2):S64−73 discussion S77−79.

79. Talley JD, Crawley IS. Transdermal nitrate, penile erection, and spousal headache. *Ann Intern Med* 1985;**103**:804.

80. Klotz L. How (not) to communicate new scientific information: a memoir of the famous Brindley lecture. *BJU Int* 2005;**96**:956−7.

81. Rosen RC, Riley A, Wagner G, Osterloh IH, Kirkpatrick J, Mishra A. The international index of erectile function (IIEF): a multidimensional scale for assessment of erectile dysfunction. *Urol* 1997;**49**:822−30.

82. Tokatli Z, Akand M, Yaman O, Gulpinar O, Anafarta K. Comparison of international index of erectile function with nocturnal penile tumescence and rigidity testing in evaluation of erectile dysfunction. *Int J Impot Res* 2006;**18**:186−9.

83. Kandeel FR, Koussa VKT, Swerdloff RS. Male sexual function and its disorders: physiology, pathophysiology, clinical investigation, and treatment. *Endocrinol Rev* 2001;**22**: 342−88.

84. Yaman O, Tokatli Z, Ozdiler E, Anafarta K. Effect of aging on quality of nocturnal erections: evaluation with NPTR testing. *Int J Impot Res* 2004;**16**:150−3.

85. Morales A, Condra M, Heaton JP, Johnston B, Fenemore J. Diurnal penile tumescence recording in the etiological diagnosis of erectile dysfunction. *J Urol* 1994;**152**:1111−4.

86. Hatzichristou DG, Hatzimouratidis K, Ioannides E, Yannakoyorgos K, Dimitriadis G, Kalinderis A. Nocturnal penile tumescence and rigidity monitoring in young potent volunteers: reproducibility, evaluation criteria and the effect of sexual intercourse. *J Urol* 1998;**159**:1921−6.

87. <http://www.uroan.com/productosing.htm>2013 [accessed 15.11.13].

88. Hirshkowitz M, Schmidt MH. Sleep-related erections: clinical perspectives and neural mechanisms. *Sleep Med Rev* 2005;**9**:311−29.

89. Basar MM, Atan A, Tekdogan UY. New concept parameters of RigiScan in differentiation of vascular erectile dysfunction: is it a useful test? *Int J Urol* 2001;**8**:686−91.

90. Melman A, Fogarty J, Hafron J. Can self-administered questionnaires supplant objective testing of erectile function? A comparison between the international index of erectile function and objective studies. *Int J Impot Res* 2006;**18**:126−9.

91. McMahon CG, Touma K. Predictive value of patient history and correlation of nocturnal penile tumescence, colour duplex Doppler ultrasonography and dynamic cavernosometry and cavernosography in the evaluation of erectile dysfunction. *Int J Impot Res* 1999;**11**:47−51.

92. Mancini M, Bartolini M, Maggi M, Innocenti P, Villari N, Forti G. Duplex ultrasound evaluation of cavernosal peak systolic velocity and waveform acceleration in the penile flaccid state: clinical significance in the assessment of the arterial supply in patients with erectile dysfunction. *Int J Androl* 2000;**23**:199−204.

93. Sattar AA, Wery D, Golzarian J, Louis L, Raviv G, Schulman CC, et al. Erectile dysfunction: the role of Rigiscan in the diagnosis. [Article in French.]. *Acta Urol Belg* 1996;**64**:43−5.

94. Jannini EA, Granata AM, Hatzimouratidis K, Goldstein I. Use and abuse of Rigiscan in the diagnosis of erectile dysfunction. *J Sex Med* 2009;**6**:1820−9.

95. Oates CP, Pickard RS, Powell PH, Murthy LNS, Whittingham TAW. The use of duplex ultrasound in the assessment of arterial supply to the penis in vasculogenic impotence. *J Urol* 1995;**153**:354−7.

96. Speel TGW, van Langen H, Wijkstra H, Meuleman EJH. Penile duplex pharmaco-ultrasonography revisited: revalidation of the parameters of the cavernous arterial response. *J Urol* 2003;216−20.

97. Stoner L, Young JM, Fryer S. Assessments of arterial stiffness and endothelial function using pulse-wave analysis. *Int J Vasc Med* 2012.10.1155/2012/903107 2012: 903107. Published online 2012 May 14.

98. Wu HT, Lee CH, Chen CJ, Sun CK. Penile arterial waveform analyzer for assessing penile vascular function in young adults. *Ann Biomed Eng* 2011;**39**:2857−68.

99. Wu HT, Lee CH, Chen CJ, Sun CK. Penile arterial waveform analyzing system for early identification of young adults with high risk of erectile dysfunction. *J Sex Med* 2012;**9**:1094−105.

100. Wu HT, Lee CH, Chen CJ, Tsai IT, Sun CK. A simplified approach to assessing penile endothelial function in young individuals at risk of erectile dysfunction. *J Androl* 2012; **May 17.** [Epub ahead of print].

Chapter 3

Cardiovascular Health Hazards Impairing Sexual Vitality

3.1 INTRODUCTION

The Merriam-Webster dictionary defines "calamity" as "a disastrous event marked by great loss and lasting distress and suffering." By all accounts, hypertension, atherosclerosis, cardiovascular and heart disease, diabetes, and metabolic syndrome are calamities ... now linked to yet another calamity, erectile dysfunction (ED). For the most part, these calamities are not altogether self-inflicted, although we are told that we may contribute to their severity by the Standard American Diet (SAD), consumed against a background of general inactivity.

Media reports suggest that it is a level playing field out there, and what brings us down is what and how much we eat and how little we walk daily, assuming also that we drink alcohol immoderately all the while we chain-smoke. As noted previously, genetic factors play a significant role in determining the slope of the age-related progressive decline timeline. Our individual task is to avoid accelerating it.

3.2 HYPERTENSION

Several variant forms of a gene (*GRK4*) interacting with other genes raise the risk of hypertension by altering the body's ability to eliminate excess salt. The *GRK4* variants appear to cause *dopamine*, a neurotransmitter principally involved in regulating mood, to cause salt retention in the kidneys.[1]

3.3 ATHEROSCLEROSIS

Researchers were able to directly compare carotid artery wall thickness of many individuals with gene mutations in several factors affecting cholesterol metabolism, particularly the so-called apoA-I mutation. Their results showed that carriers of an apoA-I mutation exhibit the most accelerated

Robert Fried: Erectile Dysfunction as a Cardiovascular Impairment.
DOI: http://dx.doi.org/10.1016/B978-0-12-420046-3.00003-2

atherosclerosis compared with those carrying mutations in others (ABCA1 and LCAT). Intima-media wall thickness studies have provided evidence that the presence of abnormally high levels of high-density lipoproteins (HDLs) due to mutations are associated with increased progression of atherosclerosis.[2]

3.4 CARDIOMYOPATHY AND ARRHYTHMIAS

The Department of Cardiovascular Medicine, Oxford University, UK, informs us that deterioration in the function of the heart muscle, cardiomyopathy, and arrhythmias are usually familial and they are caused by a change in a single gene, which is enough on its own to cause the disorder. This is different from more common forms of heart diseases such as coronary artery disease, which are caused by many different genes and are influenced as well by environmental factors such as diet and lifestyle.

Familial cardiomyopathies and arrhythmias are usually inherited from one parent, but they can arise for the first time in an individual. When this happens, there is no previous family history of the disorder, but the person in whom the disorder first occurs can pass it on to his/her children.[3,4]

3.5 CORONARY ARTERY DISEASE

The *Lancet* reported recently that the human Y chromosome is associated with risk of coronary artery disease in men of European ancestry,[5] and the *World Heart Federation* offers the following information about family history and cardiovascular disease (CVD):

- If a first-degree male relative has suffered a heart attack before the age of 55, or if a first-degree female relative has suffered one before the age of 65, one is at greater risk of developing heart disease.
- If both parents have heart disease before the age of 55, one's risk of developing heart disease can rise to 50 percent compared to the general population. However, one can take precautions as the development of CVD involves many different factors, not just family history.[6]

3.6 TYPE-2 DIABETES

Free fatty acid (FFA) metabolism and insulin resistance are known to predict type-2 diabetes. The journal *Diabetes* reports that variation in the *CAPN10* gene influences FFA levels and insulin resistance, and these variants may explain how the *CAPN10* gene increases susceptibility to type-2 diabetes.[7] It is pretty clear that for many, the tossed dice are loaded, so how does each calamity result in loss of sexual vigor?

3.7 HOW DOES HYPERTENSION CAUSE ERECTILE DYSFUNCTION?

Most authorities are beginning to agree that ED is linked to hypertension. The evidence is that hypertension damages the endothelium and so reduces control of blood vessel tone by one of the two main blood pressure-control mechanisms, the endothelium-dependent nitric oxide/cyclic guanosine monophosphate (NO/cGMP) mechanism. On the other hand, there is nothing published to disprove the hypothesis that something yet unknown damages the endothelium and that's how we develop hypertension. It could actually go either way.

In treating hypertension, medicine does not typically address what is known about enhancing the control of blood vessel function by enhancing the endothelium-dependent mechanism (NO/cGMP). Current methods favor non-endothelium-dependent methods including β-blockers and kidney diuresis hormone controls. Perhaps it should treat both simultaneously. This might entail recommendations for nitrate rich foods and supplementing with the amino acid L-arginine to pave the Acetylcholine (ACh)/NO/cGMP pathway.

3.8 BLOOD PRESSURE

During each heartbeat, blood pressure varies between maximum in systole when the heart contracts, and a minimum in diastole when it is "relaxed" between contractions. In fact, blood vessels are never totally relaxed, but always partly constricted. That is called *tone*. Hypertension is chronic blood pressure elevation above "tone" pressure during rest. Tone also fluctuates slightly, so carefully measured blood pressure is never exactly the same. Loss of blood vessel tone is devastating ... fatal.

3.9 HOW IS BLOOD PRESSURE REGULATED?

There are three principal modes of blood pressure regulation: non-endothelium-dependent, endothelium-dependent, and as noted in Chapter 2, flow-mediated. The non-endothelium-dependent blood pressure regulation is controlled by kidney/hormone reflexes; but despite its name, it is not actually independent of endothelium influence, though it is not principally regulated by the ACh/NO/cGMP pathway. Because this control mechanism is not of primary concern here it will be described only briefly below.

3.10 THE LYMPHATIC SYSTEM AND NITRIC OXIDE BIOSYNTHESIS

An additional factor in blood pressure regulation that is rarely considered is the role of the lymphatic system. Lymphatics, the third component of the circulation, may promote high blood pressure due to malfunction caused by the same factors as in veins and arteries: overload or dyscrasia.[8]

The lymphatic system capillary has an endothelium somewhat different from vascular endothelium, nevertheless, a study published in the *Anatomical Record* aimed to investigate the possible occurrence and distribution of endothelin (ET) and endothelial nitric oxide synthase (eNOS; the enzyme that generates NO from L-arginine) in (bovine) lymphatic vessels and primary culture of lymphatic endothelium by using immunocytochemistry.

The investigators concluded that the endothelium is a major source of ET and NO in lymphatic vessels. The occurrence of ET and NOs (assessed by immunoreactivity) in lymphatic vessel endothelium suggests that lymphatic endothelium may play an important role in the regulation of lymphatic vascular tone and in the production of vascular contractile activity promoting lymph flow and immune function.[9]

Lymphatics and sodium balance: A study published in *Medical Science Monitor* proposed that lymphatic vessels regulate sodium and fluid homeostasis: It was hypothesized that because vascular endothelial growth factor C (VEGF-C) plays a vital role in lymphatic capillary hyperplasia, that VEGF-C was involved in salt-sensitive hypertension. Therefore, the authors investigated its plasma concentration in salt-sensitive individuals.

Participants from a rural community of northern China, ranging in age between 25 and 50 years, whose blood pressure (BP) was less than or equal to 160/100 mm Hg, were monitored for 3 days. This period was followed by a low-salt diet for 7 days (3 g/day, NaCl) and a high-salt diet for 7 days (18 g/day, NaCl). Individuals whose BP rose by at least 10% above low-salt period to high-salt period were labeled salt-sensitive. The concentration of plasma VEGF-C was measured by an immunoenzyme method (ELISA).

High salt intake significantly increased the plasma VEGF-C level: It was significantly higher in the salt-sensitive individuals (3642.2 ± 406.1 pg/mL) than in those salt-resistant subjects (2249.8 ± 214.6 pg/mL). The difference in VEGF-C levels between the two groups was also statistically significant. The authors concluded that VEGF-C could serve as a biomarker of salt sensitivity.[10]

3.11 NON-ENDOTHELIUM-DEPENDENT BLOOD PRESSURE CONTROL

Non-endothelium-dependent regulation of arterial pressure relies on the following factors:

- Baroreceptors in the aortic arch and carotid sinus that detect changes in arterial pressure and send signals to the medulla of the brain stem. From the medulla, the autonomic nervous system adjusts the pressure by signaling alteration of both the force and speed of the heart's contractions, as well as the diameter of the blood vessels throughout the body.

- The renin–angiotensin system (RAS) affects long-term adjustment of arterial pressure. It makes the kidneys compensate for loss in blood volume or drops in arterial pressure by activating an endogenous vasoconstrictor, angiotensin II.
- The steroid hormone aldosterone is released from the adrenal cortex in response to angiotensin II or high serum potassium levels. Aldosterone stimulates sodium retention and potassium excretion by the kidneys. Because sodium mainly determines the amount of fluid in the blood vessels, aldosterone will increase fluid retention and indirectly, arterial blood pressure.
- There are several other mechanisms, including those that result in feedback by regulating the secretion of antidiuretic hormone (ADH/Vasopressin), renin, and aldosterone.

The journal *Current Pharmaceutical Design* offers a tantalizing theory tying the endothelium-independent hormones control mechanisms to the endothelium-dependent NO/cGMP blood vessel control mechanism.

3.12 ALDOSTERONE ACTIVATION AND ERECTILE DYSFUNCTION

An article appearing in that journal addresses an endothelium pro-inflammatory process involving aldosterone, a steroid hormone that controls blood pressure by binding to a receptor (mineralocorticoid receptor, MR) that regulates genes that play a role in salt and water balance by the kidneys. This MR system has been shown to play a crucial role in various human diseases including hypertension, atherosclerosis, and cardiac failure. In patients with cardiovascular risk factors and disease, including diabetes, hypertension, and/or congestive heart failure, an excess of MR activation has been shown to be pro-inflammatory and to impair endothelial function, thus altering the balance between vasoconstriction and vasodilation.

The MR activation-induced imbalance may be involved in ED, a condition that occurs frequently in patients with increased cardiovascular risk and involves endothelial impairment of blood vessel relaxation. The authors of the study propose the potential benefits of selective MR blockade in treating subsets of ED patients, such as those with congestive heart failure and hypertension, where the MR system may be over activated.[11]

3.13 ENDOTHELIUM-DEPENDENT BLOOD PRESSURE REGULATION

In 1989, an article titled "Role of endothelium-derived nitric oxide in the regulation of blood pressure (L-arginine/hypertension/vascular endothelium/endothelium-derived relaxing factor)" appeared in the *Proceedings of the*

National Academy of Sciences, USA: Three British scientists, one of them Dr. S. Moncada, stated that, "... nitric oxide formation from L-arginine by the vascular endothelium plays a role in the regulation of blood pressure and in the hypotensive actions of acetylcholine."[12]

In 1992, the *Postgraduate Medical Journal* further reported on "Endothelial regulation of vascular tone." Here is the essence of the report: There is a single layer of endothelial cells coating the intimal surface of all blood vessels in the entire circulatory system. These highly specialized cells translate signals from the passing blood cells in the lumen, including ACh, into messages for the smooth muscle cells of the vessel. The smooth muscle cells can also respond directly to physical stimuli, allowing calcium to flow in and potassium to flow out. The author then continues with emphasis on the importance of the work of Furchgott and Zawadzki (1989):

The vasodilation action of acetylcholine is mediated indirectly through release of an intermediary substance arising from the endothelium. Initially called endothelium-derived relaxing factor (EDRF), it is now known to be a simple gas, nitric oxide (NO). Once synthesized by the help of the enzyme NO synthase (NOS) from the amino acid L-arginine, NO diffuses from the endothelium to the underlying smooth muscle, where it activates an enzyme that causes a rise in intracellular cyclic GMP (cGMP) and relaxation of the vessel.

The NOS in healthy endothelial cells is calcium-dependent and responds rapidly to changes in the concentration of intracellular free calcium. The increase in calcium that follows stimulation of the acetylcholine receptor leads to activation of NOS, and the NO generated as a result can only act locally, because it has a chemical half-life of only a few seconds in biological solutions, and it is rapidly inactivated on contact with hemoglobin. In terms of controlling blood circulation, the ACh/NO/cGMP pathway is clearly the first signal system.[13]

So that's how it works and you've read it all before. But it doesn't answer the question: Why is it that it can fail and lead to ED? There are a number of attempts to explain that. For instance, a current WebMD Internet page informs us that understanding how elevated blood pressure can cause ED, one must first "understand how erections works." It goes on to inform us that: "High blood pressure keeps the arteries that carry blood into the penis from dilating the way they're supposed to. It also makes the smooth muscle in the penis lose its ability to relax. As a result, not enough blood flows into the penis to make it erect".[14]

This explanation leaves the question, "How ...?" unanswered, and it is part of the impetus to write this monograph because it illustrates the fact that the ACh/NO/cGMP pathway is still not well understood by the medical community.

- "... dilating from the way they're supposed to ....." Exactly how are they to dilate the way they're supposed to? What fails?

- "... makes the smooth muscle(s) in the penis lose its ability to relax ...." What causes that?
- "As a result, not enough blood flows into the penis to make it erect." As a result of what?

An article in *Acta Cardiologica* points out that different conditions of reduced NO activity have been shown both in hypertensive states and several CVDs, and endothelial dysfunction is likely to occur prior to vascular dysfunction: Endothelial dysfunction leading to reduced NO bioavailability is likely to precede vascular dysfunction leading to hypertension.[15]

This conclusion proposes that (a) something(s) unknown damages the endothelium leading to hypertension, rather than (b) hypertension damages the endothelium. There is presently more evidence that it is the first rather than the latter conclusion, but there are also publications that support the latter.

What the authors seem to propose is that some forms of hypertension may not constitute a distinct medical disorder, *per se*—the theme of this monograph also—but a symptom of another underlying medical condition. In fact, some investigators question whether hypertension causes endothelial damage or whether previously damaged endothelium causes hypertension. See also "Endothelial dysfunction and hypertension. Cause or effect?"[16]

Some studies suggest that alterations in endothelial function in hypertension are a consequence rather than a cause of high blood pressure. The bulk of research on oxidative stress damage to the endothelium supports the proposition that endothelial dysfunction contributes to higher blood pressure at later stages of hypertension, most likely due to the development of other cardiovascular complications.[17] In later reports, the finger points to atherosclerosis, perhaps because the development of high blood pressure appears to parallel the development of atherosclerosis.

The authors of a study appearing in the journal *Circulation* contend that high blood pressure is independently associated with aortic atherosclerosis. Among patients with atherosclerosis, high blood pressure is also associated with complex atherosclerosis involving more than simply plaque formation, but ulceration and other characteristics as well.[18] The comparison of the incidence of hypertension with aging (Figure 3.1), along with the average course of atherosclerosis (Figure 3.2) as we age, show the parallel development of both.

In this report, the same age-related elevation trend is evident for blood pressure in parallel with increase in age-related atherosclerotic coronary artery disease. Again, the comparison does not answer any question about causality.

But now, compare this to the data about NO production with aging detailed in a previous chapter: It was shown that the relationship is negative and near-linear: As NO production declines with aging, blood pressure rises; as NO production declines with aging, atherosclerosis rises.

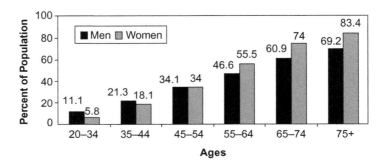

FIGURE 3.1 Prevalence of high blood pressure in Americans by age and sex (NHANES: 1999–2002). Source: *CDC/NCHS and National Heart, Lung, and Blood Institute; National Institutes of Health; US Department of Health and Human Services.*

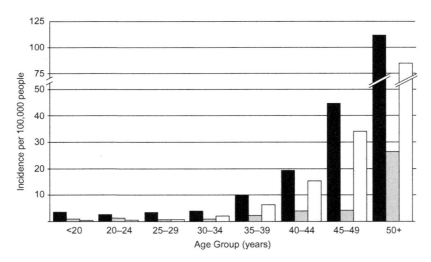

FIGURE 3.2 Mean Incidence of atherosclerotic coronary artery disease (open bars) as a function of age imbedded in data about sudden cardiac death (SCD) as function of age, stratified by sex and other data. Source: *Eckart, R. E., Shry, E. A., Burke, A. P., et al. (2011). Sudden death in young adults: An autopsy-based series of a population undergoing active surveillance.* J Am Coll Card, *58, 1254–1261.*

We can now add to the mix the data about age-related rise in prevalence of ED in Figure 3.3 to the mix, and compare it to the data in Figure 3.4, declining NO production as we age and we get the following: When NO production goes down, blood pressure, atherosclerosis, and ED go up. Not a pretty picture, as they say. What is common to all these conditions, bottom line, is impairment of the endothelium—cause *or* effect.

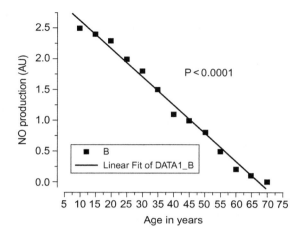

FIGURE 3.3 Prevalence of age-related ED. Source: *Prins J., Blanker M.H., Bohnen A. M.,* et al. *(2002). Prevalence of erectile dysfunction: a systematic review of population based studies.* Int J Impot Res, *14:422–32. Reproduced with kind permission from the* International Journal of Impotence Research.

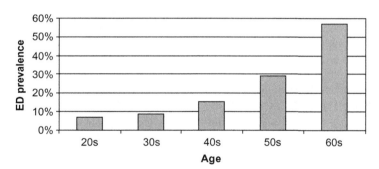

FIGURE 3.4 Aging progressively impairs endothelium-dependent vasodilation. Source: *Gerhard M., Roddy M.A., Creager S.J.,* et al. *(1996). Aging progressively impairs endothelium-dependent vasodilation in forearm resistance vessels of humans.* Hypertension, *27, 849–53. Reproduced with kind permission from* Hypertension.

Investigators reported in the journal *Zhonghua Nan Ke Xue* (*National Journal of Andrology*) that, based on the standard International Index of Erectile Dysfunction 5 (IIEF-5), and after controlling for age, nationality, occupation, education, income, smoking, alcohol consumption, exercise, obesity, fatty liver, blood lipids, blood glucose, and blood uric acid, almost 29 percent of men in their study with high-normal blood pressure reported ED. That was double the number with normal blood pressure.[19] Pre-hypertension?

Finally, according to the journal *Postgraduate Medicine*, it should be noted that "Erectile dysfunction may improve by blood pressure control in patients with high-risk hypertension": In patients at high risk of hypertension who are treated with beta-blockers, blood pressure control resulted in a lower prevalence of ED, independently of age, CVD, and medical treatments. The effect of BP control was higher in older patients.[20]

3.14 ATHEROSCLEROSIS

Another study in the *International Journal of Andrology* reported that "subclinical" endothelial dysfunction and low-grade inflammation play a role in the development of ED in young men with low risk of coronary heart disease. Their patients with ED had significantly higher levels of systolic blood pressure (SBP), total cholesterol and triglyceride, high sensitivity C-reactive protein (hs-CRP), greater carotid intima-media thickness (CIMT) and higher Framingham risk score (FRS) than the control group, though *all of these values were within the respective normal range* (my italics).

In other words, it could not be said that they had anything other than very slightly elevated risk factors with modest markers of inflammation of unknown origin—not an uncommon finding in otherwise asymptomatic individuals. However, the brachial artery FMD values were significantly lower in ED patients and correlated positively with the severity of ED. These values are used as an index of endothelium viability. In this case, it indicated impairment.

To the authors of this study, the results suggested subclinical endothelial dysfunction and low-grade inflammation as the underlying cause of ED. In their conclusion, they stated that young patients complaining of ED should be screened for cardiovascular risk factors and possible subclinical atherosclerosis.

In regard to "subclinical atherosclerosis," they propose measuring FMD as a possible clue to early "subclinical cardiovascular disease."[21] What they are saying here is that ED may not, in many cases, be the end result of CVD, but that in fact it may be an early warning sign." That has also been said by others who called it a "harbinger" of cardiovascular "events."[22]

The diseases that precede the illness are said to be "subclinical cardiovascular disease." One may be expected to have ED accompanying hypertension, but it may also be the case with *almost*-hypertension. Need one have atherosclerosis to get ED, or does *almost*-atherosclerosis qualify? It may seem hyperbole, but it begs the question, how much damage to the endothelium forebodes ED... even before clinical signs emerge? The answer to this question has practical implications: It may suggest that we should consider protecting the endothelium from damage *before* symptoms draw our attention to the damage.

3.15 PLAQUE GROWTH AND VASCULAR REMODELING

Remodeling is a euphemism for changing basic form/shape. Atherosclerotic lesions do not occur in a random fashion. Atheromas are typically seen where blood vessels branch out, where there are irregularities, and where blood suddenly changes course in direction and/or velocity. The earliest form of atherosclerosis is the fatty streaks in the aorta and coronary arteries of most individuals by about age 20 years. The fatty streak is the result of focal accumulation of oxidized serum lipoproteins within the intima of the vessel wall. Microscopy reveals lipid-laden macrophages, T-lymphocytes, and smooth muscle cells in varying proportions. The fatty streak may progress to form a fibrous plaque, the result of progressive lipid accumulation and the migration and proliferation of arterial smooth muscle cells (SMC).

The subsequent increasing deficiency of endothelium-derived NO further potentiates this developing stage of plaque maturation. The SMCs form a connective tissue matrix, a fibrous cap that overlies a core of lipid-laden foam cells, extracellular lipid, and cellular debris. As endothelial injury and inflammation progress, plaques develop resulting in two kinds of "remodeling": In "positive remodeling," the blood vessel wall bulges outward and the vessel cavity (lumen) remains unobstructed, and in "negative remodeling" the vessel wall bulges inward, causing narrowing of the lumen.[23]

3.16 TYPICAL AGE-RELATED PROGRESS OF ATHEROSCLEROSIS

Figure 3.5 shows the degree of atherosclerosis in three sample blood vessels in adult men in four age ranges from 35 to 69 years. These data validate the basis for the belief, in connection with FMD, that endothelium impairment noted in one vessel reflects that in others, albeit perhaps not to the same degree. Aorta and carotid arteries seem to bear the brunt.

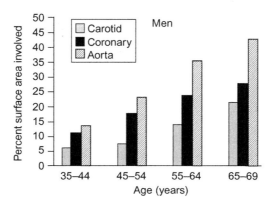

FIGURE 3.5   Age-related surface area of lesions in coronary and peripheral blood vessels. The values represent the average of the three coronary arteries, and the carotid values are the average of the common and internal carotid arteries. Source: *Hodis, H. W. & Mack, W. J. (2002). Atherosclerosis imaging methods: assessing cardiovascular disease and evaluating the role of estrogen in the prevention of atherosclerosis.* Am J Cardiol, *89, Suppl. 1, 19-27. Reproduced with kind permission from the* American Journal of Cardiology.

The age-related progression follows the same pattern: On average, it doesn't get better with age. Compare the data in Figure 3.4 to those in Figure 3.5: It is the same pattern—as atherosclerosis worsens, so does ED. While this may well be true, it is, *per se*, no clue to causality.

3.17 WHAT IS THE CONNECTION BETWEEN ATHEROSCLEROSIS AND ERECTILE DYSFUNCTION?

The standard answer is dysfunctional endothelium. But, what about the endothelium, or rather, why the endothelium? The journal *Atherosclerosis* reported a study that found that men with idiopathic ED have evidence of endothelial dysfunction in forearm resistance vessels, increased pulse pressure, and impaired heart rate variability. The authors concluded that ED is a predictor of cardiovascular dysfunction and a precursor of clinical CVD.[24]

In a study published in the *Journal of Sexual Medicine*, the authors tell us that increased cavernous intima-media blood vessel wall thickness (IMT) might predict ED. There are other aspects to the study, but they are not to the point here. The assumption underlying the measure of intima-media blood vessel wall thickness is that what thickens it is atherosclerosis. Indeed, cavernosa parameters were significantly different between ED sufferers and controls. IMT was the only predicting factor for ED, and it showed a strong direct correlation with carotid and femoral artery IMT.[25]

In other words, if there is atherosclerosis in blood vessel A, it is likely to be found also in blood vessels B, and in C, etc. This is clearly evident in Figure 3.5. Get it in one vessel and you'll get it in the others—albeit perhaps not to the same degree.

What about "pre-atherosclerosis," accepting that atherosclerosis, at least, damages the endothelium: The study above points out that atherosclerosis, once found in one vessel—perhaps the carotid artery—is typically also found in other blood vessels. But the study does not tell us whether atherosclerosis affects the cavernosae endothelium—of concern in ED—to the same extent, or whether cavernosae endothelial impairment occurs before, at the same time, or later than that in the blood vessels shown in Figure 3.5.

However, the *Journal of Urology* reported a study where the degree of impairment of the (rabbit) penis cavernosae appears to be proportional to that in another blood vessels, the iliac artery.[26] From all these data, we may conclude that the cavernosae are by no means exempt from the process of impairment by encroaching atherosclerosis. Parenthetically, rabbit blood vessels are the conventional experimental model for diet, and atherosclerosis experiments that lead to common cholesterol control prescription meds such as the statins.

Complicating the issue of causality even further, investigators report in *European Urology*, that they measured baseline blood flow in both the forearm and the penis and calculated the corresponding vascular resistance in a given region of blood vessels. They found that forearm blood flow was similar in the men with ED and those without ED, but the penile blood flow was significantly lower in men with ED compared with that in the men without ED.

The FMD studies of endothelial function focus primarily on the endothelium-dependent NO/cGMP relaxation mechanism. They could not simply extricate the action of other cavernosae relaxation mechanisms such as that relating to action of dopamine receptors in central nervous centers participating in the initiation of erection. These have also been targeted for the treatment of ED—apomorphine, administered sublingually, is the first of such drugs.[27]

The authors of the *European Urology* study concluded that: "This is the first study that provides evidence of impaired penile endothelial function without the presence of a significant peripheral endothelial dysfunction. Furthermore, these results provide further support for the notion that the development of ED could predict the future onset of cardiovascular disease."[28] It has been said to be, "The canary in the coal mine."

There is much that can be learned about the association between atherosclerosis and ED because there is much research that has gone into an attempt to find some common factor or cause. So far to no avail. There is virtually no equivocation about the meaning of these clinical research reports, but it may or it may not be significant that few address nutrition *per se*, even though it is common knowledge that the medical attempt to minimize atherosclerosis by prescription meds picks up where diet leaves off.

One notable exception is the report by Azadzoi et al. (2005)[26] cited above, that centered on "oxidative stress in atherogenic erectile dysfunction," that is, free radical damage to the endothelium (see next chapter).

Many studies of ED have found hypertension and/or atherosclerosis before any of the latter conditions were detected. Nevertheless, the study cited above that found that there are those in whom ED appears in the absence of either of the other conditions is not an isolated case.

The *Journal of the American College of Cardiology* reported a study employing brachial artery FMD. It was found that patients with ED but no clinical CVD may have a vascular system that is dysfunctional also in both endothelium-dependent and -independent blood pressure control that occurs before the development of any overt functional or blood vessel disease, and is independent of other traditional cardiovascular risk factors, including atherosclerosis.[29]

Another study from the *Journal of Sexual Medicine* reported evaluation of endothelial function with brachial artery ultrasound in ED and non-ED patients classified as "intermediate risk" based on the Framingham score. "Intermediate risk" means that there is no clinical evidence of significant atherosclerotic disease. The conclusion of the authors is unembellished: "Men without established atherosclerotic disease presenting with ED demonstrated a worse endothelial function."[30]

Again, in the *International Journal of Cardiology*, we are presented with evidence that did not support the initial theory of at least some degree of atherosclerosis underlying ED: "Because the silent or documented atherosclerosis or vascular risk factors are very frequent, the possibility of endothelial dysfunction in ED patients is expected to be increased." The usual FMD procedures were employed along with nitroglycerin challenge. Conclusion: "... ED patients have more markedly impaired endothelial and blood vessel smooth muscle functions compared with patients with similar risk factors but no ED."[31]

This is what we learn from clinical research:

- ED accompanies hypertension.
- ED accompanies atherosclerosis.
- ED is caused by hypertension.
- ED is caused by atherosclerosis.
- ED is caused by almost-hypertension, or *maybe* by almost-hypertension.
- ED is caused by almost-atherosclerosis, or *maybe* by almost-atherosclerosis.
- ED is not caused by atherosclerosis, but it could be; it predicts it, and if one has ED, he could develop atherosclerosis, and maybe even hypertension.

There are at least two other theories about what can happen to blood vessels that can damage the endothelium, resulting in ED. They deserve attention.

3.18 THEORIES OF ATHEROSCLEROSIS

The common perception is that all authorities agree that atherosclerosis is an inflammatory condition resulting from invasion of oxidized low-density lipoprotein cholesterol-laden (LDL-C) macrophages into blood vessel walls where there occurs a series of immune and other reactions choreographed in a fairly predictable way, the upshot being spiraling plaque formation and endothelium damage. This *oxidation theory* of atherosclerosis currently prevails.

Reporting in the journal *Biochemical Society Transactions*, researchers predicted that appropriate antioxidants should protect against atherosclerosis. They cite vitamin E, but aver that to date it has shown little evidence of anti-atherosclerotic potential. However, they also hold that lack of knowledge of the oxidant(s) role in lesion formation, as these can be quite complex, and our incomplete understanding of the anti- and pro-oxidant properties of vitamin E, may explain its failures to date.[32]

Studies on the impact of atherosclerosis on erectile function were presented so far in a sort of generic form, as though all authorities agree about what atherosclerosis is, how and why we get it, and what its health consequences are. In fact, nothing could be further from truth.

H. H. Wang, Fellow, Wyss Institute for Biologically Inspired Engineering, Harvard University, detailed currently favored theories of atherosclerosis in the journal *Atherosclerosis*:

- the *lipid theory* and its genetic implications, where atherosclerosis is regarded mainly as the consequence of a defect in the metabolism of cholesterol and other lipids;
- the *flow theory*, that emphasizes the association between the predominant localization of atherosclerotic lesions in regions of perturbed blood flow;
- the *response to injury theory*, that suggests that an early atherosclerotic lesion is an injury or alteration to the endothelium;
- the *fatigue theory*, that views atherosclerosis as a "wear-and-tear" disease as we age and develop hypertension;
- the *thrombogenic theory*, suggesting that atherosclerotic lesions are incorporated within the intima and become atheroma.[33]

Whichever theory turns out to be most predictive is not immaterial as treatment, and indeed prevention, depends on that. But it is well to keep in mind that prevention of ED related to atherosclerosis—whatever the theory of its cause—requires that one consider all possibilities because each one has its own preventatives. Here's yet another theory:

3.19 THE HOMOCYSTEINE THEORY OF ATHEROSCLEROSIS AND ERECTILE DYSFUNCTION

In 1990, Dr. K. S. McCully reported in the *American Journal of Medical Science* that he examined the outcome of 194 consecutive autopsies to determine the proportion of cases of atherosclerosis without elevated serum cholesterol, diabetes mellitus (DM), or hypertension. In 66% of the cases with severe atherosclerosis, the disease developed without evidence of elevated serum cholesterol, diabetes, or hypertension. He proposed reevaluating the role of blood homocysteine, which had been shown by other studies to be an independent risk factor for atherosclerosis.[34] What is homocysteine?

Metabolism of dietary proteins results in the formation of homocysteine from the amino acid methionine. Ordinarily, most of the homocysteine is recycled as other amino acids by folate, vitamin B_{12}, and B_6. Absent sufficient levels of these vitamins, blood levels of homocysteine rise, and that increase has been implicated in a number of medical disorders, though not all medical authorities agree. While the severe form, homocystinuria, is a relatively rare disease, it has been shown that many people have mildly or moderately elevated homocysteine levels (see also Chapter 10).

It was first reported in 1962 that the rare genetic condition, homocystinuria, made some people more likely to develop severe CVD in their teens and 20s. A defective enzyme was found to cause the accumulation of homocysteine in the blood, raising their risk of early development of athero-sclerosis and blood clots.

Any one of several gene mutations, such as the CBS, MTHFR, MTR, MTRR, and MMADHC, are implicated in homocystinuria. Mutations in the CBS gene cause the most common form of homocystinuria. The CBS gene controls production of an enzyme, cystathionine beta-synthase, that acts in the pathway responsible for converting homocysteine to cystathionine, as well as other amino acids, including methionine. Mutations in the CBS gene disrupt the function of cystathionine beta-synthase, preventing homocysteine from being used properly, and causing this amino acid and toxic byproducts substances to build up in the blood.

Enzymes made by the MTHFR, MTR, MTRR, and MMADHC genes act in converting homocysteine to methionine. Mutations in any of them leads to a buildup of homocysteine in the body. More about MTHFR is in Chapter 10.

Because folic acid, vitamin B_6, and vitamin B_{12} are needed to eliminate excess homocysteine in the blood, it was then recommended that one have adequate dietary intake of these vitamins.

In 2006, the journal *Metabolism* reported a study titled "Hyperhomocys-teinemia: a novel risk factor for erectile dysfunction." The report contends that elevated levels of homocysteine have an adverse impact on the forma-tion of eNOS by the endothelium. This enzyme is one of the regulators of NO production, and thus endothelium function. In fact, as detailed in Chapter 10, consistently elevated homocysteine tends to uncouple the enzyme eNOS from NO biosynthesis.

Patients with ED had higher levels of homocysteine, and also higher levels of fasting plasma glucose, total cholesterol, and LDL-C. The higher the homocysteine blood levels, the worse the International Index of Erectile Function (IIEF) domain score. Objectively, by means of penile Doppler ultrasound-measurements of blood flow through the penis, it was shown that the higher the homocysteine blood levels, the lower the 1st, 5th, and 10th minute peak-systolic blood pressure velocity.[35]

A study in the journal *Urology*, reported that plasma homocysteine was significantly higher in men with type-2 diabetes compared with controls. Patients with elevated homocysteine had 5.2 times greater odds of ED than men with normal levels. High plasma homocysteine appears linked to ED in patients with adult-onset DM.[36]

Similar findings in the *Journal of Diabetes Complications* reported that elevated homocysteine is an additional risk factor in diabetic patients with ED. Fasting plasma glucose, postprandial plasma glucose, and glycated hemoglobin (HbA1c) levels were significantly higher in diabetic patients with ED, as were homocysteine levels in diabetic patients with ED. High levels of HbA1c and homocysteine level were the main determinants of the presence of ED in the diabetic population.[37]

The question is how do elevated levels of homocysteine cause ED or contribute to it? It seems the consensus that elevated homocysteine levels damage the endothelium and therefore impair NO/cGMP formation in sexual arousal. In fact, a report in the *Journal of Biological Chemistry* asserts that homocysteine induces apoptosis in human vascular endothelial cells, and it does that by promoting thromboembolisms that damage endothelium folds.[38]

Another factor in endothelial impairment by homocysteine was proposed in the *Journal of Clinical Investigation*: Brief (15 min) exposure of endothelial cells stimulated to secrete NO to homocysteine resulted in formation of a potent anti-platelet compound and vasodilator. Exposure lasting more than 3 hours resulted in impaired NO/cGMP response.[39]

A retrospective study involving nearly one thousand men that appeared in the *International Journal of Impotence Research* reported that ED was common in patients with biomarkers of atherosclerosis. Yet, a number of factors including homocysteine level were found not to be associated with the degree of erectile function.[40]

As noted above, elevated levels of homocysteine has been strongly implicated as a cardiovascular risk and has been shown to be causally linked to endothelial impairment, and ED. However, it should be noted that not all sources agree both on the hypothetical cause, suggesting vitamin B_6, B_{12}, and folate deficiency, and treatment with these vitamins. For that reason, the controversy is revisited and examined in greater detail in Chapter 10.

3.20 THE *VASA VASORUM* THEORY OF ATHEROSCLEROSIS AND ERECTILE DYSFUNCTION

Vasa vasorum is now said to be a major player in the development of atherosclerosis. Before examining the theory of how it might explain the cause of ED, it is important to keep in mind the meaning of the two terms below because they figure prominently in the way aspects of atherosclerosis are labeled.

Biochemical signals trigger new growth of small blood vessels both in healthy and disease tissues. The signals are different in each case, and the ensuing blood vessels are likewise different:

- *Angiogenesis* typically consists of protrusion and outgrowth of capillary buds and sprouts from pre-existing blood vessels. It is a normal and vital process in growth and development as well as in wound healing.
- *Neovascularization* is the formation of functional new microvessel networks that can transport red blood cells. It is held to be set in motion by inflammatory signals to the immune system that result in these microvessel growth spurts. Neovascularization in the *vasa vasorum* plays an important role in the development of atherosclerosis lesion and plaque.

The *vasa vasorum* are a network of microvessels mostly originating in the adventitia of large arteries (see Chapter 1). They are end-arteries that do not redirect blood flow. Type 1 course through the vessels supplying the outer layer of arterial walls with oxygen and nutrients. Additionally, there are the venous *vasa vasorum* that drain the arterial wall in associated veins. According to the *European Journal of Vascular Medicine*, venous *vasa vasorum* can also be found in lymphatics.[41]

There is an expansion of *vasa vasorum*—type 2, the "second order—due to neovascularization resulting from the buildup of atherosclerosis: factors relating to changes in artery walls signal the *vasa vasorum* in ways that worsen the disease process. However, it is thought that increased density of *vasa vasorum* is essentially vasoprotective, but the thickening wall hampering oxygen delivery to the vessel is considered to be a major factor in development of atherosclerosis.[42]

Neovascularization may be a mechanism that keeps the wall oxygenated and fed in order to counter hypoxia, a condition thought to figure prominently in the development of atherosclerosis.[43–45] Likewise, studies of human tissue clearly indicate that angiogenic *vasa vasorum* also contribute to plaque growth and progression. However, what causes this cascade of events is not known.

3.21 HOW THE *VASA VASORUM* THEORY OF ATHEROSCLEROSIS DIFFERS FROM PREVIOUS THEORIES

The original theory is an *inside-out* theory where the process begins in the intima and works its way outward into the adventitia: Various risk factors stimulate leukocytes to infiltrate the blood vessel wall causing inflammation, which results in the formation of oxygen free radicals that oxidize lipids. Accumulation of oxidized lipids causes injury to the endothelium and initiates an inflammatory response.[46] Endothelium impairment occurs early in atherosclerosis due to multiple factors that include elevated and modified low-density lipoproteins (LDLs), free radicals, and hypertension.

An *outside-in* theory proposes that vessel wall inflammation begins in the adventitia and works its way into the media and intima. Increased *vasa vasorum* neovascularization and increased number of white blood cells in the adventitia support this theory.[47]

The adventitia *vasa vasorum* are thought to be the way that inflammation signals and cells enter into the vessel wall and eventually lead to the developing plaque where they contribute significantly to disease progression.[48,49] This is made possible by a *vascular permeability factor* (VEGF). Both vasculogenesis and angiogenesis rely on this protein. *Vascular permeability factor* has been shown to stimulate endothelial cell replication and cell migration, and to enhance microvessel wall permeability. The permeability leads to edema that attracts inflammatory cells to the permeable vessels.

A report in the *American Journal of Physiology* in part concludes that angiogenesis and inflammation are major contributors to the diseased vessel wall and that the adventitial *vasa vasorum* facilitate both processes. However, the sequence of events and the factors that initiate the neovascularization to promote disease progression are unknown. The human plaque data have improved understanding of plaque progression, however, as the tissue came from autopsied specimens; controlled experiments to follow the order of disease events could not be undertaken."[50]

A "functional hypoxia" theory is proposed in the journal *Medical Hypotheses*: The *vasa vasorum* are end arteries that do not lead to veins and may curtail blood flow to the cells of the intima or media of the arterial wall. The sites most vulnerable to hypoxia are at arterial branching, also known to be the most common sites for formation of atherosclerosis. The known risk factors for atherosclerosis, such as high blood pressure, reduce the blood flow in the end branches of the *vasa vasorum*. The local reduced blood flow will cause inflammation affecting the endothelium by rendering it permeable to large particles such as bacteria, LDL-lipoproteins, and fatty acids conveyed by macrophage foam cells.[51] For more detailed information on the function of *vasa vasorum* and cardiovascular disorders. See also Ritman and Lerman (2007).[52]

3.22 THE *VASA VASORUM* THEORY COULD TAKE THE ONUS OFF DIETARY CHOLESTEROL

Investigators from the Department of Clinical Chemistry, Tempere Medical School, Finland, put it this way in the title of their report, "*Vasa vasorum* hypoxia, but not cholesterol, might be the cause of atherosclerosis." If endothelial dysfunction initiates cholesterol accumulation by scavengers, that cholesterol accumulation and inflammation should first occur superficially under the endothelium. Only their theory on *vasa vasorum* hypoxia provides a logical explanation for the early development of atherosclerosis deep in the arterial wall, between the media and intima at the branching sites of arteries.[53]

The investigators applied the theory to prevention of atherosclerosis using niacin or statins, and they concluded that a peripheral vasodilatation of *vasa vasorum* might be more important in the primary prevention of atherosclerosis than decreasing cholesterol.[54]

Parenthetically, a report in *Current Opinion in Lipidology* affirms that, "Well designed, randomized, placebo-controlled studies show that niacin prevents cardiovascular disease and death." The authors conclude that niacin alters lipoprotein metabolism in novel ways, and mediates other beneficial non-lipid changes that may be atheroprotective. This information forms the rationale for the use of niacin in combination with agents possessing complementary mechanisms of action (e.g. statins) for cardiovascular risk reduction beyond that observed with monotherapy.[55]

On the other hand, not all sources are so enthusiastic about niacin in treatment of atherosclerosis and heart disease: The *Cleveland Clinic Journal of Medicine* reports that the AIM-HIGH trial (Atherothrombosis Intervention in Metabolic Syndrome With Low HDL/High Triglycerides: Impact on Global Health Outcomes) found no cardiovascular benefit from taking extended-release niacin (Niaspan). In fact, there was a statistically non-significant greater risk of ischemic stroke. But questions remain about this complex trial, which included intensive statin therapy in the active-treatment group and the control group."[56]

Tuohimaa and Järvilehto[53] contend that hypoxia of *vasa vasorum* develops gradually in response to a constriction of the peripheral small arteries and hypertension compressing small arteries in the vessel wall and capillaries of the wall of large arteries. Their *vasa vasorum* hypoxia hypothesis is that cholesterol or microbes are not the initial cause of the atherosclerosis, but that vasoconstriction, and consequent hypertension, begins the fatal process. Accumulation of cholesterol, microbes, and inflammatory cells are *consequences* of damage to capillaries in the arterial wall. They hold that niacin is exceptionally beneficial in increasing HDL-cholesterol (HDL-C), and the flush can be prevented. What's more, it is possible that the side effect, vasodilatation, could be an important factor in the prevention of atherosclerosis.

This leads one to wonder whether the failure of the recent Merck niacin trials may not be due to inhibition of flushing.[57]

Finally, they point out that whereas all the known risk factors of atherosclerosis are *vasoconstrictive* and *hypertensive*, all the protective factors are *vasodilatative*, but not all of them influence cholesterol levels, suggesting that cholesterol cannot explain the development of atherosclerosis.[53] It must be said that the issue of the therapeutic potential of niacin in treatment of atherosclerosis is controversial at the very least, and it is revisited in Chapter 10 from a somewhat different perspective.

3.23 THE CARDIOVASCULAR SYSTEM AND THE HEART

"Cardiovascular and heart disease" is a catch-all category of medical conditions that affect the circulatory system, consisting of the blood vessels and the heart. These conditions include peripheral artery disease, coronary heart disease, heart attack, congestive heart failure, and congenital heart disease. Heart disease is the leading cause of death for men and women in the US Because it has been repeatedly linked to ED, the focus of this section is on coronary heart disease (CHD), a narrowing of the small blood vessels that supply blood and oxygen to the heart.

3.24 THE HEART

It took several years after the discovery of the obligatory role of NO in erectile function for the idea to sink in that ED is basically a medical disorder and that in most cases it is actually vascular disease. Once that idea was accepted, medical research publications on the nature and character of that issue that began as a trickle became a virtual deluge. As Figure 3.6 shows, the data are unequivocal. The point of this section is not so much to show that CVD is predicted by ED—that is now well known—but rather to show why that may be so. What actually turns out to be the case, is that everyone now accepts the fact that these entities go hand-in-hand, but that perhaps none can say why that is so.

In 2005, the *Journal of the American Medical Association* (JAMA) reported on "Erectile Dysfunction and Subsequent Cardiovascular Disease." That is the title of a study conducted by authors with various affiliations ranging from the Department of Urology, University of Texas Health Science Center at San Antonio, and Southwest Oncology Group, San Antonio; and Fred Hutchinson Cancer Research Center, and the University of Washington, Seattle.

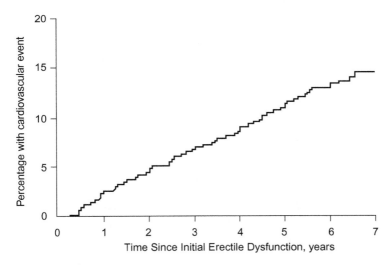

FIGURE 3.6 ED and subsequent CVD. Time to any cardiovascular event from initial report of ED for those with incident ED and no previous cardiovascular event. At risk, n = 2,495; number of cardiovascular events, 255; 5-year estimate of cardiovascular events, 11%. Source: *Thompson, I. M., Tangen, C. M., Goodman, P. J., et al. (2005). Erectile dysfunction and subsequent cardiovascular disease.* JAMA, 294, 2996–3002. *Reproduced with permission from the* Journal of the American Medical Association.

Figure 3.6 tells us that the incidence of CVD rises with time from the initial reported ED. Something must be happening during that time. Note that the acceleration of the trend is exactly opposite to that for age-related decrease in NO formation by the endothelium shown in Figure 3.3: As NO formation declines with age, the incidence of ED rises. We've seen that before, but now it bodes ill for developing CVD as well.

In the clinical study, men aged 55 years or older in a Prostate Cancer Prevention Trial at 221 US centers were evaluated every 3 months for CVD and ED between 1994 and 2003.[58] Men with ED and low libido were excluded from the analysis.

It is not reasonable to assume that endothelium aging began at the time in the patients' life (averaging age 55) when the investigators decided to study them. Therefore, regardless of the age where ED is first manifest, the process must have already been ongoing, and so it really doesn't matter at what age ED is first detected, as each individual would have the same proportional *chance* of CVD as time progresses.

The evidence that the study caught the patients mid-stream, as it were, is clearly shown in Figure 3.7. The incidence of ED had an effect equal to or greater than family history of myocardial infarction, cigarette smoking, or hyperlipidemia, on subsequent cardiovascular events.

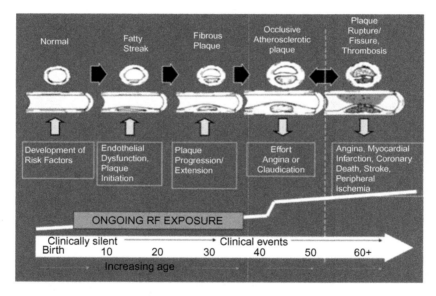

FIGURE 3.7 The pathologic progression of atherosclerosis with aging from no visible atherosclerosis at birth to the development of complex plaques with potential rupture and thrombosis in mid-to-late adulthood. The process begins in the first decade of life when initial risk exposures occur. The progression of atherosclerosis is exacerbated and intensified by the presence of risk factors. The solid white line indicates clinical events as shown. Except in rare circumstances, atherosclerotic disease is subclinical for the first two to three decades of life. National Heart, Lung, and Blood Institute Clinical Practice Guidelines. State of the Science: Cardiovascular Risk Factors and the Development of Atherosclerosis in Childhood. Source: *National Heart, Lung, and Blood Institute; National Institutes of Health; US Department of Health and Human Services.*

Epidemiological studies have demonstrated a close association between ED and vascular disease. The shared etiological factor is endothelial dysfunction. According to the *International Journal of Clinical Practice*, ED tends to precede the onset of symptoms of other vascular diseases because blood vessels in the penis are narrower in diameter than elsewhere in the body, so blood flow is restricted sooner by atherosclerosis. This means that it can be used as a "window" on vascular health.[59] It is not clear, but it almost seems as though the authors may be alluding to penis conduit blood vessel blood flow measures as an early predictor of CVD.

3.25 DEPRESSION, CARDIOVASCULAR DISEASE, AND ERECTILE DYSFUNCTION

In a study titled "The mutually reinforcing triad of depressive symptoms, cardiovascular disease, and erectile dysfunction," The *American Journal of Cardiology* summarized data from population-based epidemiologic studies considering age, heart disease, hypertension, sedentary behavior, related medications, cigarette smoking, and abnormal lipids. All these were found to be highly associated with depressive symptoms, CVD, and ED. The study concluded that all three medical conditions must share many of the same risk and causal factors. Patients with sexual dysfunction are also likely to have CVD and depression, as well as the potential increased risk for heart disease and heart attack mortality.[60]

How might endothelium dysfunction feature in depression? Endothelium dysfunction is related to measurably decreased vasodilator NO/cGMP action. It is not obvious that systemic deficit means cerebral deficit because nerve and brain cells respond to NO formed by a different isoform of NOS. Nevertheless, a decrease in oxygen in the body adversely alters the response of dopamine receptors. Decreased dopamine activity known to depress mood can also depress sexual function.[61]

It has also been shown that endothelial cells synthesize and release dopamine, and they do so in response to ischemia.[62] This causes a shortage of oxygen and glucose needed for cellular metabolism.

3.26 DOPAMINE LEVELS DECLINE WITH AGE

Dopamine levels also progressively decline with age. In fact, that decline may be causally linked to the relatively high incidence of elderly persons falling and sustaining severe injury.[63]

3.27 DIABETES IS STRONGLY LINKED TO ERECTILE DYSFUNCTION

Adult-onset type-2 diabetes is a metabolic disorder causing high blood glucose levels as a result of either insulin resistance, and/or insulin deficiency. Some hold obesity to be the primary cause of type-2 diabetes.[64] However, a recent report on weight reduction in type-2 diabetes sponsored by the Department of Health and Human Services of the NIH concluded that an intensive diet and exercise program causing weight loss does not reduce cardiovascular events such as heart attack and stroke in people with longstanding type-2 diabetes. Intervention was terminated early in this NIH-funded study of weight loss in overweight and obese adults with type-2 diabetes after "finding no harm, but no cardiovascular benefits".[65]

Rates of diabetes increased markedly in the US over the last 50 years in parallel with obesity. As of 2010 there are approximately 285 million people with the condition, compared to around 30 million in 1985. Long-term complications from high blood sugar can include cardiovascular and heart disease and ED, but it is the role that diabetes type-2 plays in ED that is central to the concerns of this book.

3.28 DIAGNOSTIC CLINICAL CRITERIA

Table 3.1 shows the cut-off limits criteria for normal and abnormal values of blood sugar levels. These are shown here because they are also the conventional criteria used in clinical trials cited in this monograph.

TABLE 3.1 Normal Blood Sugar Levels and Abnormal Blood Sugar Levels Reflecting Hyperglycemia

	2-Hour Glucose mmol/L (mg/dL)	Fasting Glucose mmol/L (mg/dL)
Normal	<7.8 (<140)	<6.1 (<110)
Impaired fasting glycemia	<7.8 (<140)	≥6.1 (≥110) and <7.0 (<126)
Impaired glucose tolerance	≥7.8 (≥140)	<7.0 (<126)
Diabetes mellitus	≥11.1 (≥200)	≥7.0 (≥126)

3.29 GLYCATED (GLYCOSYLATED) HEMOGLOBIN HbA1c

Glycated hemoglobin (HbA1c) in red cells builds up in blood because these cells remain glycated for their life duration. HbA1c level reflects average blood glucose concentration over the four weeks to three months prior to testing and the major proportion of its value may be weighted toward the most recent 2 to 4 weeks.[66] The higher the HbA1c level, the higher the risk of diabetes-related medical disorders. If HbA1c is more than 7%, an urgent need to consult a health care provider is recommended.

3.30 PREVALENCE OF DIAGNOSED AND COMMONLY UNDIAGNOSED TYPE-2 DIABETES

Investigators at the Diabetes Initiative at Detroit Receiving Hospital, Detroit, MI, looked at the prevalence of type-2 diabetes in men with ED: Diabetes and pre-diabetes were found to be highly prevalent in the adult population in the Detroit area among those who sought medical help at the Emergency Department: over 43% tested positive. The authors of the report stated that "What is most troublesome [is] that these cases were previously undiagnosed."[67]

3.31 TYPE-2 DIABETES AND ERECTILE DYSFUNCTION

The *Journal of Diabetes and its Complications* reported a study that aimed to find out whether assessing ED on the basis of a single YES/NO question could be as reliable as the use of a questionnaire because questionnaires are too elaborate for general medical practice. During routine annual check-ups, men were asked, "Do you have erection problems? Yes/No." Age, medication, and other known factors associated with ED and/or CVD were likewise analyzed.[68]

The data in Figure 3.8 show the trend in prevalence of ED in the patients sampled. Prevalence rises in parallel with age consonant with all the other risk factors age-related to ED, except for NO formation, which has an exact inverse relationship. Here, prevalence of ED rises to the 70 to

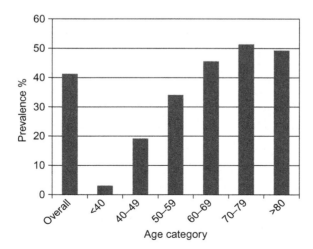

FIGURE 3.8 Prevalence of ED in men with diabetes type-2 by 10-year age category. Source: *Cleveringa, F. G. W., Meulenberg, M. G. G., Gorter,* et al. *(2009). The association between erectile dysfunction and cardiovascular risk in men with Type 2 diabetes in primary care: it is a matter of age.* J Diab Complic, *23, 153–159. Reprinted with kind permission from the* Journal of Diabetes Complications.

79 years-range and then declines slightly in the 80-years-or-older group. This is most likely due to death-related attrition bias. Because type-2 diabetes entails a greater risk of death, decreased prevalence of ED might reflect a loss in the population itself.

3.32 "THE CANARY IN THE COAL MINE"

Logic dictates that no decrease here is likely due to chance sampling factors because one would expect a further increase in data over time with such a strong and consistent trend. Trend reversals in biological observations invariably raise questions about how many variables are actually being analyzed.

One possibility is to look at prevalence of *erectile function* as compared to prevalence of ED. Prevalence of erectile function is 100% minus percent prevalence of dysfunction: In the men sampled in routine medical checkups, there is a higher prevalence of unimpaired erectile function in the 80 + age group (ca. 51%) than in the younger, 70 to 79 years age group (ca. 49%).

If the thesis of this monograph is correct, that ED is tied to all the cardiovascular and heart disease risks (including diabetes) that kill and maim by the common denominator, endothelial impairment, then one explanation is that those who do not have, or only partly have, those risks may be those who both survive longer, and by the same token have a somewhat lower incidence of ED ... conversely, a higher incidence of unimpaired erectile function. Death-related attrition bias is the most likely explanation and the obverse is known as *survival bias.*

A good portion of men with serious cardiovascular and heart disease and diabetes would have died by age 80. Only that group 80 years old or older may also have a larger proportion still alive because presumably they do not suffer from cardiovascular heart disease and/or diabetes all related to significant endothelium impairment. Thus, erectile strength becomes yet another test of endothelial health.

Indeed, a report in the *American Journal of Medicine* refers to type-2 diabetes as a coronary heart disease (CHD) "risk equivalent." The study aimed to determine whether in an older adult population where both glucose disorders and pre-existing atherosclerosis are common, cardiovascular and all-cause mortality rates would be similar in participants with prevalent CHD, versus those with diabetes. The investigators concluded that among older adults, diabetes alone confers a similar risk for cardiovascular mortality as established clinical CHD.[69]

Writing in the *Journal of the American College of Cardiology*, investigators concluded that in men under 60 years old, and in men with diabetes or hypertension, ED can be a critical warning sign for existing or impending CVD and risk of death. Their article is titled "The Link Between Erectile and Cardiovascular Health: The Canary in the Coal Mine."[70]

The *Journal of the American College of Cardiology* reported a study where a cohort of men between the ages of 55 and 88 years with type-2 diabetes were enrolled in the Action in Diabetes and Vascular Disease: Preterax and Diamicron Modified-Release Controlled Evaluation (ADVANCE) program. Baseline medical examination included inquiries about ED. Over 5 years of follow-up during which participants attended repeat clinical examinations, the occurrence of fatal and nonfatal CVD episodes, cognitive decline, and dementia were ascertained.

After adjusting for a range of cofactors such as existing illness, psychological health, and classic CVD risk factors relative to those who were free of the condition, baseline ED was associated with an elevated risk of all CVD events, CHD, and stroke and dementia. Men who experienced ED at baseline and at 2-year follow-up were at the highest risk for these events[71,72] in the journal *Clinical Cardiology*: "Erectile dysfunction is common among patients at high risk for cardiovascular disease because of diabetes and/or [hypertension]. Diabetic men are affected earlier than those with [hypertension] …."

3.33 HOW DOES DIABETES LEAD TO ERECTILE DYSFUNCTION?

According to a report in the *American Journal of Physiology—Heart and Circulatory Physiology*, the endothelial cell dysfunction caused by elevated glucose is due to the activity of free radicals. The authors propose treatment with antioxidants to protect against impaired endothelium-dependent relaxations caused by elevated glucose.[73]

A report titled "Reactive oxygen-derived free radicals are key to the endothelial dysfunction of diabetes" appeared in the *Journal of Diabetes*. Elevated levels of oxygen-derived free radicals are the initial source of endothelial dysfunction in diabetes. Oxygen-derived free radicals not only reduce NO bioavailability, but also facilitate the production and/or action of the antagonist—endothelial-derived contracting factors (EDCFs) causing the endothelial balance to tip towards vasoconstrictor responses over the course of diabetes.[74]

According to the *International Journal of Vascular Medicine*, type-2 diabetes is characterized by hyperglycemia due to lack of, or resistance to, insulin. Patients with diabetes frequently suffer from poor blood circulation and/or impaired wound healing. Type-2 diabetes accelerates atherosclerosis and endothelial cell dysfunction.[75]

3.34 CAUTIONARY NOTE

An article in the *British Journal of Pharmacology* reviews studies on the dysfunction of endothelium-dependent vasodilatation in diabetes and highlights the disparity in the clinical and experimental findings and cautions that the susceptibility of tissues to the damaging effects of elevated serum glucose may vary. The article further cautions that the process of endothelium-dependent vasodilatation may be quite different in blood vessels of different size and in different anatomical locations. Thus, it may be hazardous to extrapolate conclusions drawn from one vessel type or diabetes model to another. It is said to be important to select clinically relevant models for future studies on endothelial dysfunction[76]

Type-2 diabetes is characterized by oxidative stress, which in turn causes endothelial dysfunction. The *Journal of Diabetes Complications* reports that antioxidant treatment with glicazide, an antidiabetic drug with antioxidant properties, improves both antioxidant status and NO-mediated vasodilation in diabetic patients.[77]

A review appearing in the journal *Cardiovascular Research* holds that damaged endothelium may be the key factor in diabetic impairments of endothelium- and NO-dependent microvascular function. This may contribute to several other clinical aspects of diabetes such as hypertension, abnormal amounts of blood lipids, and in insulin resistance. There are now several reports describing elevations in specific oxidant stress markers in both insulin resistance syndrome (IRS) and diabetes, together with determinations of reduced total antioxidant defense and depletions in individual antioxidants.

The pro-oxidant environment in diabetes may disrupt endothelial function through the inactivation of NO formation, resulting in the reduction of basic anti-atherogenic and normal glucose concentration. One clinical study showed that the supplementation of insulin-resistant or diabetic states with antioxidants such as vitamin E can reduce oxidant stress and improve both endothelium-dependent vasodilation and insulin sensitivity.[78]

Endocrinological Reviews published a very comprehensive review titled "Diabetes and Endothelial Dysfunction: A Clinical Perspective"[79]: The effects of aging, elevated serum lipids, hypertension, and other factors add to the complexity of the role of endothelial dysfunction in type-2 diabetes, making it more complicated than type-1. For one thing, markers of endothelial dysfunction are often elevated years before any evidence of damage to small blood vessels becomes evident.

A major element of type-2 diabetes is insulin resistance. As a result, research efforts have focused on defining the possible contribution of insulin resistance to endothelial dysfunction. In patients with type-2 diabetes, endothelial dysfunction occurs with insulin resistance and precedes elevated serum glucose. It follows that the metabolic abnormalities of insulin resistance may lead to endothelial dysfunction.

Endothelial cells are easily damaged by oxidative stress, a characteristic of the diabetic state where there is often an increased tendency for oxidative stress and high levels of oxidized lipoproteins, especially the small, dense, LDL-C. High levels of fatty acids and elevated serum glucose have both been shown to induce an increased level of oxidation of phospholipids as well as proteins. Type-2 diabetes is also associated with increased blood platelet aggregation and a prothrombotic tendency that may be related to diminished NO formation.

The report also addresses L-arginine supplementation and antioxidants— one may reasonably assume that L-arginine supplementation would activate eNOS and produce more NO with greater vasodilation because L-arginine is a substrate for eNOS. In fact, this idea has been implemented in different cases, including reversal of endothelial dysfunction associated with chronic heart failure and diabetes.

The *Arginine Paradox* (see Chapter 7) is proposed as an explanation for how elevated L-arginine plasma levels could produce more eNOS activation: In the normal state, L-arginine is present at concentrations that would support maximum enzymatic turnover of arginine to citrulline and NO. It is possible that in the cases where L-arginine supplementation restored normal endothelial function, eNOS activation was lower to begin with. However, it is difficult to know what intracellular L-arginine levels actually are, and there is some evidence that arginine transporters can concentrate arginine against a concentration gradient. Therefore, even measuring arginine levels would not necessarily result in accurate conclusions about intracellular L-arginine levels.

Consumers have been told that chronic supplementation of antioxidants and/or vitamins may be beneficial in reducing cardiovascular risk. There is some evidence that vitamin E or C and/or other antioxidant therapy may improve vascular function in patients with diabetes. However . . .

3.35 CAVEAT

The consumer public is rarely warned to exercise caution in supplementing vitamins such as vitamin E known to have anticoagulant activity, particularly if one is taking prescription anticoagulant medication. The general message is that because vitamins are "natural," they must be safe.

A meta-analysis conducted by investigators at the Division of Preventive Medicine, Brigham and Women's Hospital, Harvard Medical School, Boston, MA, reported that while vitamin E may lessen the risk of ischemic stroke, it raises the risk of hemorrhagic stroke, thus lessening the benefits of its supplementation: While vitamin E did not affect the total risk of stroke, the risk for hemorrhagic stroke rose by 22%, while the risk of ischemic stroke declined by 10%. The results project one additional hemorrhagic stroke for every 1,250 individuals and one ischemic stroke prevented for 476 individuals. The authors declare that because there is a relatively small risk reduction of ischemic stroke, and considering the generally more severe outcome of hemorrhagic stroke, they do not recommend indiscriminate widespread use of vitamin E.[80]

In one study the effects of short-term dietary supplementation with tomato juice (source of lycopene), vitamin E, and vitamin C on susceptibility of LDL to oxidation and circulating levels of C-reactive protein (CRP) and cell adhesion molecules were measured in patients with type-2 diabetes. Well-controlled type-2 diabetes patients received tomato juice (500 mL/day), vitamin E (800 U/day), vitamin C (500 mg/day), or placebo treatment for 4 weeks.

Lycopene (in tomato juice) and vitamin E were both associated with resistance of LDL to oxidation, but only the latter group showed a decrease in CRP. Levels of plasma glucose did not change significantly during the study. These authors suggested that their findings may be relevant to strategies aimed at reducing risk of myocardial infarction in patients with diabetes.[81] More about Lycopene in Chapter 9.

A clinical study was undertaken to determine whether the antioxidant vitamin C could improve endothelium-dependent vasodilation in forearm resistance vessels of patients with type-2 diabetes. Participants with diabetes and age-matched, non-diabetic control subjects were assessed during concomitant intra-arterial administration of vitamin C (24 mg/min). In diabetic subjects, endothelium-dependent vasodilation was augmented by simultaneous infusion of vitamin C. In contrast, in non-diabetic subjects, vitamin C administration did not alter endothelium-dependent vasodilation. These data show that administration of vitamin C improves the endothelial function associated with the diabetic state, presumably due to antioxidant activity.

The authors concluded that no insight can be drawn about the specificity of the effect of vitamin C because no other vitamin or drug was tested. Despite an intensive search, they could not find any long-term trial of vitamin C supplementation in patients with diabetes (type-1 or -2), and therefore could come to no conclusion about the possible use of this vitamin to prevent atherosclerosis and/or microvascular disease in patients with diabetes.[82] They concluded that in type-2 diabetes, endothelial cell dysfunction is detectable very early in the course of the disease, even before there is overt elevated serum glucose.[79]

The preponderance of the evidence is that elevated serum glucose and the ensuing free radical formation in type-2 diabetes does some serious harm to the endothelium, in turn impairing NO formation and thus causing further harm to the cardiovascular system and the heart and promoting atherosclerosis. This is clearly a scenario that could lead to ED.

Oxidative stress has been causally implicated in the cardiovascular disorders that are linked to ED. The next chapter details the common features of oxygen radicals that do harm to the endothelium.

REFERENCES

1. Felder RA, Sanada H, Xu J, Yu PY, Wang Z, Watanabe H, et al. G protein-coupled receptor kinase 4 gene variants in human essential hypertension. *Proc Natl Acad Sci USA* 2002;**99**:3872–7.

2. Hovingh GK, de Groot E, van der Steeg W, Boekholdt SM, Hutten BA, Kuivenhoven JA, et al. Inherited disorders of HDL metabolism and atherosclerosis. *Curr Opin Lipidol* 2005;**16**:139—45.

3. <www.geneticseducation.nhs.uk>; 2013 [accessed 15.11.13].

4. <http://www.cardiov.ox.ac.uk/inherited-heart-disease/inheritance>; 2013 [accessed 15.11.13].

5. Charchar FJ, Bloomer LD, Barnes TA, Cowley MJ, Nelson CP, Wang Y, et al. Inheritance of coronary artery disease in men: an analysis of the role of the Y chromosome. *Lancet* 2012;**379**:915—22.

6. <http://www.world-heart-federation.org/cardiovascular-health/cardiovascular-disease-risk-factors/family-history/>; 2013 [accessed 15.11.13].

7. Orho-Melander M, Klannemark M, Svensson MK, Ridderstråle M, Lindgren CM, Groop L. Variants in the calpain-10 gene predispose to insulin resistance and elevated free fatty acid levels. *Diabetes* 2002;**51**:2658—64.

8. Mekarski JE. Essential hypertension is lymphatic: a working hypothesis. *Med Hypotheses* 1998;**51**:101—3.

9. Marchetti C, Casasco A, Di Nucci A, Reguzzoni M, Rosso S, Piovella F, et al. Endothelin and nitric oxide synthase in lymphatic endothelial cells: immunolocalization in vivo and in vitro. *Anat Rec* 1997;**248**(4):490—7.

10. Liu F, Mu J, Yuan Z, Lian Q, Zheng S, Wu G, et al. Involvement of the lymphatic system in salt-sensitive hypertension in humans. *Med Sci Monit* 2011;**17**:CR542—6.

11. Caprio M, Mammi C, Jaffe IZ, Zennaro MC, Aversa A, Mendelsohn ME, et al. The mineralocorticoid receptor in endothelial physiology and disease: novel concepts in the understanding of erectile dysfunction. *Curr Pharm Des* 2008;**14**:3749—57.

12. Rees DD, Palmer RMJ, Moncada S. Role of endothelium-derived nitric oxide in the regulation of blood pressure (L-arginine/hypertension/vascular endothelium/endothelium-derived relaxing factor). *Proc. Nati. Acad. Sci. USA* 1989;**86**:3375—8.

13. Furchgott RF, Zawadzki JV. The obligatory role of endothelial cells in the relaxation of arterial smooth muscle by acetylcholine. *Nature* 1980;**288**:373—6.

14. <http://www.webmd.com/hypertension-high-blood-pressure/guide/high-blood-pressure-erectile-dysfunction>; 2013 [accessed 15.11.13].

15. Puddu P, Puddu GM, Zaca F, Muscari A. Endothelial dysfunction in hypertension. *Acta Cardiol* 2000;**55**:221—32.

16. Quyyumi AA, Patel RS. Endothelial dysfunction and hypertension. Cause or effect? *Hypertension* 2010;**55**:1092—4.

17. Lüscher TF. The endothelium in hypertension: bystander, target or mediator? *J Hypertens Suppl.* 1994;**12**:S105—16.

18. Agmon Y, Khandheria BK, Meissner I, Schwartz GL, Petterson TM, O'Fallon WM, et al. Independent association of high blood pressure and aortic atherosclerosis. A population-based study. *Circul* 2000;**102**:2087—93.

19. Wu XR, Wu WL, Feng ZC. [Erectile dysfunction in men with high-normal blood pressure. Article in Chinese]. *Zhonghua Nan Ke Xue* [*National Journal of Andrology*] 2012;**18**:44—7.

20. Cordero A, Bertomeu-Martínez V, Mazón P, Fácila L, González-Juanatey JR. Erectile dysfunction may improve by blood pressure control in patients with high-risk hypertension. *Postgrad Med* 2010;**122**:51—6.

21. Yao F, Huang Y, Zhang Y, Dong Y, Ma H, Deng C, et al. Subclinical endothelial dysfunction and low-grade inflammation play roles in the development of erectile dysfunction with no well defined cause in young men with low risk of coronary heart disease. *Int J Androl* 2012;**35**:653—9.

22. Miner MM. Erectile dysfunction and the "window of curability": a harbinger of cardiovascular events. *Mayo Clin Proc* 2009;**84**:102—4.

23. <http://emedicine.medscape.com/article/153647-overview#a0104>; 2013 [accessed 15.11.13].

24. Stuckey BG, Walsh JP, Ching HL, Stuckey AW, Palmer NR, Thompson PL, et al. Erectile dysfunction predicts generalised cardiovascular disease: evidence from a case-control study. *Atherosclerosis* 2007;**194**:458—64. [Epub 2006 Sep 20].

25. Caretta N, Palego P, Schipilliti M, Ferlin A, Di Mambro A, Foresta C. Cavernous artery intima-media thickness: a new parameter in the diagnosis of vascular erectile dysfunction. *J Sex Med* 2009;**6**:1117—26.

26. Azadzoi KM, Schulman RN, Aviram M, Siroky MB. Oxidative stress in atherogenic erectile dysfunction. *J Urol* 2005;**174**:386—93.

27. Andersson KE. Pharmacology of penile erection. *Pharm Rev* 2001;**53**:417—50.

28. Vardi Y, Dayan L, Apple B, Gruenwald I, Ofer Y, Jacob G. Penile and systemic endothelial function in men with and without erectile dysfunction. *Eur Urol* 2009;**55**:979—85.

29. Kaiser DR, Billups K, Mason C, Wetterling R, Lundberg JL, Bank AJ. Impaired brachial artery endothelium-dependent and -independent vasodilation in men with erectile dysfunction and no other clinical cardiovascular disease. *J Am Coll Cardiol* 2004;**43**:179—84.

30. Averbeck MA, Colares C, de Lira GH, Selbach T, Rhoden EL. Evaluation of endothelial function with brachial artery ultrasound in men with or without erectile dysfunction and classified as intermediate risk according to the Framingham. *J Sex Med* 2012;**9**:849—56.

31. Yavuzgil O, Altay B, Zoghi M, Gürgün C, Kayikçioğlu M, Kültürsay. H. Endothelial function in patients with vasculogenic erectile dysfunction. *Int J Cardiol* 2005;**103**:19—26.

32. Jessup W, Kritharides L, Stocker R. Lipid oxidation in atherogenesis: an overview. *Biochem Soc Trans* 2004;**32**(Pt 1):134—8.

33. Wang HH. Analytical modeling of atherosclerotic lesions. *Atherosclerosis* 2001;**159**:1—7.

34. McCully KS. Atherosclerosis, serum cholesterol and the homocysteine theory: a study of 194 consecutive autopsies. *Am J Med Sci* 1990;**299**:217—21.

35. Demir T, Comlekçi A, Demir O, Gülcü A, Calýpkan S, Argun L, et al. Hyperhomocysteinemia: a novel risk factor for erectile dysfunction. *Metabolism* 2006;**55**:1564—8.

36. Al-Hunayan A, Thalib L, Kehinde EO, Asfar S. Hyperhomocysteinemia is a risk factor for erectile dysfunction in men with adult-onset diabetes mellitus. *Urol* 2008;**71**:897—900.

37. Demir T, Cömlekci A, Demir O, Gülcü A, Caliskan S, Argun L, et al. A possible new risk factor in diabetic patients with erectile dysfunction: homocysteinemia. *J Diab Comp* 2008;**22**:395—9.

38. Zhang C, Cai Y, Adachi MT, Oshiro S, Teijiro Aso T, Kaufman RJ, et al. Homocysteine induces programmed cell death in human vascular endothelial cells through activation of the unfolded protein response. *J Biol Chem* 2001;**276**:35867—74.

39. Stamler JS, Osborne JA, Jaraki O, Rabbani LE, Mullins M, Singel D, et al. Adverse vascular effects of Homocysteine are modulated by Endothelium-derived relaxing factor and related oxides of nitrogen. *J Clin Invest* 1993;**91**:308—18.

40. Eaton CB, Liu YL, Mittleman MA, Miner M, Glasser DB, Rimm EB. A retrospective study of the relationship between biomarkers of atherosclerosis and erectile dysfunction in 988 men. *Int J Impot Res* 2007;**19**:218—25.

41. Schäfer K. Vasa vasorum of subcutaneous veins and lymph vessels. *Vasa* 1990;**19**:237—41.

42. de Muinck ED, Simons M. Re-evaluating therapeutic neovascularization. *J Mol Cell Cardiol* 2004;**36**:25—32.

43. Barger AC, Beeuwkes 3rd R, Lainey LL, Silverman KJ. Hypothesis: *vasa vasorum* and neovascularization of human coronary arteries. A possible role in the pathophysiology of atherosclerosis. *N Engl J Med* 1984;**310**:175—7.

44. O'Brien ER, Garvin MR, Dev R, Stewart DK, Hinohara T, Simpson JB, et al. Angiogenesis in human coronary atherosclerotic plaques. *Am J Pathol* 1994;**145**: 883−94.

45. Wilens SL, Plair CM. Blood cholesterol, nutrition, atherosclerosis. A necropsy study. *Arch Intern Med* 1965;**116**:373−80.

46. Maiellaro K, Taylor WR. The role of the adventitia in vascular inflammation. *Cardiovasc Res* 2007;**75**:640−8.

47. Galkina E, Ley K. Immune and inflammatory mechanisms of atherosclerosis. *Annu Rev Immunol* 2005;**27**:165−97.

48. de Boer OJ, van der Wal AC, Teeling P, Becker AE. Leucocyte recruitment in rupture prone regions of lipid-rich plaques: a prominent role for neovascularization? *Cardiovasc Res* 1999;**41**:443−9.

49. Kaartinen M, Penttila A, Kovanen PT. Accumulation of activated mast cells in the shoulder region of human coronary atheroma, the predilection site of atheromatous rupture. *Circulation* 1994;**90**:1669−78.

50. Mulligan-Kehoe MJ. The *vasa vasorum* in diseased and nondiseased arteries. *Amer J Physiol* 2009;**298**(2):H295−305.

51. Järvilehto M, Tuohimaa P. Vasa vasorum hypoxia: initiation of atherosclerosis. *Med Hypotheses* 2009;**73**:40−1.

52. Ritman EL, Lerman A. The dynamic *vasa vasorum*. *Cardiovasc Res* 2007;**75**:649−58 [Published online 2007 June 29].

53. Tuohimaa P, Järvilehto M. Niacin in the prevention of atherosclerosis: significance of vasodilatation. *Med Hypotheses, May* 2010;6. [Epub ahead of print].

54. Tuohimaa P, Järvilehto M. "Vasa Vasorum Hypoxia, But Not Cholesterol, Might Be the Cause of Atherosclerosis" (Commentary). <http://www.athero.org/commentaries/comm983. asp>; 2013 [accessed 15.11.13].

55. Meyers CD, Kamann VS, Kashyap ML. Niacin therapy in atherosclerosis. *Curr Opin Lipidol* 2004;**15**:659−65.

56. Nicholls SJ. Is niacin ineffective? Or did AIM-HIGH miss its target? *Clev Clin J Med* 2012;**79**:38−43.

57. Thomas K. "Merck Says Niacin Drug Has Failed Large Trial," *New York Times*. December 20, 2012. <http://www.nytimes.com/2012/12/21/business/merck-says-niacin-combination-drug-failed-in-trial.html>; 2013 [accessed 15.11.13].

58. Thompson IM, Tangen CM, Goodman PJ, Probstfield JL, Moinpour CM, Coltman CA. Erectile dysfunction and subsequent cardiovascular disease. *JAMA* 2005;**294**: 2996−3002.

59. Kirby M, Jackson G, Simonsen U. Endothelial dysfunction links erectile dysfunction to heart disease. *Int J Clin Pract* 2005;**59**:225−9.

60. Goldstein I. The mutually reinforcing triad of depressive symptoms, cardiovascular disease, and erectile dysfunction. *Am J Cardiol* 2000;**86**(Suppl. 1):41−5.

61. Giuliano F, Allard J. Dopamine and sexual function. *Int J Impot Res* 2001;**13**(Suppl. 3): S18−28.

62. Sorriento D, Santulli G, Del Giudice C, Anastasio A, Trimarco B, Iaccarino G. Endothelial cells are able to synthesize and release catecholamines both in vitro and in vivo. *Hyperten* 2012;**60**:129−36. [Epub 2012, Jun 4].

63. Bohnen NI, Muller MLTM, Kuwabara H, Cham R, Constantine GM, Studenski SA. Age-associated striatal dopaminergic denervation and falls in community-dwelling subjects. *J Rehabil Res Dev* 2009;**46**:1045−52.

64. Micic D, Cvijovic G. Abdominal obesity and type 2 diabetes. *Eur Endocrinol* 2008;**4**:26—8 <http://www.touchendocrinology.com/articles/abdominal-obesity-and-type-2-diabetes>; 2013 [accessed 15.11.13].

65. <http://www.nih.gov/news/health/oct2012/niddk-19.htm>; 2013 [accessed 15.11.13].

66. University of Michigan Health System, Hemoglobin A1c Fact Sheet: <http://www.med. umich.edu/mdrtc/cores/ChemCore/hemoa1c.htm>; 2013 [accessed 15.11.13].

67. Araia M, Roxas R, Abou-Samra AB, Seyoum B. Diabetes Initiative at Detroit Receiving Hospital: Startling Results (<http://www.drhuhc.org/articles/diabetes_and_endocrinology. htm)>; 2005.

68. Cleveringa FGW, Meulenberg MGG, Gorter KJ, van den Donk M, Rutten GEHM. The association between erectile dysfunction and cardiovascular risk in men with Type 2 diabetes in primary care: it is a matter of age. *J Diab Complic* 2009;**23**:153—9.

69. Carnethon MR, Biggs ML, Barzilay J, Kuller LH, Mozaffarian D, Mukamal K, et al. Diabetes and coronary heart disease as risk factors for mortality in older adults. *Am J Med* 2010;**123**:556.e1—9.

70. Meldrum DR, Gambone JC, Morris MA, Meldrum DAN, Esposito K, Ignarro LJ. The link between erectile and cardiovascular health: the canary in the coal mine. *Am J Cardiol* 2011;**108**:599—606.

71. Batty GD, Li Q, Czernichow S, Neal B, Zoungas S, Huxley R, et al. Erectile dysfunction and later cardiovascular disease in men with type 2 diabetes: prospective cohort study based on the ADVANCE (Action in Diabetes and Vascular Disease: Preterax and Diamicron Modified-Release Controlled Evaluation) trial. *J Am Coll Cardiol* 2010;**56**:1908—13.

72. Roth A, Kalter-Leibovici O, Kerbis Y, Tenenbaum-Koren E, Chen J, Sobol T, et al. Prevalence and risk factors for erectile dysfunction in men with diabetes, hypertension, or both diseases: a community survey among 1,412 Israeli men. *Clin Cardiol* 2003;**26**:25—30.

73. Tesfamariam B, Cohen RA. Free radicals mediate endothelial cell dysfunction caused by elevated glucose. AJP — Heart 1992;**263**:H321—6.

74. Shi Y, Vanhoutte PM. Reactive oxygen-derived free radicals are key to the endothelial dysfunction of diabetes. *J Diabetes* 2009;**1**:151—62.

75. Kolluru GK, Bir SC, Christopher G, Kevil CG. Endothelial dysfunction and diabetes: effects on angiogenesis, vascular remodeling, and wound healing. *Int J Vasc Med* 2012. [Article ID 918267, 30 pages; PMID 22611498].

76. De Vriese AS, Verbeuren TJ, Van de Voorde J, Lameire NH, Vanhoutte PM. Endothelial dysfunction in diabetes. *Br J Pharm* 2000;**130**:963—74.

77. De Mattia G, Laurenti O, Fava D. Diabetic endothelial dysfunction: effect of free radical scavenging in Type 2 diabetic patients. *J Diabetes Complications* 2003;**17**(2 Suppl):30—5.

78. Laight DW, Carrier MJ, Anggard EE. Antioxidants, diabetes and endothelial dysfunction. *Cardiovasc Res* 2000;**47**:457—64.

79. Calles-Escandon J, Cipolla M. Diabetes and endothelial dysfunction: a clinical perspective. *Endocr Rev* 2001;**22**:36—52.

80. Schürks M, Glynn RJ, Rist PM, Tzourio C, Kurth T. Effects of vitamin E on stroke subtypes: meta-analysis of randomised controlled trials. *Brit Med J* 2010;**341**:c5702.

81. Upritchard JE, Sutherland WH, Mann JI. Effect of supplementation with tomato juice vitamin E, and vitamin C on LDL oxidation and products of inflammatory activity in type 2 diabetes. *Diab Care* 2000;**23**:733—8.

82. Ting HH, Timimi FK, Boles KS, Creager SJ, Ganz P, Creager MA. Vitamin C improves endothelium-dependent vasodilation in patients with non-insulin-dependent diabetes mellitus. *J Clin Invest* 1996;**97**:22—8.

Chapter 4

Oxidative Stress Damages the Endothelium

4.1 INTRODUCTION

Ripening apple peel gradually increases in concentration of bacteria and fungus-fighting compounds protecting them from destructive microorganisms in pursuit of rising concentration of sugars in the fruit. Protective compounds are found in the chemistry of the color in the peel, and they make up what is essentially the apple immune system. They include antioxidant polyphenols, flavonoids (including quercetin, epicatechin, and procyanidin), and vitamin C. What these compounds have in common is that we can, and indeed we do, also consume them to our benefit as antioxidants.

"Antioxidant" is to oppose chemical combination with oxygen (O_2). Oxygen is essential to life, but to oxidize is to burn. A number of body mechanisms protect us from excess exposure to O_2: One of these is the ability of hemoglobin to limit O_2 blood saturation, and another consists of endogenous antioxidants that can limit tissue and cell exposure to O_2.

Oxygen can take a form, or combine with other atoms to take a form, we call a *radical*—commonly called "*free radical*," although all radicals are free. A radical is a cluster of atoms containing one that has an unpaired electron in its outermost electron shell. This is an extremely unstable chemical configuration, and radicals quickly react with other molecules or radicals to stabilize the electron configuration in their outermost shell. In the process of stabilizing, free radicals combining with other atoms or molecules in the body form new molecules often also toxic to life. For example, one combination of unstable O_2 with hydrogen (H) forms hydrogen peroxide (H_2O_2). The oxidizing capacity of H_2O_2 is so strong that it is considered a highly "reactive oxygen species" (ROSs).

We naturally produce H_2O_2 as a byproduct of oxidative metabolism, but we are protected from its corrosive effects by *catalase peroxidase* enzymes that reduce low concentrations of it to water and oxygen (Figure 4.1).

Robert Fried: Erectile Dysfunction as a Cardiovascular Impairment.
DOI: http://dx.doi.org/10.1016/B978-0-12-420046-3.00004-4
111

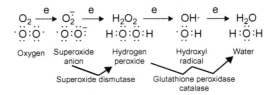

FIGURE 4.1 Diagram of ROS formation. Oxygen (O_2) plays a major role in the formation of ROSs because O_2 has unpaired electrons (represented by single dots). When O_2 picks up an electron, it becomes superoxide, an extremely reactive anion. SOD catalyzes the dismutation reaction of superoxide to H_2O_2, which is further catalyzed to the highly reactive hydroxyl radical and ultimately to water by GPx and catalase enzymes. Superoxide, H_2O_2, and OH radicals are considered to be ROSs. Source: *Racila, D., & Bickenback, J. R. (2009). Are epidermal cells unique with respect to aging.* Aging, *1, 746–750.*

It is doubtful that most would knowingly pump H_2O_2 into their blood stream and yet, there it is, often as a result of the foods we consume and digest. Take LDL cholesterol (LDL-C), for instance: LDL-C is a minor player in health until it combines with O_2 free radicals to form oxidized LDL (oxLDL) cholesterol, the form implicated in atherosclerosis.

4.2 SOURCES OF FREE RADICALS

There are many sources of free radicals; here are just some of them:

- Respiration: With every breath we take, about 2% of unused oxygen would be generated as free radicals.
- Cellular respiration.
- Air pollution, especially diesel exhaust, automobile exhaust, and industrial pollutants.
- Bacterial or viral infection and inflammation, physical injury.
- Radiation, for instance sunlight, X-rays, etc.
- Digestion of foods, consuming alcohol, smoking and second hand smoke. (Smoking also reduces nitric oxide in the blood vessels.)
- Sustained vigorous physical exercise.
- Exposure to plastics, polychlorinated biphenyls (PCBs), pesticides.
- Arsenic and chlorine (which can sometimes be found in drinking water), and many more.

4.3 REACTIVE OXYGEN SPECIES AND ANTIOXIDANTS

As body cells come under free radical attack, other free radicals, as well as non-radical but reactive oxygen derivatives, may form. These chemicals are collectively called ROSs. The term "reactive oxygen species" tells us that, strictly speaking, there are different forms that oxygen free radicals can take, different "species," as it were. ROSs are formed in many ways, including the interaction of ionizing radiation with biological molecules and as an unavoid-able byproducts of cellular respiration. Some electrons passing in the electron transport chain leak away from the main path and directly convert oxygen molecules to superoxide. Some ROSs are synthesized by certain enzymes in immune system cells such as macrophages and neutrophils (Table 4.1).

TABLE 4.1 Common Reactive Oxygen Species (ROSs)

ROS (Radical)	ROS (Non-Radical)
Hydroxyl ($^-$OH)	Peroxinitrite ($ONOO^-$)
Superoxide ($O_2\bullet^-$)	Hypochloric acid (HOCl)
Nitric oxide (NO)	Hydrogen peroxide (H_2O_2)
Peroxyl (RO_2)	Ozone (O_3)
Lipid peroxyl (LOO)	Lipid peroxide (LOOH)

4.4 SCAVENGING ANTIOXIDANTS

Scavenging antioxidants seek out ROSs and remove them. They include the water-soluble vitamin C and glutathione (GSH). Some are fat-soluble, including vitamin E, carotenoids, lipoic acid, and coenzyme Q10 (CoQ10). Scavenging antioxidant enzymes include superoxide dismutase (SOD), cata-lase (CAT), and glutathione peroxidase (GPx).

4.5 COENZYME Q10

Endothelium impairment has been attributed to the destructive action of free radicals. The antioxidant activity of CoQ10 has been investigated in a study designed to determine whether oral CoQ10 supplementation could also improve body SOD activity and endothelium-dependent nitric oxide/cyclic guanosine monophosphate (NO/cGMP) vasodilation in patients with coronary artery dis-ease (CAD). On entry and after one month, all participants in the study pub-lished in the *European Heart Journal* underwent brachial artery relaxation assessment as well as other procedures, including the measurement of endothelium-bound SOD activity. One group of participants received CoQ10 orally at doses of 300 mg/day for one month and the other group, placebo.

In the CoQ10-treated group, endothelium-dependent relaxation and other measures were significantly greater than those in the placebo group. Improvements due to CoQ10 supplementation were "remarkable" in those with low initial endothelium-bound SOD, and thus more prone to oxidative stress. It was concluded that improvement in the endothelium-dependent blood vessel relaxation and endothelium-bound SOD activity might be related to CoQ10 supplementation enhancing endothelial function, perhaps by counteracting deactivation of the process of nitric oxide (NO) formation.[1]

In a similar vein, the journal *Atherosclerosis* reported a study concluding that, "Coenzyme Q10 supplementation is associated with significant improvement in endothelial function. That study supported a role for CoQ10 supplementation in patients with endothelial dysfunction."[2]

CAD is strongly associated with erectile dysfunction (ED), and both are causally linked to endothelial impairment: In the studies cited above, the supply of the antioxidant enzyme, CoQ10, was found deficient in blood vessel walls in CAD. Oral supplementation was found to result in body absorption of the enzyme to benefit the clinical outcome of the trial. It is not unreasonable to extrapolate the outcome of this study to endothelium function in ED.

Both of the studies cited recommend supplementing CoQ10 in mending the endothelium. However, individuals with atherosclerosis and cardiovascular and heart disease in whom one might expect low CoQ10 to contribute to the endothelial impairment at the root of the disease might also be routinely prescribed statin drugs, but these medications are known to deplete CoQ10 tissue stores:

In a study titled "Atorvastatin decreases the Coenzyme Q10 level in the blood of patients at risk for cardiovascular disease and stroke," appearing in the journal *Archives of Neurology*, the authors state that, "Even brief exposure to atorvastatin causes a marked decrease in blood CoQ10 concentration. Widespread inhibition of CoQ10 synthesis could explain the most commonly reported adverse effects of statins...."[3] More on statins and CoQ10 depletion later.

On the bright side, statins have also been shown to ameliorate oxidative stress by their antioxidant activity. Writing in the journal *Biological Trace Elements Research*, researchers went almost so far as to suggest that antioxidant activity is the principal action of statins. They analyzed plasma lipid levels in treatment groups given fluvastatin over a three-month trial period. They ascribed significant improvement in endothelium function to fluvastatin and recommend it in treatment of dyslipidemia.[4]

4.6 GLUTATHIONE AND ERECTILE DYSFUNCTION

GSH has been shown to protect erectile function just as insufficiency impairs it: The *International Journal of Andrology* reported a study titled "Glutathione levels in patients with erectile dysfunction, with or without diabetes mellitus." As noted in the previous chapter, type-2 diabetes is a major cause of ED.

The reduced form of GSH is a very important cell antioxidant essential for forming endothelial nitric oxide synthase (eNOS), the enzyme that synthesizes NO from L-arginine. In diabetes, reduced levels of GSH are due in large measure to a high blood sugar-induced increase of free radical production that can hamper endothelial cell functions. In fact elevated blood sugar hampers natural antioxidant activity (serum albumin, in this case) in diabetes.[5] This condition may play an important role in the cause of erectile ED.

Men with ED, and men with ED *and* diabetes, were compared to a control group of participants with diabetes but not ED, and another control group of men with neither diabetes nor ED. It was found that concentration of GSH in red blood cells was significantly lower in men with ED than in those in the control group. In fact, GSH was lowest in men with both ED and diabetes compared to those with ED and no diabetes.

In men with diabetes only, GSH was significantly lower than in the control group and significantly higher than in men with ED *and* diabetes. GSH concentration was inversely correlated with fasting glucose concentrations and with the duration of diabetes: In other words, GSH was highest when fasting glucose was lowest. The authors concluded that GSH depletion can reduce NO synthesis, thus impairing vasodilation in the corpora cavernosa.[6]

It will be detailed again later that blood glucose has been shown to damage the endothelium by increasing oxidative stress. It should be noted in passing that elevated blood glucose levels are as serious a threat to cardiosexual health—even to life—as diabetes, and it has been called a "coronary heart disease equivalent": Diabetes without prior myocardial infarction (MI) and prior MI without diabetes indicate similar risk for CHD death in men and women. However, diabetes without prior evidence of MI, or angina pectoris, or ischemic ECG changes, indicates a higher risk than prior evidence of CHD in non-diabetic subjects.[7]

SOD detoxifies superoxide, while CAT deals with the H_2O_2 radical, and GPx detoxifies cellular peroxides.

4.7 PREVENTIVE ANTIOXIDANTS

There are proteins that bind with ROSs and prevent ROS formation to protect essential proteins. They include albumin, metallothionein, transferrin, ceruloplasmin, myglobin, and ferritin.[8] This brief description is, of course, mostly an outline. A comprehensive description of free radicals and antioxidants is beyond the scope of this monograph. Antioxidants basically are molecules that "scavenge," "mop up," in effect, neutralize free radicals. In fact, antioxidants are themselves radicals, and that's why they work. In addition, although they generally get bad press, free radicals, naturally produced by some systems in the body, also have beneficial effects that should not be overlooked.

4.8 ANTIOXIDANTS—CAVEAT

The immune system is the main body system that utilizes free radicals: Invading microorganisms or damaged tissue are marked with free radicals by the immune system. This signals immune system cells to identify what needs to be removed from the body. In addition, free radicals perform essential functions in cell signaling, including regulation of blood vessel dilation/constriction.[9] Therefore, some sources hold that because antioxidant supplementation can in some circumstances also thwart immune system function, their supplementation has to be considered in context.[10]

In fact, the body can be said to show evidence of a free radical/antioxidant homeostasis system that adjusts antioxidant activity to counter free radical formation. For that reason, one must consider judicious antioxidant supplementation, as excess supplementation may inadvertently subvert endogenous antioxidant production. This theme is also detailed later in this chapter.

How do we rid the body of excess free radicals or neutralize them for health and safety? The answer... but first, back to the apple: Damaging the peel compromises the apple immune system, resulting in oxidation of the flesh, and ultimately, bacterial and mold action and decay. Microbes could easily penetrate the peel of the apple just as they can easily penetrate us. But the immune system that protects apples, as well as our immune system, depends on a checks-and-balances system of free radicals and antioxidants.

Apple peel—especially red apples—holds a reasonable quantity of the flavonoid quercetin (about 440 mg/kg, depending on the cultivar). The title of a report in the journal *Fundamental & Clinical Pharmacology*, "Protective effect of quercetin, a polyphenolic compound, on mouse corpus cavernosum," just about says it all. The authors concluded that quercetin acts as a protective agent in mouse corpus cavernosum, increasing the bioavailability of NO by protecting it from superoxide.[11] One could use lemon juice to prevent browning. The antioxidant vitamin C in lemon juice prevents the apple flesh from oxidizing by itself oxidizing instead.

The body is constantly exposed to environmental stressors that cause free radical formation, but it forms antioxidant means to neutralize them—at least the three forms of the enzyme SOD to combat a free radical form of oxygen, *superoxide*. The damage caused by free radicals, and the energy the body expends in neutralizing them, as well as that required to repair that damage is, in the aggregate, oxidative stress. *Oxidative stress* may be a very serious threat to erectile function. The cavernosa endothelium in particular is especially vulnerable to oxidative stress.

4.9 OXIDATIVE STRESS

Oxidative stress results from the physiological cost of the interplay between types of reactive oxygen or oxygen-combination molecules and antioxidant molecules. It originates mainly in cell mitochondria from reactive oxygen and reactive nitrogen species (ROS/RNS).[12] Oxidative stress is said to cause damage to body cells, tissues, and even DNA.

Oxidative stress damage also leads to vasculogenic ED, and many authorities hold it to be in some measure due to the Standard American Diet (SAD). In a study published in *Circulation*, it was shown that a rigorous medically supervised short-term, low-fat, high-fiber diet and exercise for three weeks resulted in dramatic improvements in blood pressure, lowering of markers of oxidative stress, increased NO availability, and amelioration of the metabolic profile within 3 weeks, thus lowering the risk of atherosclerosis.

The authors report that systolic and diastolic blood pressure decreased, whereas urinary markers of NO increased. There was a significant reduction in fasting insulin, and positive correlation between decreased serum insulin and increased urinary NO. All fasting lipids decreased significantly, and the total cholesterol to high-density lipoprotein cholesterol (HDL-C) ratio improved. Body weight and body mass index decreased, but obesity was still present in some participants, and there were no correlations between change in body mass index and change in insulin, blood pressure, or urinary NO.[13] It should be noted that these findings, that insulin resistance may rise even in the face of obesity, are consistent with recent findings that obesity may not in most cases cause metabolic syndrome and type-2 diabetes.

4.10 AGE-RELATED GLUTATHIONE ACTIVITY DECLINE

GSH, as noted previously, declines with age just as evidence of oxidative stress rises.[14] We may fail to appreciate the age-related progressive decline in antioxidant activity that may contribute to all the cardio-sexual impairment. These are linked, and their age-related statistics as seen previously hardly seem random: The near linear regression of age-related endogenous antioxidant activity to cardiovascular health risks begs the conclusion that there are at least common causal factors. As we well know, correlation is not causality, but it suffices to say that the prediction is statistically significant regardless of whether the regression has positive or negative coefficients.

The three panels in Figure 4.2 (B, C, and D) come from a study appearing in *Mutation Research/Fundamental and Molecular Mechanisms of Mutagenesis*. In all three instances, activity of singularly crucial antioxidants formed in the body, SOD (B), CAT (C), and GSH (D), are seen to decline with age[15] just as cardio-sexual impairment (as detailed in a previous chapter) rise with age. The authors point out that decline in free radical scavenging enzymes elevates oxidative stress, causing damage and mutations on mitochondrial DNA (mtDNA) over aging.

One may argue that the regression line in panel B appears curvilinear and that in panel C it is hyperbolic. This may or it may not be based largely on non-homoscedasticity, but it should not prevent some degree of reliable prediction in this case with the caveat that the *reliability* of prediction may vary with age as the *standard error of estimate* may vary along the regression line. Curiously, that is not the case with GSH.

Other investigators reporting on "Superoxide dismutase and catalase levels in diseases of the aged" have also found similar decline in antioxidant activity with aging: SOD and CAT activities were measured in blood

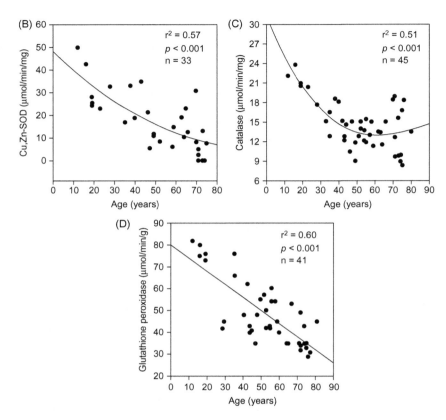

FIGURE 4.2   Age-dependent change of the activities of free radical scavenging enzymes in the fibroblasts from normal human subjects. Three of four panels: (B) Cu,Zn-SOD; (C) CAT; and (D) GSH. *Reproduced with kind permission from Mutation Research/Fundamental and Molecular Mechanisms of Mutagenesis.*

from men and women (and controls) aged 50 to 93 years, with no relevant diseases; disease-group participants also ranged in age from 50 to 93 years. Disease included those in the cardiovascular system and in the osteoarticular system, myoma, prostatic pathologies, chronic obstructive pulmonary disease (EPOC), and acute cerebral vascular accident (ACVA) AKA stroke.

It was found that levels of activity for SOD and CAT were increased for women in a control population, whereas the level of activity for SOD and CAT decreased with aging.[16] By all accounts, not only does enzyme CoQ10 decrease with age as well, as shown in Figure 4.3, but it seems to decrease faster and most profoundly in endocardium, heart tissue lining, much like the endothelium.

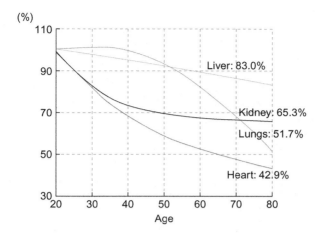

FIGURE 4.3   Progressive age-related decline in CoQ10. Source: *Kalen, A., Appelkvist, E. L., & Daliner, G. (1989). Age-related change in the lipid composition of rat and human tissue.* Lipids, *24, 579–584. Reproduced with kind permission from* Lipids.

4.11 FREE RADICAL ENDOTHELIUM DAMAGE IN DIABETES

Writing in the *Journal of Diabetes*, investigators report that, "Reactive oxygen-derived free radicals are the key to the endothelial dysfunction of diabetes." They hold that elevated levels of oxygen-derived free radicals are the initial source of endothelial dysfunction in diabetes: Oxygen-derived free radicals may reduce NO bioavailability, thus also facilitating the formation of opposing endothelium-dependent contracting factors (EDCFs). This results in the long run in favoring the endothelial balance towards vasoconstriction.[17]

According to a report published in the *American Journal of Physiology— Heart and Circulation*, investigators subjected sections of aorta from laboratory-induced diabetic rabbits, some fed an antioxidant (probucol), to elevated concentrations of glucose. Impairment of NO/cGMP relaxation to acetylcholine (ACh) caused by exposure to elevated glucose was prevented by SOD and by CAT; that impairment did not occur in aortas from rabbits fed probucol. It was concluded that the endothelial cell dysfunction caused by elevated glucose is mediated by free radicals, and treatment with antioxidants protects against that impairment.[18]

4.12 POSTPRANDIAL OXIDATIVE STRESS IN TYPE-2 DIABETES

Persons with diabetes seem much more vulnerable to after-meal oxidative stress than those who do not have diabetes. According to a study published in the journal *Diabetes and Metabolism*, dietary fructose is often recommended for diabetic patients, as this form of carbohydrate is thought to lead to a lower postprandial rise in plasma glucose and insulin.

4.13 FRUCTOSE RECONSIDERED

Fructose contributes to the generation of free radicals. The aim of the study cited next was to investigate the acute effects on several metabolic oxidative biomarkers, particularly plasma *15-F2t isoprostanes* (15-F2t isoPs), of a fructose load in patients with type-2 diabetes compared to those in healthy control participants.

Patients with type-2 diabetes, and healthy participants, underwent a single fructose tolerance test (75 g of anhydrous fructose). Plasma 15-F2t isoPs concentration, plasma total antioxidant capacity (TAS), and other biomarkers were measured at baseline and at 60, 120, 180, and 240 min after fructose absorption.

Significant increases occurred in plasma markers in the diabetic patients but not in the non-diabetic controls. In view of the potentially detrimental effect of plasma 15-F2t isoPs—in particular, vascular lesions—the investigators propose that fructose consumption in type-2 diabetes patients should be reconsidered.[19]

Much evidence from epidemiologic studies has linked fructose intake to the metabolic syndrome, and it has also been linked to inflammatory response in the kidney. For that reason, a study in the *Journal of the American Society of Nephrology* proposed to determine whether fructose directly stimulates endothelial inflammatory processes by upregulating the inflammatory molecule intercellular adhesion molecule-1 (ICAM-1, upregulated at sites prone to atherosclerosis development. These adhesion molecules allow the attachment of leukocytes to the endothelium and may permit their subsequent migration into other tissue.)

Human aortic endothelial cells were stimulated with physiologic concentrations of fructose: Fructose reduced endothelial NO levels and caused a transient reduction in endothelial NO synthase expression. The administration of an NO donor inhibited fructose-induced ICAM-1 expression, whereas blocking NO synthase enhanced it, suggesting that NO inhibits endothelial ICAM-1 expression.[20]

"Postprandial hyperglycemia in type-2 diabetic patients challenged with carbohydrates load" is the title of a study appearing in the *International Journal of Diabetes Research*: On the other hand, quercetin (rich supply in apple peel) effectively suppresses postprandial hyperglycemia in patients with type-2 diabetes after maltose load.[21]

4.14 DIABETES, ADVANCED GLYCATION END-PRODUCTS, THE ENDOTHELIUM, AND THE PENIS

Excess blood sugar damages blood vessels when it settles into blood vessel cells and forms a substance called *advanced glycation end-products* (AGEs) that accumulate over time. AGEs are proteins or lipids that become glycated after exposure to sugars. AGEs are prevalent in the diabetic vasculature and contribute to the development of atherosclerosis.[22] AGEs block NO activity in the endothelium and cause the production of ROSs. Of concern here is that AGEs are also elevated in diabetic human penile tissue, localized to the collagen of the penile tunica and corpus cavernosae.[23]

4.15 COOKING AFFECTS ADVANCED GLYCATION END-PRODUCTS

How a meal is prepared can impact postprandial oxidative stress: The difference between a low-AGE and a high-AGE meal on post-meal oxidative stress was reported in a study published in the *American Journal of Clinical Nutrition.* Flow-Mediated Dilation (FMD) and Laser-Doppler Flowmetry (LDF) were used to measure the effects of serum markers of endothelial dysfunction, oxidative stress, and serum AGE. The meals had identical ingredients but different AGE amounts achieved by varying the cooking temperature and time. The measurements were performed at baseline and 2, 4, and 6 hours after each meal.

After the high-AGE meal, FMD decreased by 36.2%, whereas after the low-AGE meal, FMD decreased by 20.9%. Impairment of vascular function after the high-AGE meal was accompanied by increased concentrations of serum AGE and markers of endothelial dysfunction and oxidative stress. The investigators concluded that in patients with type-2 diabetes, a high-AGE meal induces a more pronounced impairment of vascular function than does an otherwise identical low-AGE meal. Therefore, chemical modifications of food by means of cooking play a major role in influencing the extent of post-prandial vascular dysfunction.[24]

4.16 CAVEAT

Increased dietary formation of AGEs aggravates the vascular complications of diabetes. Foods rich in both protein and fat, and cooked at high heat, tend to be the richest dietary sources of AGEs, whereas low-fat carbohydrate-rich foods tend to be relatively low in AGEs. The so-called AGEs in the diet are not primarily generated by glycation reactions, but by interactions between oxidized lipids and protein. A low-AGE, genuinely low-fat vegan diet may be beneficial for diabetics.

However, plasma AGE content of healthy vegetarians has been reported to be higher than that of omnivores, suggesting that something about vegetarian diets may promote endogenous AGE production. Some researchers have proposed that the relatively high-fructose content of vegetarian diets may explain this phenomenon, but there is so far no clinical evidence that normal intake of fructose has an important impact on AGE production.

It is possible that the relatively poor taurine intake of vegetarians upregulates the physiological role of certain enzyme-derived oxidants in the generation of AGEs. In that case, taurine supplementation could lower AGE production.

The author of a study titled "The low-AGE content of low-fat vegan diets could benefit diabetics — though concurrent taurine supplementation may be needed to minimize endogenous AGE production," published in *Medical Hypotheses*, recommends taurine supplementation for a low-fat vegan diet to minimize AGE-mediated complications in diabetics.[25]

In another postprandial oxidative stress study reported in the journal *Metabolism*, investigators evaluated the effects of two different standard meals designed to produce different levels of postprandial hyperglycemia on the plasma oxidative status and LDL oxidation. The meals were administered in randomized order to patients with type-2 diabetes. Blood samples were collected at baseline and at 60 and 120 minutes after meals, and were analyzed for plasma levels of glucose, insulin, cholesterol, triglycerides, and total radical-trapping antioxidant parameter (TRAP). LDL susceptibility to oxidation was evaluated at baseline and after 120 minutes.

The variations in plasma glucose and TRAP were significantly greater, and LDL was more susceptible to oxidation after the meal that produced a significantly higher degree of hyperglycemia. Postprandial hyperglycemia may contribute to oxidative stress in diabetic patients.[26]

In another study titled "Effects of oxidative stress on endothelial function after a high-fat meal," published in the journal *Clinical Science*, investigators reported that endothelial dysfunction was observed after consuming a high-fat meal, and that it was associated with augmented oxidative stress as manifested by the depletion of serum antioxidant enzymes and increased excretion of "oxidative modification products" that are mostly oxidative modification of LDL-C particles. Before, and 2, 4, and 6 hours after a standard high-fat meal (50 g of fat), serum triglyceride rose and FMD decreased significantly. Plasma GSH dropped two hours after the meal, and urinary excretion of a marker of oxidative stress significantly increased four hours after the meal.[27]

It seems that avoiding oxidative stress is even more essential in persons with type-2 diabetes than in those not so afflicted. These clinical studies are simply examples of the many dozens of similar studies that report the same outcomes.

4.17 OXIDATIVE STRESS: *NEMESIS* OF THE ENDOTHELIUM

The *Journal of Sexual Medicine* reported a review titled "The endothelial-erectile dysfunction connection: an essential update." The authors tell us that the purpose of their review was to identify up-to-date information on the basic physiological mechanisms underlying loss of cavernosal endothelial function due to vascular risk factors (VRFs), with special emphasis on evaluation by the Penile Nitric Oxide Release Test (PNORT), a variant form of the FMD test.[28] In their conclusion, the authors emphasize the importance of endothelial dysfunction assessment of NO formation to also diagnose generalized vascular disease.[29]

Investigators from the Bristol Royal Infirmary, UK, reported in the journal *Expert Opinion in Pharmacotherapy* that basic scientific studies show that oxidative stress acting through the superoxide and other oxygen free radical "species" may be central to impaired cavernosae function in ED. Superoxide inactivates NO biosynthesis, resulting in impaired penile NO/cGMP formation and therefore impaired signaling of penile vascular smooth muscle relaxation. Chronic endothelial dysfunction caused by oxygen radicals may also result in permanent impairment of penile vascular function. They hold this process to parallel early atherogenesis, as ED and atherosclerosis are closely linked through shared risk factors, and they recommend that antioxidants be part of treatment; that may be of benefit in both the short-, and the long-term.[30]

Writing in the *British Journal of Urology International*, investigators show that oxidative stress not only damages the endothelium in the cavernosae and in the cardiovascular system as a whole, but further causes "neurodegeneration" in the penis. By laboratory means, they impaired the endothelium in the iliac artery supplying blood flow to the penis in their rabbit models. This procedure resulted in chronic ischemia to the penis.

Nerve damage mediated by free radicals was evident by upregulation of SOD, and nerve growth factor (NGF) in the ischemic erectile tissue. The body's natural antioxidant, SOD, and the NGF that maintains the viability of sensory nerves, were seen to coordinate defensive reactions to free radicals that seem to fail to prevent nerve injury in the ischemic penis.[31]

In this study published in the *Journal of Andrology*, investigators induced atherosclerosis in rabbits in order to reduce blood flow through the iliac artery to the cavernosae. They hypothesized that this atherosclerosis-induced ischemia would alter the production of the several forms of the NOS enzymes—particularly eNOS—and therefore adversely alter NO formation by the endothelium. Their study showed that while iliac artery blood flow, blood flow through the cavernosae, and blood oxygen concentration in the cavernosae were unchanged four weeks after the induction of arterial atherosclerosis, these markers were significantly diminished at weeks 8 and 16. Likewise, while erectile responses to nerve stimulation and cavernosal smooth muscle relaxation were unchanged at week 4, these markers were significantly diminished at weeks 8 and 16 after the induction of atherosclerosis.

The investigators explained atherosclerosis-related ED as a consequence of ischemia lowering production of eNOS needed to form NO by the endothelium. This is another plausible explanation of why atherosclerosis leads to ED: It impeded blood flow to the cavernosal endothelium, thereby impairing the production of eNOS needed for NO biosynthesis.[32]

4.18 IS THE ENDOTHELIUM *SEAMLESS*?

The vascular system is apparently seamless. Many clinical studies have shown that advanced endothelium impairment in any segment predicts likewise impairment in other segments. However, there is some doubt that endothelium impairment—all factors equal—begins uniformly at the same time in all parts of the vascular system. If that is the case, then the schemata for the age-related atherosclerosis timeline is misleading because it assumes that all parts of the vascular system begin to develop plaque deposits at about the same time. There may be good reason to think that the endothelium is not *functionally* seamless, and if that is so, it is not good news.

A report published in *European Urology* concerned the results of a study comparing "penile and systemic endothelial function in men with and without erectile dysfunction." The participants were divided into two groups, those with ED and those without ED, according to their International Index of Erectile Function (IIEF) ED domain scores. Blood flow measurements, penile endothelial function, and forearm endothelial function were assessed in all participants using veno-occlusive plethysmography (see Chapter 10).

General characteristics of the two groups of participants were comparable except for age. Forearm blood flow was similar in the two groups, but the penile blood flow was significantly lower in men with ED compared with that in the men without ED. Penile vascular resistance was higher in the ED group compared with the control group. The indices of forearm endothelial function were comparable in both groups. However, indices of penile endothelial function were significantly higher in the control group compared with those of the ED group.

This is the first study that shows impaired penile endothelial function "*without the presence of a significant peripheral endothelial dysfunction*" (my italics).[33] In other words, endothelial dysfunction in the penis blood vessels preceded endothelial dysfunction in other regions of the vasculature: The penis may be at risk of atherosclerosis even before other parts of the body are targeted.

4.19 FIXING IT WITH ANTIOXIDANTS

The *Journal of Urology* reported a study titled "Oxidative stress in arteriogenic erectile dysfunction: prophylactic role of antioxidants." It concerned antioxidant activity of beverages including pomegranate juice, as well as other juices and green tea. Pomegranate juice was found by special analysis to have the highest free radical scavenging capacity. The effect of long-term consumption of pomegranate juice on rabbit cavernosae blood flow and penile erection was then recorded.

It was found that pomegranate juice resulted in the highest decrease of low-density lipoprotein oxidation (oxLDL). The rabbit model of arterosclerosis-related ED demonstrated decreased intracavernous blood flow, erectile dysfunction, loss of smooth muscle relaxation, and decreased eNOS. Long-term pomegranate juice intake increased intracavernous blood flow, improved erectile response and smooth muscle relaxation in ED and control groups, while having no significant effect on NOS formation. Pomegranate juice intake prevented erectile tissue fibrosis in the ED group.

The authors concluded that oxidative stress may be a significant factor in development of vascular-related ED, and antioxidant therapy may be a useful tool in preventing penis blood vessel dysfunction.[34,35] There are numerous isolated studies that report the benefits of antioxidants. There are also reports and reviews that raise red flags. These will be detailed later.

Caveat emptor: It is not possible to determine the antioxidant value of commercially available pomegranate juices that also hold various unknown "proprietary" flavorings and offer a considerable amount of sugar per serving (8 oz. glass). On the other hand, whole fruit has fiber, lacking in juice, and thus typically contributes less to blood sugar. In the case of diabetes, a high dietary food fiber intake is recommended because it lowers blood sugar.[36] Eating fruit would be preferable to drinking juice.

4.20 WHEN EVEN VIAGRA® FAILS

Sometimes, atherosclerosis is so profound, and presumably the endothelium is so impaired, that even phosphodiesterase type 5 (PDE5) inhibitor treatment is ineffective. In a study published in the *Journal of Sexual Medicine*, investigators speculated that the lack of response to PDE5 inhibitor effects by patients with ED of arterial origin may be caused by endothelial dysfunction, resulting in reduced NO formation and increased oxidative stress.

The aim of their study was to treat patients with arterial ED, already treated with sildenafil (VIAGRA) twice a week for 8 weeks with endothelial antioxidant compounds (EAC), and note the effect on the erectile response to sildenafil. Patients with arterial ED, hypertension, and diabetes mellitus were randomly given EAC (1 dose/day) for 8 weeks, and after a wash out of 8 weeks, sildenafil (100 mg) plus EAC: EAC consisted of L-arginine (2,500 mg), propionyl-L-carnitine (250 mg), and nicotinic acid (20 mg). The administration of EAC plus sildenafil resulted in a significantly higher number of "responsive" patients compared with sildenafil alone or EAC alone. Furthermore, patients treated with EAC and sildenafil reached a successful response in a shorter length of time (3 weeks) than patients responsive to sildenafil or EAC alone. The authors contend that EAC administration to their patients improved the success rate of sildenafil. They suggest implementation of combined treatment to increase bioavailable NO and to neutralize oxygen free radicals.[37]

The same EAC compounds as cited in the study above, but in connection with tadalafil, were shown in patients with ED to significant decrease serum markers of endothelial cell apoptosis. The authors concluded that treatment with tadalafil reduces endothelial cell loss in patients with arterial ED, and EAC treatment prolongs and stabilizes the duration of the endothelial cell protective effects of tadalafil.[38]

4.21 LIFESTYLE APPROACH

The *International Journal of Impotence Research* reported, "Lifestyle and metabolic approaches to maximizing erectile and vascular health." The authors of the article contend that oxidative stress impairs NO formation and causes insulin resistance: decreased vascular NO formation has been linked to obesity, smoking, and high consumption of fat and sugar, all of which cause oxidative stress by increasing free radical formation.

Smoking: Much has been said about the role of poor diet and lack of exercise, but smoking is also a good source of free radicals. Chronic cigarette smokers have a lower antioxidant capacity (TAC) than nonsmokers, and the antioxidant capacity falls further immediately after smoking.[39]

Exercise has been found to increase NO formation, reflected in the increase in urinary excretion of nitrate and cGMP,[40] and it promotes more frequent erections even in men under 40 years of age.[41] Exercise also increases insulin sensitivity. In fact, one study found that a single session of exercise can increase insulin sensitivity for at least 16 hours in healthy as well as in non-insulin dependent diabetic (NIDDM) individuals.[42]

The effect of exercise on muscle insulin sensitivity is thought to mimic hypoxia. The post exercise increase in sensitivity of muscle glucose transport to activation is not specific for insulin, but also involves increased susceptibility to activation by a submaximal contraction/hypoxia stimulus that may be mediated by translocation of more glucose transported 4 (GLUT4) to the cell surface.[43,44]

Antioxidants raise vascular NO, improve vascular and erectile function, and they may be useful for men with ED who smoke, who are obese, or who have diabetes. Omega-3 fatty acids reduce inflammatory markers, decrease cardiac death, and increase endothelial NO production, and are therefore important supplements for men with ED under the age of 60 years, and who may have diabetes, hypertension, or coronary artery disease.

PDE5 inhibitors have recently been shown to improve antioxidant status and NO production. Some angiotensin II receptor blockers decrease oxidative stress and improve vascular and erectile function, and may therefore be preferred choices for lowering blood pressure in men with ED.

Some sources recommend lifestyle modifications, including physical exercise, weight loss, omega-3 and folic acid supplements, reduced intakes of fat and sugar, and improved antioxidant status through diet and/or supplements for any comprehensive approach to maximizing erectile function.[45]

4.22 OXYGEN RADICAL ABSORBANCE CAPACITY

It is increasingly more evident that the host of different types of oxygen radicals is to be reckoned with both in health and in disease. The questions are, can volume or concentration of free radicals be measured, and what units shall be used? How about number of free radicals produced per unit time in a given cell or tissue? That must be in the millions of trillions, even in small samples and in small time lapses. And then there are the questions: How many free radicals per unit time does it take to cause any kind of damage, or even how many per unit time do we need to mop up to avert it? Then there is also the question: What's the best mopper-upper?

Some, of course, say it is selected foods and beverages—in whatever form. If so, how much and how often are questions left largely unanswered except in the broadest, most vague terms. Many sources that are often food or supplement product-related simply list Oxygen Radical Absorbance Capacity (ORAC), and sometimes also Trolox Equivalent (TEAC) values, that seem to stumble from product Nutrient Facts/Content table to product table… not infrequently with errors.

Although there had initially been a recommendation by the US Department of Agriculture (USDA) that gave birth to ORAC for a minimum daily ORAC unit intake that may promote health (5 K units/day), the USDA reversed its position on recommending ORAC because there seemed no way to avoid its abuse in marketing health foods and related products. The stated basis for this reversal was that no systematic, peer-reviewed clinical studies had shown that a given ORAC unit value, or range of values, proffer any benefit to health or reverse illness. Here is the 2012 USDA statement on ORAC[46]:

4.23 OXYGEN RADICAL ABSORBANCE CAPACITY OF SELECTED FOODS, RELEASE 2 (2010)

Recently the USDA's Nutrient Data Laboratory (NDL) removed the USDA ORAC Database for Selected Foods from the NDL website due to mounting evidence that the values indicating antioxidant capacity have no relevance to the effects of specific bioactive compounds, including polyphenols on human health.

There are a number of bioactive compounds which are theorized to have a role in preventing or ameliorating various chronic diseases such as cancer, coronary vascular disease, Alzheimer's, and diabetes. However, the associated metabolic pathways are not completely understood and non-antioxidant mechanisms, still undefined, may be responsible. ORAC values are routinely misused by food and dietary supplement manufacturing companies to promote their products and by consumers to guide their food and dietary supplement choices.

A number of chemical techniques, of which Oxygen Radical Absorbance Capacity (ORAC) is one, were developed in an attempt to measure the antioxidant capacity of foods. The ORAC assay measures the degree of inhibition of peroxy-radical-induced oxidation by the compounds of interest in a chemical milieu. It measures the value as Trolox equivalents and includes both inhibition time and the extent of inhibition of oxidation. Some newer versions of the ORAC assay use other substrates and results among the various ORAC assays are not comparable. In addition to the ORAC assay, other measures of antioxidant capacity include ferric ion reducing antioxidant power (FRAP) and Trolox equivalence antioxidant capacity (TEAC) assays. These assays are based on discrete underlying mechanisms that use different radical or oxidant sources and therefore generate distinct values and cannot be compared directly.

There is no evidence that the beneficial effects of polyphenol-rich foods can be attributed to the antioxidant properties of these foods. The data for antioxidant capacity of foods generated by in vitro (test-tube) methods cannot be extrapolated to in vivo (human) effects and the clinical trials to test benefits of dietary antioxidants have produced mixed results. We know now that antioxidant molecules in food have a wide range of functions, many of which are unrelated to the ability to absorb free radicals.

For these reasons the ORAC table, previously available on this web site, has been withdrawn."

In all fairness, it may be said that the statements in the last paragraph do not fit the facts: As shown in this book in detailed description of clinical studies cited in prestigious peer-reviewed medical and basic related medical science reports, the benefit of antioxidants have been repeatedly reported in human clinical trials. What's more, clinical trials to test any treatment have often resulted in conflicting published reports. It is invariably the preponderance of the clinical trials evidence that rules, not that no one disagrees with it. If one picks and chooses what suits the taste in science, one will likely find it.

For instance, how many have been told that being overweight is a health hazard and that they must shed weight? According to an article recently published in the *Journal of the American Medical Association* (JAMA), persons with type-2 diabetes and who are overweight (obese) actually outlive those with normal weight. The report goes on to say that "BMI may be a terrible indicator of the health risks of obesity in Type II diabetic patients": A higher waist-to-hip ratio went with higher rates of patient death even though a higher ratio means that one has higher fat deposits in the abdominal region, and that is a better measure of both adiposity and fat localization, a major factor in health.[47]

The position reversal and conclusion of the 2012 USDA "withdrawal" appears to rely heavily on a publication in the journal *Free Radical Biology & Medicine* that questions the validity of ORAC on the grounds that bioavailability of flavonoids reach only very low concentrations in human plasma after the consumption of flavonoid-rich foods. Furthermore, most flavonoids are metabolized, and that can affect their antioxidant capacity. Fruits and vegetables contain many macro- and micronutrients, in addition to flavonoids, that may directly, or through their metabolism, affect the total antioxidant capacity of plasma.[48]

In this article, they conclude that the large increase in plasma total antioxidant capacity observed after the consumption of flavonoid-rich foods is not caused by the flavonoids themselves, but is likely the consequence of increased uric acid levels.

On the other hand, the journal *Food Chemistry* reported a study comparing ORAC and TEAC to measure antioxidant capacity of milk, orange juice, and a milk/orange juice combination. It was found that ORAC is an accurate way to measure food antioxidant capacity—better than TEAC.[49] However, there is no indication that a better assessment of antioxidant capacity of foods says anything about their contribution to better health or to combating illness.

A recent clinical study published in the *Nutrition Journal* reported that dietary total antioxidant capacity is inversely related to CRP concentration in young Japanese women.[50] In yet another recent study, the total antioxidant capacity of different strengths of coffee was compared in human volunteers over a period of four weeks: Plasma total antioxidant status (TAS), ORAC, oxidized LDL and 8-epi-prostaglandin F2α, erythrocyte SOD, GPx, and CAT activity were measured at baseline and after the interventions.

The authors reported that ORAC increased after ingestion of mild light roast coffee. However, no significant alteration in lipid peroxidation biomarkers was observed: Both coffees had antioxidant effects. Although mild light roast coffee contained more chlorogenic acids, there were similar antioxidant effects between the treatments.[51]

There seems little disagreement about the beneficial effects of certain antioxidants. It remains to be seen whether a unit-value such as ORAC or TEAC can be a helpful benchmark to health benefit. ORAC at present is no more than an *interval scale*: 1,800 ORAC units are twice 900. But, can we say that 1,800 units are twice as beneficial as 900 units? We also don't know what is the lowest ORAC units-value that has any health benefits, and what is the highest unit-value that has adverse, even toxic, effects. Parenthetically, one source lists raw sumac bran as having 312,400 ORAC units (*micromoles of Trolox Equivalent per 100 grams*)!

4.24 INHIBITION ASSAY

ORAC and TEAC assays are *inhibition* methods where a sample is added to a free radical-generating system, and inhibition of free radical action is measured. This inhibition of free radicals is related to the sample antioxidant capacity.[52–55]

4.25 HOW ADVISABLE IS SUPPLEMENTING?

Antioxidant supplementation should depend on evidence of deficiency. The progressive age-related decline in antioxidant activity is seen in the first four panels in Figure 4.4, shown previously in the chapter, that indicate declining SOD, CAT, GSH, and CoQ10 activity trends.

The first four panels in Figure 4.4 were shown previously in Figures 4.2 and 4.3. They compare age-related decline in endogenous antioxidant activity.

The next three panels show concomitant changes in health function: It can be seen that the frequency of the cardiovascular health risks, hypertension, atherosclerosis, and diabetes rise with age, while at the same time, major

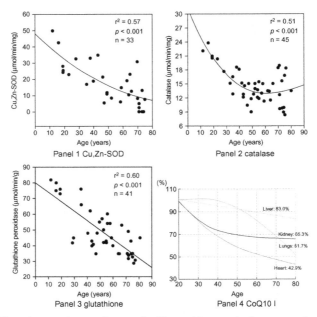

These four panels were shown earlier. The next four panels show concomitant changes in health function:

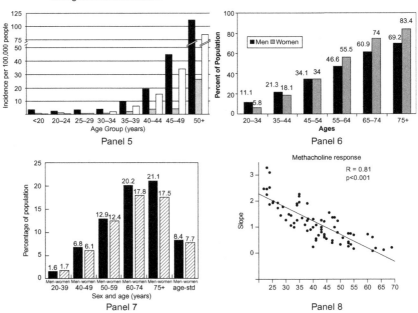

FIGURE 4.4 **Panel 5** Mean Incidence of atherosclerotic coronary artery disease (open bars) as a function of age imbedded in data about sudden cardiac death (SCD) as function of age, stratified by sex and other data. . . .[a] **Panel 6** Prevalence of cardiovascular diseases in Americans age 20 and

(Continued)

endogenous antioxidants decline with age. In summary, as endogenous anti-oxidants activity declines progressively with age:

- endothelium impairment sets in and worsens progressively with age,
- blood pressure rises progressively as NO/cGMP formation declines,
- atherosclerosis worsens progressively, and
- coronary artery diseases and diabetes become a progressively greater risk.

The incidence of cardiovascular and heart risks and diabetes increases with age, but by the 60 to 74 years of age bracket, the incidence of diabetes trend stops rising. Obviously it doesn't mean that many are well now and that they finally outgrew their diabetes. What it means is that by that age, many with diabetes already died: The prevalence of persons with diabetes—proportion of the total population—is not rising because there are fewer of them to be counted. This is commonly called "survivorship bias." As noted previously, diabetes is one of the health calamities that befall us and it may perhaps be the costliest.

Figure 4.5 shows data initially presented in the *New England Journal of Medicine* (NEJM) and adapted in a study titled "Optimizing cardiovascular outcomes in diabetes mellitus," appearing in the *American Journal of Medicine*.[56] The risk of MI is almost the same in patients with diabetes without previous MI as non-diabetic patients with MI.[57]

Back to our panels and to sum up: Antioxidant activity declines with age, endothelium-dependent NO/cGMP formation declines with age, blood pressure rises, atherosclerosis progressively worsens, and so does diabetes. However, there is presently no basis to say: "*As* antioxidant activity declines with age, *so* does endothelium NO/cGMP formation decline..., etc.," because "as" and "so" imply causality, and we don't know that to be the case. The inference is certainly not without basis, but it is only inference so far.

Figure 4.6, seen previously, shows that age-related prevalence of ED also rises. That's no surprise because much of this linkage has already been made so far in this book. There is a regular increase in incidence of ED to nearly 60% by the sixth decade: more than half of men are *impotent* by age 60. Of course, there is ED and then again, there is ED: There is no indication of degree of severity of ED in Figure 4.6, and ED does occur in degrees of severity.

---

older by age and sex (NHANES: 1999–2002). These data include CHD, CHF, stroke, and hypertension.[b] **Panel 7** shows the prevalence of diabetes in men and women in the US population age 20 years or older, based on the National Health and Nutrition Examination Survey III. Diabetes is defined by fasting plasma glucose greater than or equal to 126 mg/dL. (age-std = age-standardized).[c] **Panel 8** shows metacholine "challenge" results in progressively less NO/cGMP formation by the endothelium. (Aging impairs NO/cGMP vasodilation).[d] Source: [a]*Eckart, R. E., Shry, E. A., Burke, A. P., et al. (2011). Sudden Death in Young Adults: An Autopsy-Based Series of a Population Undergoing Active Surveillance. J Am Coll Cardiol, 58, 1254–1261. Reproduced with kind permission from* Journal of the American College of Cardiology; [b]*CDC/NCHS and NHLBI;* [c]*National Diabetes Fact Sheet, Center for Disease Control (CDC): http://www.cdc.gov/diabetes/pubs/factsheet11.htm;* [d]*Gerhard, M., Roddy, M. A, Creager, S. J., & Creager, M. A. (1996). Aging Progressively Impairs Endothelium-Dependent Vasodilation in Forearm Resistance Vessels of Humans. Hypertension, 27, 849 -853. Reproduced with kind permission from* Hypertension.

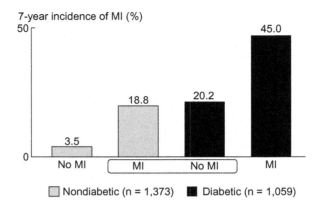

FIGURE 4.5 Patients with diabetes but without previous myocardial infarction (MI) are at the same risk for MI as patients without diabetes with previous MI. *Sobel, B. E. (2007). Optimizing cardiovascular outcomes in diabetes mellitus. Am J Med, 120, Suppl 2, S3–S11; adapted from Haffner, S. M. (2002). Lipoprotein disorders associated with type 2 diabetes mellitus and insulin resistance.* American J Cardiol, *90, 8, Suppl, 55–61. Reproduced with kind permission from the* American Journal of Medicine.

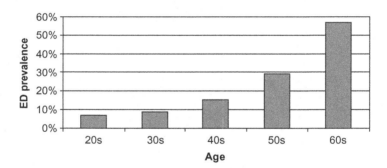

FIGURE 4.6 Prevalence of ED. Source: *Prins, J., Blanker, M. H., Bohnen, A. M., Thomas, S., et al. (2002). Prevalence of erectile dysfunction: a systematic review of population-based studies.* Int J Impot Res, *14, 422–432. Reproduced with kind permission from the* International Journal of Impotence Research.

Figure 4.7 gives a somewhat different picture of age-related ED: The left gray bars show prevalence of *minor* ED and fluctuate little over the age span; the center white bars show *moderate* ED progressively increasing over the lifespan; the right black bars show *complete* ED progressively worsening over the life span but at lesser acceleration than *moderate* ED.

Admittedly, Figure 4.6 presents a rather dim and discouraging view of the impact of age on erectile function as all categories (mild, moderate, and severe) are lumped together. However, differentiating between degrees of

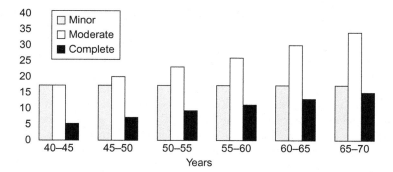

FIGURE 4.7   Age-related declining ED (by decade). *Reproduced with permission from the* Journal of Urology. Source: *Feldman, H. A., Goldstein, I., Hatzichristou, D. G., et al. (1994). Impotence and its medical and psychosocial correlates: results of the Massachusetts Male Aging Study.* J Urol, *151, 54–61.*

ED presents an altogether more encouraging picture, and it tells us that the cardiovascular hazards may not have quite so serious an impact on erectile function if one differentiates degree of ED. This is by no means intended to downgrade the hazards of the impact, for instance, of the SAD diet on health and longevity with respect to hypertension, atherosclerosis, coronary heart disease, and diabetes.

Type-2 diabetes is a separate and somewhat dire case. Figure 4.5 also shows that diabetes may be more serious a risk than the other cardiovascular and heart disorders. What's more, one source reports in *Medical Clinics of North America* that there is 95% prevalence of ED in diabetes and cautions that: "This prevalence does not represent normal aging, however, any more than hypertension is considered to be normal in an older adult, and erectile dysfunction is not an inevitable effect of aging."[58] First of all, population "prevalence" represents precisely what is normal in that population and not a quirk. One can only hope that the consistent data presented in this monograph may go a ways to dispel that idea generally held by many both in medicine and in the public forum.

There is even strong evidence of age-related eNOS uncoupling (in rats).[59] Those data could provide some explanation, at least in part, for the age related decline in NO biosynthesis and availability evidenced in panel 8.

4.26 IS THERE AN AGE-RELATED PROGRESSIVE ANTIOXIDANT DEFICIENCY SYNDROME?

Evidence strongly supports the temptation to suggest that an age-related progressive antioxidant deficiency syndrome (PADS) that impairs endothelium function is at the root of the cluster of health calamities, including hypertension, atherosclerosis, coronary heart disease, metabolic syndrome, type-2 diabetes, and ED. There is some support for such a hypothesis because for one thing, many of the conventional medical treatments for these calamities address deficiency in the NO/cGMP pathway.

However, stronger support would come from evidence that the age-related decline in endogenous antioxidants in parallel with declining NO biosynthesis and rising health calamities also parallels age-related rise in serum ROSs. It should surprise no one that this evidence exists: The journal *Mechanisms of Ageing and Development* reports a study on age-induced ROSs effect on ROS and reactive nitrogen species (RNSs) in human phagocyting granulocytes. The ROSs and RNSs were quantified by chemiluminescence assay and by measurement of nitrite production. The age-induced ROS generation was studied in healthy subjects ranging from 20 to 80 years old, divided into six age groups: The results showed the procedure to parallel age-induced generation of the ROS and RNS. A significant increase of ROS production was observed from 40 years old, while for RNS this increase was observed only from 50 years old. These data suggest an age-related increase of oxidizing species generation (ROS/RNS). The increased generation of ROS in 40-to-49-year-old participants was induced before the increase in RNS in those 50−59 years old.[60]

It is not part of the agenda of this monograph to show how the endothelium non-independent vaso-relaxation pathway may actually not be all that non-independent. However, writing in the *Journal of Cardiovascular Pharmacology*, investigators also found (in rats) that eNOS undergoes uncoupling with aging, manifested by a decrease in NO and a 3-fold increase in peroxynitrite for 16-week-old and 110-week-old rats, respectively. Metoprolol, a *beta blocker*, reversed eNOS uncoupling, and increased the production rate and concentration of NO. This effect was observed also with L-arginine and SOD, among several factors.[61]

The *Journal of Hypertension* reported in 2012 that angiotensin II receptor blockers (ARB) improve endothelial dysfunction associated with sympathetic hyperactivity in metabolic syndrome. In short, ARB type 1 treatment ameliorated impaired forearm vasodilation in response to ACh, thus improving impaired endothelial and baroreflex function.[62]

In a study titled "The antioxidant effects of statins" appearing in the journal *Coronary Artery Disease*, the authors state that statins interfere with oxidation in several ways that may contribute to reducing the atherogenic process. Statins inhibit oxidant enzyme activity and upregulate the activity of antioxidant enzymes such as catalase. They reduce endothelial dysfunction mainly by their ability to enhance endothelial NO bioavailability.[63]

The issue regarding statins is not free of controversy. There is a substantial difference between statins have antioxidant properties, and statins function principally as antioxidants: In an article titled, "Antioxidants, statins, and atherosclerosis," appearing in the *Journal of the American College of*

Cardiology, the author contends that "... despite lipoprotein oxidation's biologically plausible role in atherogenesis, several studies have reported inconsistent effects of antioxidants on clinical coronary end points, in sharp contrast with the studies of lipid modification with... statins."[64] Obviously no one to whom antihypertensive or statin meds is prescribed should be encouraged to chuck them in favor of broccoli, rose hips, and fish oil, but adding a little of those to diet "might not hurt," as they say.

This monograph aims to review recent research on cardio-sexual function and health and the possible basis for linking health hazards and ED to damage caused also by the SAD insofar as it has been shown that damage to be due to resulting oxidative stress. That said, it is not a mission of this monograph to suggest that anyone engage simply in ad lib antioxidant consumption. While cautious and guided supplementation of micronutrients—especially antioxidants—may in theory be a good idea, and clinical research cited here shows it to advantage in selected cases, there are many reasons why it cannot be encouraged in haphazard fashion. To begin with, it is not wise to supplement something for which individual deficiency has not been shown to exist as this may actually backfire. Furthermore, there are some individuals in whom antioxidant supplementation is contraindicated.

For instance, a report titled "The benefits and hazards of antioxidants: Controlling cancer cell apoptosis and other protective mechanisms in cancer patients and the human population," appeared in the *Journal of the American College of Nutrition*. The author makes a number of cogent points:

In moderate concentrations, ROSs are essential mediators of antimicrobial phagocytosis, detoxification reactions carried out by the cytochrome P-450 complex, and apoptosis which eliminates cancerous and other life-threatening cells. Excessive antioxidants could endanger these protective functions.

Exogenous antioxidants (vitamins E, C, beta-carotene, and others) could protect against cancer and other degenerative diseases in people with innate or acquired high levels of ROSs. However, excessive levels of antioxidants might suppress these protective functions, particularly in people with a low innate baseline level of ROSs.

Screening human populations for ROS levels could both help identify groups with a high level of ROSs that are at a risk of developing cancer and other degenerative diseases, and also identify groups with a low level of ROSs that are at a risk of downregulating ROS-dependent anti-cancer and other protective reactions.

Finally, the author proposes screening populations to provide a basis for antioxidant supplementation.[65]

In essence, ROSs are a key element in the immune system defenses against microbes and cancer. It might be a good idea to raise antioxidant levels if it can be ascertained that there is a deficiency that threatens health and sexual function. However, assessment of body antioxidant capacity is not presently a part of routine medical practice. One might argue that, as it has been shown earlier, antioxidant capacity decreases with age, and because that correlates with ED, why not simply supplement as we age. One good reason is that the age-related decline in antioxidants is an epidemiological population-generated statistic, and that it tells you nothing about antioxidant functional capacity and cancer in any particular individual.

On the other hand, considerable evidence to the contrary notwithstanding, some sources actually suggest that antioxidant supplementation is useless at best. A report in *Circulation* concluded that scientific data do not justify the use of antioxidant vitamin *supplements* (my italics) for CVD risk reduction. The authors aver that CVD risk reduction can be achieved by the long-term consumption of diets consistent with the AHA Dietary Guidelines; the long-term maintenance of a healthy body weight through balancing energy intake with regular physical activity; and the attainment of desirable blood cholesterol and lipoprotein profiles and blood pressure levels: "No consistent data suggest that consuming micronutrients at levels exceeding those provided by a dietary pattern consistent with AHA Dietary Guidelines will confer additional benefit with regard to CVD risk reduction."[66] This conclusion is not uncorroborated.

In a review titled "Oxidative stress and inflammation in heart disease: Do antioxidants have a role in treatment and/or prevention?" Dr. Pashkow levels a number of pertinent valuable criticisms at the outcomes of many studies on the benefits of antioxidants. He avers that the results of research on the effectiveness of antioxidants have been largely disappointing due to a number of naive or erroneous assumptions and poor study design procedures. Here are some:

- Nutritional compounds such as vitamin E, vitamin C, and β-carotene are mainly responsible for the benefit of increased intake of dietary fruits and vegetables.
- All antioxidants are essentially the same.
- Misunderstanding of the mode of action of different antioxidants, leading to erroneous clinical designs and patient selection.
- Frequently absent dose−response documentation.

Dr. Pashkow further holds that antioxidants, because of their provenance as "natural products" or "nutritional supplements," and their presumption of safety and efficacy generated from the results of epidemiologic and observational studies early-on, have not been subjected to the same stringent developmental requirements that are applied to new pharmaceutical drug candidates. "Biologically active compounds, formulated properly, administered in appropriate amounts for an appropriate duration to the right patients will be required to achieve all the requirements that truly define therapeutic success."[12]

On the other hand, many of the studies cited previously, as well as those cited in later chapters, all drawn from journals just as prestigious, resorted to antioxidant formulations other than "natural antioxidant supplements" to treat conditions associated with CVD and endothelial dysfunction.

4.27 NATURE'S ANTIOXIDANT RESPONSE ELEMENT

The media make much of free radical damage and antioxidant prevention and repair. Yet it occurs to few that nature is not known for overlooking details vital to survival. For the most part, we survive many years without glaring evidence of widespread corrosion by free radicals. Furthermore, we have a protective process inside many of our cells that adjusts cell antioxidant capacity to meet free radical density. It is an endogenous antioxidant defense involving an *antioxidant response element* (ARE).

Research informs us that excess antioxidant intake can lead to reducing that ARE. By consuming more antioxidants, we become dependent on them and reduce the ability to detoxify. We now know that cells exposed to oxidative stress, or drugs and toxins, respond by triggering upregulation of genes that contain a DNA regulating element that promotes the antioxidant ARE.[67,68] It is probably unwise to spook those!

In the next chapter, free radicals, excess weight, and insulin resistance are shown to conspire in creating Syndrome X, metabolic syndrome.

REFERENCES

1. Tiano L, Belardinelli R, Carnevali P, Principil F, Seddaiu G, Littarru GP. Effect of coenzyme Q10 administration on endothelial function and extracellular superoxide dismutase in patients with ischaemic heart disease: a double-blind, randomized controlled study. *Eur Heart J* 2007;**28**:2249–55.
2. Gao L, Mao Q, Cao J, Wang Y, Zhou X, Fan L. Effects of coenzyme Q10 on vascular endothelial function in humans: a meta-analysis of randomized controlled trials. *Atherosclerosis* 2012;**221**:311–6.
3. Rundek T, Naini A, Sacco R, Coates K, DiMauro S. Atorvastatin decreases the Coenzyme Q10 level in the blood of patients at risk for cardiovascular disease and stroke. *Arch Neurol* 2004;**61**:889–92.
4. Yilmaz MI, Baykal Y, Kilic M, Sonmez A, Bulucu F, Aydin A, et al. Effects of statins on oxidative stress. *Biol Trace Elem Res* 2004;**98**:119–27.
5. Bourdon E, Loreau N, Blache D. Glucose and free radicals impair the antioxidant properties of serum albumin. *FASEB J* 1999;**13**:233–44.
6. Tagliabue M, Pinach S, Di Bisceglie C, Brocato L, Cassader M, Bertagna A, et al. Glutathione levels in patients with erectile dysfunction, with or without diabetes mellitus. *Int J Androl* 2005;**28**:156–62.
7. Juutilainen A, Lehto S, Rönnemaa T, Pyörälä K, Laakso M. Type 2 diabetes as a "coronary heart disease equivalent": an 18-year prospective population-based study in Finnish subjects. *Diab Care* 2005;**28**:2901–7.
8. <http://www.glutathione-for-health.com/index.html>; 2013 [accessed 10.11.13].
9. Droge W. Free radicals in the physiological control of cell function. *Physiol Rev* 2002;**82**:47–95.
10. Schroecksnadel K, Fischer B, Schennach H, Weiss G, Fuchs D. Antioxidants suppress Th1-type immune response in vitro. *Drug Metab Lett* 2007;**1**:166–71.

11. Ertuğ PU, Olguner AA, Oğülener N, Singirik E. Protective effect of quercetin, a polyphenolic compound, on mouse corpus cavernosum. *Fundam Clin Pharmacol* 2009;**24**:223−32.
12. Pashkow FJ. Oxidative stress and inflammation in heart disease: do antioxidants have a role in treatment and/or prevention? *Int J Inflam* 2011;**2011**.
13. Roberts CK, Vaziri ND, Barnard RJ. Effect of diet and exercise intervention on blood pressure, insulin, oxidative stress, and nitric oxide availability. *Circulation* 2002;**106**:2530−2.
14. Samiec PS, Drews-Botsch C, Flagg EW, Kurtz JC, Sternberg P. Glutathione in human plasma: decline in association with aging, age-related macular degeneration, and diabetes. *Free Rad Biol Med* 1998;**24**:699−704.
15. Lu C-Y, Lee H-C, Fahn H-J, Wei Y-H. Oxidative damage elicited by imbalance of free radical scavenging enzymes is associated with large-scale mtDNA deletions in aging human skin. *Mutat Res Fundam Mol Mech Mutag* 1999;**423**:11−21.
16. Casado A, de la Torre R, López-Fernández E, Carrascosa D, Venarucci D. [Superoxide dismutase and catalase levels in diseases of the aged] [Article in Spanish]. *Gac Med Mex* 1998;**134**:539−44.
17. Shi Y, Vanhoutte PM. Reactive oxygen-derived free radicals are key to the endothelial dysfunction of diabetes. *J Diab* 2009;**1**:151−62.
18. Tesfamariam B, Cohen RA. Free radicals mediate endothelial cell dysfunction caused by elevated glucose. *Am J Physiol − Heart* 1992;**263**:H321−6.
19. Faure P, Polge C, Monneret D, Favier A, Halimi S. Plasma 15-F2t isoprostane concentrations are increased during acute fructose loading in type 2 diabetes. *Diab Metab* 2008;**34**:148−54.
20. Glushakova O, Kosugi T, Roncal C, Mu W, Heinig M, Cirillo P, et al. Fructose induces the inflammatory molecule ICAM-1 in endothelial cells. *J Am Soc Nephrol* 2008;**19**:1712−20.
21. Hussain SA, Ahmed ZA, Mahwi TO, Aziz TA. Quercetin dampens postprandial hyperglycemia in type 2 diabetic patients challenged with carbohydrates load. *Int J Diabet Res* 2012;**1**:32−5.
22. Goldin A, Beckman JA, Schmidt AM, Creager MA. Advanced glycation end products—sparking the development of diabetic vascular injury. *Circulation* 2006;**114**:597−605.
23. Seftel AD, Vaziri ND, Ni Z, Razmjouei K, Fogarty J, Hampel N, et al. Advanced glycation end products in human penis: elevation in diabetic tissue, site of deposition, and possible effect through iNOS or eNOS. *Urol* 1997;**50**:1016−26.
24. Negrean M, Stirban A, Stratmann B, Gawlowski T, Horstmann T, Götting C, et al. Effects of low- and high-advanced glycation endproduct meals on macro- and microvascular endothelial function and oxidative stress in patients with type 2 diabetes mellitus. *Am J Clin Nutr* 2007;**85**:1236−43.
25. McCarty MF. The low-AGE content of low-fat vegan diets could benefit diabetics − though concurrent taurine supplementation may be needed to minimize endogenous AGE production. *Med Hypotheses* 2005;**64**:394−8.
26. Ceriello A, Bortolotti N, Motz E, Pieri C, Marra M, Tonutti L, et al. Meal-induced oxidative stress and low-density lipoprotein oxidation in diabetes: the possible role of hyperglycemia. *Metabolism* 1999;**48**:1503−8.
27. Tsai W-C, Li Y-H, Lin CC, ChaO T-H, Chen JH. Effects of oxidative stress on endothelial function after a high-fat meal. *Clin Sci* 2004;**106**:315−9.
28. Virag R, Floresco J, Richard C. Impairment of shear-stress-mediated vasodilation of cavernous arteries in erectile dysfunction. *Int J Impot Res* 2004;**16**:39−42.
29. Costa C, Virag R. The endothelial−erectile dysfunction connection: an essential update. *J Sex Med* 2009;**6**:2390−404.

30. Jones RW, Rees RW, Minhas S, Ralph D, Persad RA, Jeremy JY. Oxygen free radicals and the penis. *Expert Opin Pharmacother* 1992;**3**:889−97.

31. Azadzoi KM, Golabek T, Radisavljevic ZM, Yalla SV, Siroky MB. Oxidative stress and neurodegeneration in penile ischaemia. *BJU Int* 2010;**105**:404−10.

32. Azadzoi KM, Master TA, Siroky MB. Effect of chronic ischemia on constitutive and inducible nitric oxide synthase expression in erectile tissue. *J Androl* 2004;**25**:382−8.

33. Vardi Y, Dayan L, Apple B, Gruenwald I, Ofer Y, Jacob G. Penile and systemic endothelial function in men with and without erectile dysfunction. *Eur Urol* 2009;**55**:979−85.

34. Azadzoi KM, Schulman RN, Aviram M, Siroky MB. Oxidative stress in arteriogenic erectile dysfunction: prophylactic role of antioxidants. *J Urol* 2005;**174**:386−93.

35. Zhang Q, Radisavljevic ZM, Siroky MB, Azadzoi KM. Dietary antioxidants improve arteriogenic erectile dysfunction. *Int J Androl* 2011;**34**:225−35.

36. Riccardi G, Rivellese AA. Effects of dietary fiber and carbohydrate on glucose and lipoprotein metabolism in diabetic patients. *Diab Care* 1991;**14**:1115−25.

37. Vicari E, La Vignera S, Condorelli R, Calogero AE. Endothelial antioxidant administration ameliorates the erectile response to PDE5 regardless of the extension of the atherosclerotic process. *J Sex Med* 2010;**7**:1247−53.

38. La Vignera S, Condorelli R, Vicari E, D'Agata R, Calogero AE. Endothelial antioxidant compound prolonged the endothelial antiapoptotic effects registered after tadalafil treatment in patients with arterial erectile dysfunction. *J Androl* 2012;**33**:170−5.

39. MacNee W. Oxidants/antioxidants and COPD. *CHEST* 2000;**117**(5_suppl_1):303S−17S.

40. Bode-Boger SM, Boger RH, Schroder EP, Frolich JC. Exercise increases systemic nitric oxide production in men. *J Cardiovasc Risk* 1994;**1**:173−8.

41. Hsiao W, Shrewsberry AB, Moses KA, Johnson TV, Cai AW, Stuhldreher P, et al. Exercise is associated with better erectile function in men under 40 as evaluated by the international index of erectile function. *J Sex Med* 2012;**9**:524−30.

42. Borghouts LB, Keizer HA. Exercise and insulin sensitivity: a review. *Int J Sports Med* 2000;**21**:1−12.

43. Holloszy JO. Exercise-induced increase in muscle insulin sensitivity. *Appl Physiol* 2005;**99**:338−43.

44. Watson RT, Pessin JE. Bridging the GAP between insulin signaling and GLUT4 translocation. *Trends Biochem Sci* 2006;**31**:215−22.

45. Meldrum DR, Gambone JC, Morris MA, Esposito K, Giugliano D, Ignarro LJ. Lifestyle and metabolic approaches to maximizing erectile and vascular health. *Int J Impot Res* 2012;**24**:61−8.

46. USDA: <http://www.ars.usda.gov/services/docs.htm?docid = 15866>;   2013   [accessed 10.11.13].

47. Carnethon MR, De Chavez PJD, Biggs ML, Lewis CE, Pankow JS, Bertoni AG, et al. Association of weight status with mortality in adults with incident diabetes. *JAMA* 2012;**308**:581−90.

48. Lotito SB, Frei B. Consumption of flavonoid-rich foods and increased plasma antioxidant capacity in humans: cause, consequence, or epiphenomenon? *Free Radic Biol Med* 2006;**41**:1727−46.

49. Zulueta A, Esteve MJ, Frígola A. ORAC and TEAC assays comparison to measure the antioxidant capacity of food products. *Food Chem* 2009;**114**:l10−316.

50. Kobayashi. S, Murakami K, Sasaki S, Uenishi K, Yamasaki M, Hayabuchi H, et al. Dietary total antioxidant capacity from different assays in relation to serum C-reactive protein among young Japanese women. *Nutr J* 2012;**11**:91.

51. Corrêa TA, Monteiro MP, Mendes TM, Oliveira DM, Rogero MM, Benites CI, et al. Medium light and medium roast paper-filtered coffee increased antioxidant capacity in healthy volunteers: results of a randomized trial. *Plant Foods Hum Nutr* 2012;**67**:277−82.

52. Cao G, Prior RL. Comparison of different analytical methods for assessing total antioxidant capacity of human serum. *Clin Chem* 1998;**44**:1309−15.

53. Cao G, Prior RL. Measurement of oxygen radical absorbance capacity in biological samples. *Methods Enzymol* 1999;**299**:50−62.

54. Lightbody JH, Stevenson LM, Jones DG, Donaldson K. Standardization of a spectrophotometric assay for plasma total antioxidant capacity. *Anal Biochem* 1998;**258**:369−72.

55. Wang CC, Chu CY, Chu KO, Choy KW, Khaw KS, Rogers MS, et al. Trolox-equivalent antioxidant capacity assay versus oxygen radical absorbance capacity assay in plasma. *Clin Chem* 2004;**50**:952−4.

56. Sobel BE. Optimizing cardiovascular outcomes in diabetes mellitus. *Am J Med* 2007;**120** (Suppl 2):S3−11.

57. Haffner SM, Lehto S, Rönnemaa T, Pyörälä K, Laakso M. Mortality from coronary heart disease in subjects with type 2 diabetes and in nondiabetic subjects with and without prior myocardial infarction. *N Engl J Med* 1998;**339**:229−34.

58. Kaiser FE. Erectile dysfunction in the aging man. *Med Clin North Am* 1999;**83** (5):1267−78 1 September 1999.

59. Johnson JM, Bivalacqua TJ, Lagoda GA, Burnett AL, Musicki B. eNOS-uncoupling in age-related erectile dysfunction. *Int J Impot Res* 2011;**23**:43−8.

60. Martins Chaves M, Rocha-Vieira E, Pereira dos Reis A, de Lima e Silva R, Gerzstein NC, Nogueira-Machado JA. Increase of reactive oxygen (ROS) and nitrogen (RNS) species generated by phagocyting granulocytes related to age. *Mech Ageing Dev* 2000;**119**:1−8.

61. Funovic P, Korda M, Kubant R, Barlag RE, Jacob RF, Mason RP, et al. Effect of beta-blockers on endothelial function during biological aging: a nanotechnological approach. *J Cardiovasc Pharmacol* 2008;**51**:208−15.

62. Kishi T, Hirooka Y, Konno S, Sunagawa K. Angiotensin II receptor blockers improve endothelial dysfunction associated with sympathetic hyperactivity in metabolic syndrome. *J Hypertens* 2012;**30**:1646−55.

63. Davignon J, Jacob RF, Mason RP. The antioxidant effects of statins. *Coron Artery Dis* 2004;**15**:251−8.

64. Gotto Jr. AM. Antioxidants, statins, and atherosclerosis. *J Am Coll Cardiol* 2003;**41**:1205−10.

65. Salganik RI. The benefits and hazards of antioxidants: controlling apoptosis and other protective mechanisms in cancer patients and the human population. *J Am Coll Nutr* 2001;**20** (Suppl 5):464S−72S.

66. Kris-Etherton PM, Lichtenstein AH, Howard BV, Steinberg D, Witztum JL. Antioxidant vitamin supplements and cardiovascular disease. *Circulation* 2004;**110**:637−41.

67. Nerland DE. The antioxidant/electrophile response element motif. *Drug Metabol Rev* 2007;**39**:235−48.

68. Nguyen T, Sherratt PJ, Pickett CB. Regulatory mechanisms controlling gene expression mediated by the antioxidant response element. *Ann Review Pharmacol Toxicol* 2003;**43**:233−60.

Chapter 5

Metabolic Syndrome Impairs Erectile Function

5.1 INTRODUCTION

A study titled "High proportions of erectile dysfunction (ED) in men with the metabolic syndrome," published in *Diabetes Care*, informs us of a predictable rise in the inflammation marker C-reactive protein (CRP) level and in impairment of endothelium function score as the number of components of metabolic syndrome increase. The link between the International Index of Erectile Function (IEEF) scores and endothelial function scores found in their participants suggested the presence of a vascular basis to both metabolic syndrome and ED.[1]

The authors proposed that the *mechanism* underlying both metabolic syndrome and ED is reduced NO availability: Impaired endothelium dependent relaxation (NO/cGMP) has been demonstrated in the laboratory in isolated corpus cavernosum strips from patients with ED.[2]

It will become clear that metabolic syndrome may be any combination of otherwise distinct clinical entities, including hypertension, atherosclerosis, coronary artery disease (CAD), visceral obesity, endothelial dysfunction, insulin resistance (IR) or insufficiency, and testosterone insufficiency. Depending on where an investigator focuses attention, metabolic syndrome yields signs and, bottom line, it features ED.

In any event, health authorities commonly consider metabolic syndrome an increasingly prevalent risk of cardiovascular and heart disease—which, of course, it is—and rarely see its connection to ED, despite its known prevalence. In part, this may be due to the fact that the link between ED and hypertension, and ED and atherosclerosis, has captured the limelight as an alternative to "psychogenic" theories of ED.

5.2 WHAT IS METABOLIC SYNDROME?

Conventional diagnostic syndrome criteria include:

- Abdominal obesity—waist circumference exceeding 40 inches in men, and 35 in women.

Robert Fried: Erectile Dysfunction as a Cardiovascular Impairment.
DOI: http://dx.doi.org/10.1016/B978-0-12-420046-3.00005-6
141

- Triglycerides exceeding 150 mg/dL.
- Low high-density lipoprotein cholesterol (HDL-C)—greater than 40 mg/dL in men, and greater than 50 mg/dL in women.
- High blood pressure (BP)—equal to or greater than 130/85 mm Hg
- High fasting glucose.[3]

The prevalence of metabolic syndrome is on the rise as shown in Figure 5.1, drawn from a report in the *Journal of the American Medical Association* (JAMA) titled "Prevalence of the metabolic syndrome among US adults: findings from the Third National Health and Nutrition Examination Survey."[4]

By the 60 to 70 years age bracket, just under half the US population qualifies for the diagnosis "metabolic syndrome." Similar to age-related prevalence of type-2 diabetes encountered earlier, the decline in prevalence in the greater than 70-years-old male population (as noted in Figure 5.1) is most likely due to survivorship bias—attrition by death. On that score, it resembles type-2 diabetes. Note that more men than women are fatally affected by the condition.

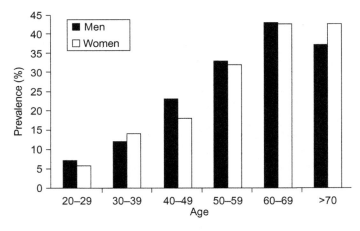

FIGURE 5.1 Age-specific prevalence of metabolic syndrome. Source: *Ford, E. S., Giles, W. H., & Dietz, W. H. (2002). Prevalence of the metabolic syndrome among US adults: findings from the Third National Health and Nutrition Examination Survey.* JAMA, *287, 356–359. Reproduced with permission from the* Journal of the American Medical Association.

5.3 METABOLIC SYNDROME MAY ENTAIL CARDIO-SEXUAL RISK EQUAL TO DIABETES AND CORONARY HEART DISEASE

The prevalence of metabolic syndrome rises with age just like the previously encountered age-related cardiovascular and heart disease, and diabetes risk data ... and ED. The same trend in men—a steady near-linear rise, then reversal—as seen in prevalence of diabetes in a previous chapter, certainly gives pause. If epidemiological data allow inferences about clinical data, then it may be concluded that metabolic syndrome is riskier to life than hypertension. Figure 5.2 illustrates the point.

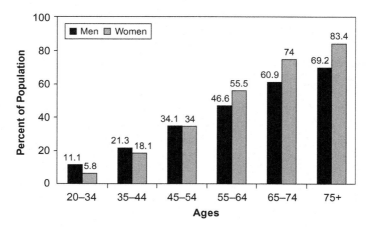

FIGURE 5.2 Prevalence of high BP in Americans by age and sex, NHANES 1999–2002. Source: *Catravas, J. D. (2006). The 8th International Conference on the Vascular Endothelium: Translating Discoveries into Public Health Practice − Part I.* Vasc Pharm, *45, 302–307. Reproduced with kind permission from* Vascular Pharmacology.

The trend for hypertension in men rises well into the seventies,[5] whereas it dips for metabolic syndrome by that age as seen in Figure 5.2. Unlike "uncomplicated" hypertension and even "uncomplicated" atherosclerosis in men, metabolic syndrome may exact a cost somewhat greater even than ED. There is a nearly 10% loss of the population with metabolic syndrome as they pass the end of their sixties.

Note: *Prevalence* depends on number of individuals in a population counted and constitutes a percentage of the population having a given property—metabolic syndrome in this case. If the incidence of death in metabolic syndrome sufferers is greater in those in their mid-to-late sixties than in the population sampled, there will be fewer of them sampled, and that will result in a smaller prevalence percentage of the total population counted. That is attrition due to (non-) survivor bias.

5.4 BLAME IT ON DIET

Circulation reported a study titled "Dietary intake and the development of the metabolic syndrome—The Atherosclerosis Risk in Communities Study." What is the evidence? The goal of the study was to find the link between diet and the onset of metabolic syndrome. They relied on prospective data from participants aged 45 to 64 years old enrolled in the Atherosclerosis Risk in Communities (ARIC) Study, the data assessing dietary frequency intake of 66 food items. Metabolic syndrome was defined by American Heart Association (AHA) guidelines.

The analysis aimed to determine "Western" and "prudent" dietary patterns from 32 food groups, plus 10 food groups used in previous studies of the ARIC cohort.[6] Adjusting for demographic factors, smoking, physical activity, and energy intake, it was found that consumption of a Western dietary pattern was associated with incidence of metabolic syndrome.

Further analysis of food consumption revealed that meat, fried foods, and diet soda were strongly linked to incidence of metabolic syndrome. Conversely, dairy consumption was found to be beneficial. There seemed to be no links of metabolic syndrome to a prudent dietary pattern or intakes of whole grains, refined grains, fruits and vegetables, nuts, coffee, or sweetened beverages. No attempt was made to look for a connection between metabolic syndrome and diet soda.[7]

The evidence of a link between certain foods and metabolic syndrome is supportive of the theory, but it is not overwhelming. If foods X, Y, Z are causally linked to metabolic syndrome, then shouldn't withdrawal reverse the condition, unless there is evidence that consuming X, Y, and Z causes irreversible harm? None of that exists at present. Furthermore, it is clearly the case that not everyone who consumes X, Y, and Z develops the condition, even if it causes weight gain. According to the American Diabetes Association (ADA), most overweight people never develop type-2 diabetes, and many people with type-2 diabetes are at a normal weight or only moderately overweight.[8] What proffers immunity?

There is a certain logic, albeit probably erroneous, to the argument that metabolic syndrome is largely due to what is typically consumed in the US: colossal servings of salty, saturated fat-rich fast foods, washed down with soda pop. This argument does not take into account all the known facts: for instance, the role of endogenous antioxidant activity in maintaining the endothelial function in the face of near-universal, age-related, critical life function changes, including rising IR. This may eventually prevent us from looking for other culprits such as serum-available testosterone levels that impact the usual-suspects risks.

Then, as consumption of the Standard American Diet (SAD) diet increases, the prevalence of metabolic syndrome rises, and that further strengthens the notion that we need look no further than the lunch counter and the dinner plate to explain it. Additional weight is given the argument by clinical studies that report amelioration of the syndrome by diet change. To wit: A clinical study titled "Effect of a Mediterranean-style diet on endothelial dysfunction and markers of vascular inflammation in the Metabolic Syndrome: A randomized trial" appeared in the *Journal of the American Medical Association* (JAMA). Contending that the role of diet in the cause and development of metabolic syndrome "is poorly understood," the aim of the study was to assess the effect of a Mediterranean-style diet on endothelial function and vascular inflammatory markers in patients with the disorder.

The study was conducted on patients with metabolic syndrome as defined by the Adult Treatment Panel III (see below: Expert Panel on Detection, Evaluation, and Treatment of High Blood Cholesterol in Adults Treatment Panel III; ATP III) over a period of about two-and-one-half years at a university hospital in Italy. Participants in the intervention group were to follow a Mediterranean-style diet and were told how to increase daily consumption of whole grains, fruits, vegetables, nuts, and olive oil. A control group followed a *prudent diet* (50 to 60% carbohydrates, 15 to 20% proteins, and not more than 30% total fat).

Blood pressure, platelet aggregation, as well as arterial blood vessel relaxation response to L-arginine were taken as endothelial function scores. So were lipid and glucose values, insulin sensitivity, circulating levels of CRP (hs-CRP), and interleukins (IL-6, 7, and 18). After two years follow-up, patients on the Mediterranean-style diet (more foods rich in monounsaturated fat, polyunsaturated fat, and fiber) had a lower ratio of omega-6 to omega-3 fatty acids. Total fruit, vegetable, and nuts intake, whole grain intake, and olive oil consumption were also significantly higher in the intervention group. Mean body weight decreased more in the intervention group than in the control group.

Compared with patients on the control diet, those in the intervention diet group had significantly reduced serum concentrations of inflammatory markers (hs-CRP, IL-6, IL-7, and IL-18), as well as decreased IR. Endothelial function scores improved and remained stable. At the two-years follow-up also, a smaller number of patients in the intervention group still had features of metabolic syndrome as compared to patients in the control group. The authors concluded that a Mediterranean-style diet reduces the prevalence of metabolic syndrome and related cardiovascular risk.[9]

To what nutrition or other factors do these investigators attribute the benefits of the Mediterranean Diet? Among those cited are anti-inflammatory, anti-oxidative stress, and antioxidant activity of dietary fiber, antioxidant vitamins, and anti-inflammatory effect of omega-3 fatty acids. Likewise, the authors contend that there was a limitation in the study due to the inability to determine whether individual components of the diet accounted for the changes observed, or whether the changes in metabolic risk factors resulted from all the dietary changes.

There is no question but that such studies are helpful guides to reducing health hazards. But they fail us by suggesting that *correlation* is *cause* and that guesses—even educated guesses—are akin to knowing: What does what? Why? And, How does it do it? The results of this study and similar ones, though they may ultimately serve a purpose, fail to answer questions about the very basics of the physiology of the disorder(s) they are studying. What exactly are the deficiencies induced by "unhealthy" diets.

In the end, the difference between medicine and physiology is that medicine looks at outcome and guesses at cause, whereas physiology looks at cause and guesses at outcome. The danger in correlation studies is that they seduce us into believing that correlation is cause.

5.5 NOTE ON RELEVANT TYPES OF DIET

"Western pattern diet" is common in developed countries and consists mostly of high intakes of red meat, sugary desserts, high-fat foods, and refined grains. It also contains high-fat dairy products, high-sugar drinks, and higher intakes of processed meat. Prospective study findings suggest that consumption of a Western dietary pattern may promote the incidence of metabolic syndrome.

5.6 STANDARD AMERICAN DIET

The SAD is similar to the Western pattern diet with about 50% carbohydrate, 15% protein, and 35% fat. The SAD exceeds dietary guidelines for fat (below 30%), is below the guidelines for carbohydrate (above 55%), and is at the upper end of the guidelines for recommended for protein (below 15%). The quality of the carbohydrate, protein, and fat is at least as important as the quantity: Complex carbohydrates such as starch are believed to be healthier than sugar so frequently consumed in the SAD.

5.7 PRUDENT DIET

The "Prudent Diet" is associated with enhanced insulin sensitivity and a lower risk of type-2 diabetes.[10] The diet emphasizes fruits and vegetables as the mainstay of a prudent heart diet: at least 4-1/2 cups of fruits and vegetables every day, fiber-rich whole grains (three 1-oz. servings each day), and reduced salt consumption.

Go for colorful products, reducing consumption of potatoes (high starch content), and instead dark, leafy greens such as kale and collard greens, cooked tomatoes and colorful bell peppers. Fish is an excellent protein source, and the AHA recommends at least two, 3-1/2 oz. servings of fish, preferably oily fish, each week: oily fish are high in omega-3 fatty acids. Nuts, legumes, and seeds are also a good source of protein. Avoid processed meats and limit saturated fat to no more than 7% of overall calorie intake.

Data in Table 5.1 come from "Prudent diet and preventive nutrition from pediatrics to geriatrics: Current knowledge and practical Recommendations."[11] This is one of the most detailed Prudent Diet food

TABLE 5.1 Recommended Daily Energy Intake and Major Sources of Dietary Fats

Fat Sources	Details
Total Fat	Total Fat: 30%−35%; 34% dairy products and meat are the major sources of dietary fat; the (40−75 g) former contribute more than the latter.
Cholesterol	Less than 200 mg 270 mg egg yolk, brain, organ meat, beef, lamb, pork, poultry (thigh and skin), shellfish, shrimp, prawn, full-fat milk, and high-fat dairy products (cream, ice cream, milk shake, cheese, curd).
SAFA*	Less than 7%; 12% beef, lamb, pork, bacon, sausage, ribs, poultry with skin, butter, ghee, (10−15 g) vanaspathi (vegetable ghee), desserts, bakery products (cakes, biscuits, cookies, donut), cheese, ice cream, full-fat milk, and tropical oils (coconut, palm kernel, and palm oils).
TRAFA*	Avoid, or less than 2% hard margarine, vegetable shortening, frying fats (especially those used repeatedly), bakery products (cakes, Danish pastry, donuts, crackers, rusk, biscuits, cookies, white bread), French fries, fried chicken, peanut butter, nondairy creamer, tortillas, pizza, and virtually all "crispy and crunchy foods."
MUFA*	Up to 20%; 14% olive oil, canola oil, mustard oil, peanut oil, macadamia nuts, hazelnuts, pecans, peanuts, almonds, cashew nuts, pistachio nuts, avocado, dairy products, beef, lamb, mutton, poultry.
n-6 PUFA*	7%−8 %; 6%−8% vegetable oils (soybean, corn, safflower, sunflower, and cottonseed).
n-3 PUFA*	2%−3% less than 1% fatty fish (mackerel, halibut, lake trout, herring, sardines, albacore tuna, salmon), meat, poultry, vegetables (tofu, soybeans, pinto beans, flax seeds), vegetable oils (soybean, canola), salad dressing, whole grains, DHA-enriched egg, and nuts (walnuts, butternut, flax seeds, pecans).

*Abbreviations: SAFA, saturated fat; MUFA, monounsaturated fat; PUFA, polyunsaturated fat; and TRAFA, transunsaturated fat.
Source: Enas, E. A., Senthilkumar, A., Chennikkara, H., *et al.* (2003). Prudent diet and preventive nutrition from pediatrics to geriatrics: Current knowledge and practical recommendations. *Indian Heart J*, 55, 310−338. Reprinted with permission from the *Indian Heart Journal.*

plans readily accessible:Additional recommendations include that the main form of carbohydrates should ideally be whole grains and various types of fruits and vegetables, and that one should avoid prolonged cooking of vegetables.

Recommendations: The vegetarian diet is healthy only when it is low in saturated fat (SAFA), the predominant energy comes from foods with a low glycemic index, and when substituting full-fat dairy products for low-fat dairy products. Cooking oils containing high SAFA should be replaced with those containing high MUFA. Deep-frying, especially with previously used oils, should be discouraged. Nuts are healthy, wholesome foods, and make an excellent replacement for unhealthy calories. A diet rich in fish raises

HDL and lowers triglyceride levels. Consumption of fish is preferable to taking fish oil capsules.[11]

5.8 METABOLIC SYNDROME AND HYPERTENSION

The *Journal of Internal Medicine* reported a lengthy and detailed account of the prominent features of metabolic syndrome in a study titled "Influence of metabolic syndrome on hypertension-related target organ damage." Metabolic syndrome entails structural damage to heart muscle and kidneys.

Even after correction for age, 24-hour BPs, duration of hypertension, previous antihypertensive therapy, and gender distribution, hypertensive patients with metabolic syndrome have greater left ventricular (LV) mass (Figure 5.3), higher myocardial relative wall thickness, higher albumin excretion rate (AER), and a greater prevalence of left ventricular hypertrophy (LVH—more than double—among other signs, than patients without metabolic syndrome. This is not simply a matter of "pot-belly."

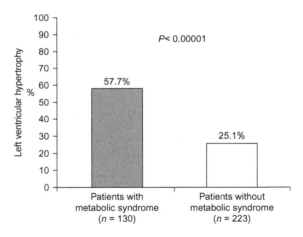

FIGURE 5.3 Prevalence of left ventricular hypertrophy in hypertensive patients with metabolic syndrome and those without it. Source: *Mule, G., Nardi, E., Cottone, S, Cusimano, P., Volpe, V., Piazza, VG., et al. (2005). Influence of metabolic syndrome on hypertension-related target organ damage.* J Int Med, *257, 503−513. Reproduced with kind permission from the* Journal of Internal Medicine.

The authors concluded that metabolic syndrome may complicate hypertension-related cardiac and kidney changes beyond the potential contribution of each single component of this syndrome. These signs of "target organ damage" are well-known predictors of cardiovascular events, and explain the increased cardiovascular risk associated with metabolic syndrome. In addition, an increased prevalence of different degrees of hypertensive retinal blood vessel damage (retinopathy) was observed in patients with metabolic syndrome when compared with persons without that disorder.

There were many other findings that need not be specified here except for prevalence of increased antioxidant response element (ARE), associated with adverse accumulation of cell oxidation byproducts.[12]

A report titled "The metabolic syndrome in hypertension: European Society of Hypertension position statement," published in the *Journal of Hypertension*, informs us that metabolic syndrome raises the risk of cardiovascular and renal events in hypertension. Obesity and IR have been implicated in the pathogenesis of the syndrome as well as number of independent factors of hepatic, vascular, and immunologic origin with proinflammatory properties. A high prevalence of end-organ damage and a poor prognosis have emerged from cross-sectional and prospective studies.

Treatment aims to reduce the high risk of a cardiovascular or a renal event and to prevent the greater chance that metabolic syndrome patients have to develop type-2 diabetes or hypertension. Recommended treatment includes lifestyle interventions that effectively reduce visceral obesity with or without the use of drugs that oppose the development of IR or body weight gain.[13]

5.9 METABOLIC SYNDROME ENTAILS INCREASED RISK OF CORONARY HEART DISEASE AND STROKE

The *Journal of the American College of Cardiology* reported on the link between hypertension and metabolic syndrome: Patients with the syndrome had nearly double the number of cardiovascular events compared to those without it. Even after adjustment for factors including age, gender, total cholesterol, smoking, left ventricular hypertrophy, and elevated systolic BP, the risk of developing cardiovascular events was still higher in patients with metabolic syndrome.

The authors concluded that in hypertensive subjects, metabolic syndrome raises cardiovascular risk associated with hypertension irrespective of the effects of several traditional cardiovascular risk factors. Hypertensive patients with metabolic syndrome are at increased risk of coronary heart disease (CHD) and stroke.[14]

It seems reasonably clear from the sources cited above that chronically elevated BP and metabolic syndrome potentiate to form a nasty mix that bodes ill for cardio-sexual function. Entering "hypertension, metabolic syndrome" in PubMed[15] yields countless relevant entries. It is doubtful that any can be found that contradict the premise.

In a previous chapter, the authors of a report alluded to the impact of almost-atherosclerosis on endothelial (and therefore) erectile function. The following report alludes to almost-hypertension:

Age, waist circumference, fasting plasma glucose and triglyceride were independently and significantly correlated with systolic BP in a study published in the journal, *Aging Male*. Their patient group with the highest, albeit normal, BP had a significantly higher likelihood of having abnormal waist circumference, fasting plasma glucose and triglycerides. The authors concluded that in older normotensive men, the risk for having metabolic syndrome was significantly associated with higher, albeit normal, systolic BP.[16]

5.10 METABOLIC SYNDROME AND ATHEROSCLEROSIS

It has been well established that in some individuals, postprandial hyperlipidemia influences the development of atherosclerosis. Age has been thought to be a major factor in the extent of this phenomenon. To test the theory that age is an independent factor influencing postprandial hyperlipidemia, investigators compared the serum lipid response to a rich fatty meal (60% fat) of healthy young men under 30 years old to that in older metabolic syndrome patients aged over 40 years, and healthy people over 65, in fasting state and at 2nd and 4th postprandial hours.

The study published in the journal *Atherosclerosis* reports that metabolic syndrome seems to account for the differences in postprandial hyperlipidemia that have previously been attributed to age. There were no significant differences between a young population (mean age 22.6 years) and healthy people over 65-years-old absent metabolic syndrome.[17]

5.11 METABOLIC SYNDROME AND ALMOST-ATHEROSCLEROSIS

The aim of this study published in the journal *Arteriosclerosis, Thrombosis & Vascular Biology* was to examine whether small, low-density lipoprotein (LDL) particle size was associated with metabolic syndrome and with subclinical atherosclerosis. Carotid and femoral artery ultrasound provided data from a population-based sample of 58-year-old, clinically healthy men, free of treatment with cardiovascular drugs.

Participants with evident metabolic syndrome had thicker mean carotid and femoral artery wall thickness (intima-media complex, IMT), and also mean lower LDL particle size, unlike participants with no risk factors. The metabolic syndrome group also had higher mean serum cholesterol and heart rate.

In the whole study group, there were significant but weak negative correlations between small LDL particle size, increasing IMT, and increasing cross-sectional intima-media area of the carotid and femoral arteries, and also negative relationships between LDL particle size and plaque occurrence and size in the carotid and femoral arteries. This study was the first to show a relationship between the clustering of metabolic syndrome risk factors and small LDL particle size pattern suggesting *preclinical* atherosclerosis.[18]

Another study published in the journal *Diabetes Care* aimed to determine the association between the National Cholesterol Education Program Third Adult Treatment Panel Report (ATP III) definition of metabolic syndrome and cardiovascular disease (CVD), and to estimate relative risk of incident CHD and stroke among black and white middle-aged individuals in the ARIC over an average 11-year follow-up.

Symptoms of metabolic syndrome were present in approximately 23% of participants without diabetes or prevalent CVD at baseline. Among the components of metabolic syndrome, elevated BP and low levels of HDL-C exhibited the strongest associations with CHD. After adjustment for age, smoking, LDL-C, and race/ARIC center, men and women with metabolic syndrome were approximately 1.5 to 2 times more likely to develop CHD than control participants. Similar associations were found between metabolic syndrome and incident stroke. It was concluded that individuals without diabetes or CVD, but with metabolic syndrome, were at increased risk for long-term cardiovascular outcomes.[19]

It seems at times, in reviewing the relevant publications on metabolic syndrome and atherosclerosis, that metabolic syndrome was largely defined by signs and symptoms exclusion rather than inclusion. There is an almost modular quality to the syndrome: one can insert or remove various components and, bottom line, one is left with metabolic syndrome. Yet, no study leaves out one or a set of signs chosen from the "usual suspects," preexisting or existing hypertension, preexisting or existing atherosclerosis, heart disease, small-particle LDL, IR, and so on.

5.12 METABOLIC SYNDROME AND HEART DISEASE

The National Heart, Lung, and Blood Institute (NHLBI) of the National Institutes of Health (NIH) tells us that metabolic syndrome is a name given to a group of factors that raises the risk for heart disease ... and other health problems, such as diabetes and stroke. By "heart disease" they mean CHD, a condition resulting from atherosclerotic plaque accumulation in the coronary arteries.

If you possess the right *genes*, you need not wait to age, you can get CAD while you're still young: Patients with a first-ever acute coronary syndrome, and age-matched and sex-matched controls, were evaluated for metabolic syndrome. The prevalence of metabolic syndrome was found to be significantly higher in the patient group compared with the control group.

Analysis showed that smoking, *positive family history of premature CAD* (my italics), and metabolic syndrome were associated with significantly higher odds of having acute CHD, after taking into account the matching for age and sex. Moreover, a 10 mg/dL increase in total cholesterol was associated with higher odds of having an acute coronary event. It was also shown that smoking and a positive family history of premature CAD in young individuals with metabolic syndrome had an incremental effect on the odds of suffering an acute coronary event. Metabolic syndrome is highly associated with acute CHD in patients younger than 45 years of age.[20]

The study titled "Which features of the metabolic syndrome predict the prevalence and clinical outcomes of angiographic coronary artery disease?" appearing in the journal *Cardiology* was designed to determine whether a diagnosis of metabolic syndrome and its signs could predict CAD as measured by angiography. Laboratory evaluation included levels of fasting glucose (FG), triglyceride (TG), HDL-C, BP, and body mass index (BMI, used as an alternative to waist circumference). The investigators concluded that metabolic syndrome primarily predicts CAD by virtue of high fasting glucose, and secondarily by diabetes alone.[21]

5.13 METABOLIC SYNDROME, OBESITY, INSULIN RESISTANCE, AND DIABETES

In her book, *Diabesity: The Obesity—Diabetes Epidemic That Threatens America—And What We Must Do To Stop It*, Dr. Francine R. Kaufman[22] tells us that more and more of us are becoming obese because, "Our ancient genes and our modern environment have collided." In ancient times calories were hard to come by, but now our body can store excess calories as fat, and that is made all the easier by the ready availability of fast food and junk food, and an increasingly inactive lifestyle.

Both IR and defective insulin secretion appear prematurely in obese people, and both work their way to diabetes. The "hyperbolic relationship" between IR and insulin secretion and the "glucose allostasis" concept dominate the accepted theories about the relationship between "resistance" and "defective secretion."

5.14 GLUCOSE ALLOSTASIS

The term glucose allostasis first appeared in the journal *Diabetes* as an explanation for an apparent physiological adaptation to chronic IR: "Allostasis, stability through change, is thought to ensure the continued homeostatic response (stability through staying the same) to acute stress at some cumulative costs to the system. With increasing severity and over time, the allostatic load, increase in serum glucose, may have pathological consequences, such as the development of type-2 diabetes."[23] However, some sources hold that glucose allostasis does not exist.[24]

The journal *Diabetes* reported a study that first detailed the relationship between insulin sensitivity and β-cell function. Participants were relatively young and healthy with varying degrees of obesity. Insulin sensitivity index (SI) measures, body mass index (BMI), and fasting glucose levels were obtained. SI was compared to measures of body adiposity and β-cell function. BMI and SI showed a negative curvilinear relationship so that on average, an increase in BMI predicted a lower SI value.

The relationship between the SI and the β-cell measures was more clearly curvilinear and reciprocal for fasting insulin and β-cell secretory capacity (Figure 5.4). The curvilinear relationship between SI and the β-cell measures could not be mathematically distinguished from a hyperbola (i.e. SI × β-cell function = constant).

The authors state that the relationship is consistent with a "regulated feedback loop control system" such that for any difference in SI, a proportionate reciprocal difference occurs in insulin levels and responses in those with similar carbohydrate tolerance. They conclude that present normal glucose tolerance, and varying degrees of obesity, β-cell function varies quantitatively with differences in insulin sensitivity: Because the function is hyperbolic, when insulin sensitivity is high, large changes in insulin sensitivity produce relatively small changes in insulin levels and responses, whereas when insulin sensitivity is low, small changes in insulin sensitivity produce relatively large changes in insulin levels and responses."[25,26]

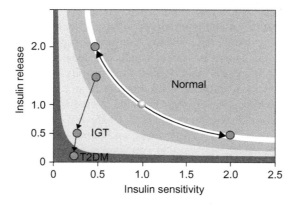

FIGURE 5.4 Hyperbolic relationship between insulin sensitivity and insulin release in health and disease: Relationship between insulin sensitivity and the β-cell insulin response is nonlinear. This hyperbolic relationship means that assessment of β-cell function requires knowledge of both insulin sensitivity and the insulin response. Hypothetical regions delineating normal glucose tolerance (▢), impaired glucose tolerance (IGT; ▢), and type 2 diabetes mellitus (T2DM; ▮) are shown. In response to changes in insulin sensitivity, IR increases or decreases reciprocally to maintain normal glucose tolerance—"moving up" or "moving down" the curve. In individuals who are at high risk of developing type-2 diabetes, the progression from normal glucose tolerance to type-2 diabetes transitions through impaired glucose tolerance and results in a "falling off the curve." Those individuals who do progress will frequently have deviated away from the curve even when they have normal glucose tolerance, in keeping with β-cell function already being decreased before the development of hyperglycemia. Source: *Kahn, S. E., Hull, R. L., & Utzschneider, K. M. (2006). Mechanisms linking obesity to insulin resistance and type 2 diabetes.* Nature, *444, 840-846. Reproduced with kind permission from* Nature.

Why is it so difficult to lose weight? The road from obesity to diabetes is paved with both a progressive defect in insulin secretion and IR, and declining serum testosterone. Both of these coupled with low testosterone are seen very early in obese patients, worsening with encroaching diabetes. There is an addition factor, perhaps at least, if not more important than, IR, detailed below.

Increase in visceral and ectopic fatness is linked to IR. To lose weight one must oxidize fat. The accumulation of fat globules in skeletal muscle may be due to reduced lipid oxidation capacity. A defect in fat oxidation capacity may be responsible for energy economy and hampered weight loss.[27] The recent discovery of the two hormones that control food intake, leptin and ghrelin (described below), also factor into increased fat storage and difficulty in shedding it.

According to a report in *Diabetologia*, normalization of both *beta*-cell function and insulin sensitivity in type-2 diabetes can be achieved just by dietary energy restriction. This was associated with decreased pancreatic and liver triacylglycerol stores. The abnormalities underlying type-2 diabetes are reversible by reducing dietary energy intake:

Among other findings, after 1 week of restricted energy intake, fasting plasma glucose normalized in the diabetic group as insulin suppression of glucose output improved and liver triacylglycerol content fell.[28]

There is also evidence that dietary L-arginine supplementation reduces adiposity in genetically, as well as diet-induced, obese human subjects with type-2 diabetes. The journal *Amino Acids* cites a study where L-arginine supplementation stimulates mitochondrial biogenesis and brown adipose tissue development, possibly through the enhanced synthesis of NO. The authors aver that L-arginine holds great promise as a safe and cost-effective nutrient to reduce adiposity, increase muscle mass, and improve the metabolic profile in animals and humans.[29] L-Arginine supplementation in type-2 diabetes is detailed in Chapter 7.

5.15 OBESITY, LEPTIN/GHRELIN, AND THE LINK TO ERECTILE DYSFUNCTION

To further underscore the previous assertion that the SAD *per se* may not necessarily be the principal villain in the complex aspects of health calamities leading to ED, obesity taken as an isolated phenomenon—which of course it is not quite that—is yet another of the factors with age-related progression. As noted below, on average, fat piles up as the years pile up.

Figure 5.5 shows that there is a fairly linear relationship between the increase in the ratio of fat accumulated in the region of the abdomen (here termed "visceral"—V) to that under the skin in other parts of the body.[30]

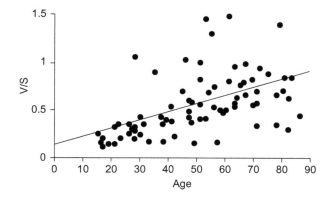

FIGURE 5.5   Visceral-to-subcutaneous fat (V/S) ratio in men with age. V/S in men increases with age. V/S ratio was measured by CT-based body composition analysis at the level of navel. Source: *Yanase, T., Fan, W-Q, Kyoya, K., Mina, L., Takayanagi, R., Kato, S.,* et al. *(2008). Androgens and metabolic syndrome: Lessons from androgen receptor knock out (ARKO) mice.* J Steroid Biochem Mol Biol, *109, 254−257. Reproduced with kind permission from* The Journal of Steroid Biochemistry and Molecular Biology.

The section below will show that testosterone is also a key player in this scenario, and it also shows a pretty linear progressive age-related decline. It is an important factor for determining body composition in men, as well as an important element in cardiovascular and heart health.

5.16 EATING, ENERGY CONSUMPTION AND STORAGE

The primary clue to understanding how obesity may also be controlled by the brain is the discovery of the hormones leptin and ghrelin, which have a profound impact on how we eat and how we marshal energy storage as fat. A first hint to their existence came in the 1950s when observations of obese rats in an otherwise normal-weight colony ultimately led to the identification in the mid-1990s of a mutant gene that controlled food intake, leptin, and another one that controlled satiation, ghrelin.

Note: This book is about health conditions that lead to ED. It is not about diet for weight loss. For that reason, the description of these hormone regulators is limited to that which may lead to understanding of the pathway to metabolic syndrome and ED.

Leptin is formed primarily in the adipocytes of white adipose tissue, fat cells derived from connective tissue cells that differentiated and became specialized in the synthesis and storage of fat. They are important to the body in maintaining proper energy balance, storing calories in the form of lipids, and mobilizing energy sources in response to hormonal stimulation. The level of circulating leptin is directly proportional to the total amount of fat in the body. In other words, fat begets fat.

A study published in the *Journal of Clinical Endocrinology & Metabolism* reported that leptin circulates in plasma at concentrations that parallel the amount of fat reserves. Their clinical trials showed that in obese men, testosterone levels decline in proportion to the degree of obesity. This indicates that excess circulating leptin may be an important contributor to reduced testosterone in male obesity.[31] In other words, fat begets fat … begets ED.

Leptin is present in blood serum in direct proportion to the amount of adipose tissue, and as fat cells become enlarged in obesity, they secrete more leptin. The hormone communicates with the hypothalamus in the brain to regulate energy intake and energy stores in the body. It also affects heart function, BP, immune function, and cognition. In fact, high leptin levels may result in left ventricular hypertrophy (LVH), a major risk factor for congestive heart failure,[32] and leptin levels are a significant predictor of fibrinogen,[33] an important clotting factor, and major risk factor for heart disease.

Leptin suppresses food intake and increases metabolic rate, mimicking the action of insulin. In most obese individuals, leptin levels are actually excessively high due to leptin resistance, a process comparable to IR. The brain gauges the levels of leptin secreted by fat cells, and if these are within normal range, the brain shuts off the signal and the body now stores extra calories as fat. Also, one no longer wishes to eat because one feels "full."

Because food is plentiful nowadays, overeating is common. Overeating causes numerous complex changes to certain brain region cells receptors so that they become desensitized, or resistant, to leptin. Other changes affect the ability of excess leptin to pass through the blood/brain barrier. In effect, even though blood levels of leptin may be very high, brain levels become insufficiently low, resulting in food cravings and weight gain: The brain tells us that we are hungry and tells our body to store fat.

5.17 A NOTE ON LEPTIN AND DOPAMINE

Dopamine is a brain neurotransmitter that affects movement, cognition, pleasure, and motivation. It is a key factor in motivation, and in affecting behaviors that lead to pleasure and a feeling of satisfaction. Feelings of satisfaction can easily become desirable, motivating a person to seek satisfaction to attain it, even if it entails severe risk taking. Such behavior can, and often does, lead to various addictions where one basically unwittingly surrenders the will to control his/her behavior. Omitting here the psychiatric strategies focusing on dopamine in such emotional disorders as depression and schizophrenia, dopamine plays a key role in the way we eat, and it is affected by what we eat.

In yet another instance where it can be shown that becoming obese is not simply a matter of changing diet or eating less, nature has endowed us with the addition of addictive pleasure to nourishment in the form of the brain messenger dopamine: For instance, a study published in the journal *Diabetes*

Care focused on the fact that dopamine neurons, involved in reward and motivation, interact with hormones such as insulin, leptin, and ghrelin that regulate food intake.

The authors hypothesized that these hormones are associated with deficits in dopamine in obesity. Fasting levels of insulin, leptin, dopamine, BMI, and SI were assessed. The results of the study confirmed the hypothesis of deficient central dopamine (DA type-2) receptor activity in obese subjects, reflecting relatively reduced dopamine level availability.[34,35] In fact, age-related decline in brain dopamine activity has been reported, as noted in Figure 5.6.

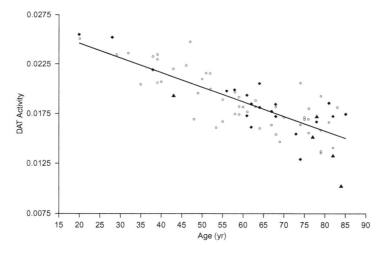

FIGURE 5.6   Age-related decline in brain dopamine activity. Source: *Bohnen, N. I., Muller, M. L. T. M., Kuwabara, H., Cham, R., Constantine, G. M., Studenski S. A., (2009). Age-associated striatal dopaminergic denervation and falls in community-dwelling subjects.* J Rehab Res Dev, *46, 1045–1052. Reproduced with kind permission from* Journal of Rehabilitation Research and Development.

These studies were conducted by the Department of Veterans Affairs in connection with the incidence of falling in the elderly.[36] An average person may lose about 33% of striatal dopaminergic innervation between the ages of 25 and 75.[37]

In another study appearing in the journal *Neuron*, direct administration of leptin to the ventral tegmental area (VTA) in mice caused decreased food intake, while genetically altering the receptor led to increased food intake, locomotor activity, and sensitivity to highly palatable food. The results pointed to a critical role for leptin in regulating feeding behavior and provided functional evidence for direct action of a metabolic signal on dopamine neuron activity.[38]

5.18 LEPTIN LEVELS TEND TO RISE WITH AGE

Age-related increase in leptin may be one reason why we have an easier time losing weight when we are less than 30 years old than do seniors. Reporting in the journal *Diabetes*, research concluded that in the rat model, adiposity increases with age and cannot be attributed simply to increased food intake.[39]

5.19 GHRELIN

The hormone ghrelin, produced mainly by cells lining the stomach, the pancreas, and the hypothalamus, stimulates hunger. Ghrelin levels rise before meals and decline after meals. It also plays a key role in cognitive adaptation to changing environments and the process of learning. Ghrelin has also been shown to activate endothelial nitric oxide synthase (eNOS).

5.20 GHRELIN LEVELS TEND TO DECLINE WITH AGE

A study appearing in the *International Journal of Molecular Medicine* examined whether ghrelin undergoes changes in the adrenal cortex in aging. A significant negative correlation was found between ghrelin and age in the adrenal cortex of patients 33 to 82 years old who had undergone surgical unilateral adrenal/kidney removal due to cancer. No significant gender differences were noted.[40]

In addition, it was shown in another study that advanced age is associated with a poorer ghrelin postprandial recuperation phase, a reduced cholecystokinin (CCK) postprandial response, and exaggerated postprandial insulin release. A loss of ghrelin prandial rhythm is present in old, frail persons. The impaired response of this hunger regulatory hormone with age is thought to contribute to the anorexia associated with advanced aging.[41]

As we age, we tend to sleep less, and that may also contribute to a problem with weight control. However, *Annals of Internal Medicine* reported that even in some healthy young men, sleep curtailment is associated with lower leptin levels, elevated ghrelin levels, and increased hunger and appetite. In that study, sleep restriction (4 hrs) relative to sleep extension (10 hrs) was associated with a 24% increase in hunger ratings and a 23% increase in appetite ratings for all food categories combined.

The increase in appetite tended to be greatest for calorie-dense foods with high carbohydrate content: sweets, salty foods, and starchy foods increase 33% to 45%. The increase in appetite for fruits and vegetables was less consistent and of lesser magnitude (increase, 17% to 21%). Appetite for protein-rich nutrients (meat, poultry, fish, eggs, and dairy foods) was not significantly affected by sleep duration. The increase in hunger was found to be proportional to the increase in ghrelin-to-leptin ratio.[42]

5.21 LEPTIN IS LINKED TO INFLAMMATION MARKERS

According to investigators reporting in *Circulation*, leptin is linked to the inflammation marker CRP in normal humans, and implicated in metabolic and inflammatory CVD mechanisms, including atherosclerosis.[43] Another article in the same journal links leptin to inflammation and atherosclerosis not only by promotion of CRP, but by the consequent reduction in eNOS expression and NO bioactivity in cultured human aortic endothelial cells (HAECs).[44]

In light of these findings, it is hard to know how to interpret the report in *Biochemical & Biophysics Research Communications* that leptin causes arterial relaxation mediated by the nitric oxide/cyclic guanosine monosphosphate (NO/cGMP) pathway.[45]

5.22 LEPTIN AND BLOOD PRESSURE

The term "resistant hypertension" (RHTN) refers to elevated BP that requires four or more classes of medications for control. Leptin and aldosterone have been associated with hypertension. A study published in the *Journal of Human Hypertension* aimed to determine the link between plasma leptin and aldosterone levels with BP in uncontrolled controlled resistant hypertension (UCRHTN) and controlled RHTN (CRHTN). Plasma leptin and aldosterone levels, BP, and heart rate were obtained from participants in the groups.

No gender, body mass, or age differences were observed between groups. Compared to well-controlled and CRHTN groups, the uncontrolled RHTN group had significantly elevated leptin levels. The same pattern held for aldosterone levels. Uncontrolled RHTN patients had higher heart rates. Plasma leptin levels correlated positively with systolic BP, diastolic BP, and aldosterone in uncontrolled RHTN versus the other groups: systolic BP, diastolic BP, and aldosterone may be predicted by leptin in the uncontrolled RHTN group.

In conclusion, UCRHTN patients had higher circulating leptin levels and increased plasma aldosterone and BP levels than the other patient groups.[46]

There is some question about whether controlled and uncontrolled RHTN patients are similar or dissimilar: In 2008, the AHA Guidelines on Resistant Hypertension included in the definition of this syndrome patients whose BP is controlled, but who require four or more classes of antihypertensive drugs to do so, as noted above. The inclusion of this "controlled RHTN" subgroup was thought justified because of the associated high cardiovascular risk level. However, the hormonal and vascular basis for this distinction is unknown. For more information, see Gordo, Faria, Barbaro *et al.*[47]

5.23 LEPTIN DAMAGES THE ENDOTHELIUM

Leptin receptors are found in coronary arteries, and it appears they affect the cardiovascular system: high leptin levels increase the risk of CVD. Obesity is associated with marked increases in plasma leptin concentration, and elevated leptin is an independent risk factor for CAD.

A study titled "Leptin receptors are expressed in coronary arteries, and hyperleptinemia causes significant coronary endothelial dysfunction," appearing in the *American Journal of Physiology, Heart & Circulation Physiology*, aimed to determine whether elevated leptin levels induce coronary endothelial dysfunction. Using a laboratory method to amplify short DNA sequences (RT-PCR analysis), it was found first that the leptin receptor gene is expressed in human coronary endothelium.

The major findings of this investigation were as follows[1]: the leptin receptor (ObRb) is present in coronary arteries and coupled to NO/cGMP vasodilation, and[2] elevated leptin (hyperleptinemia) results in considerable coronary artery endothelial dysfunction.[48]

To the extent that obesity results from elevated levels of leptin, that holds bad news for erectile function: As shown in the study above, elevated leptin impairs NO/cGMP activity required for attaining and sustaining erection. It must seem a paradox, then, that in a study published in the *Korean Journal of Urology*, it was shown that leptin enhances NO/cGMP relaxation of the clitoral corpus cavernosum—in rabbits, anyway.[49]

In another study, it was shown that elevated leptin accompanying obesity impairs endothelial NO formation and therefore NO/cGMP vasodilation: In obesity, leptin increases formation of eNOS, decreasing intracellular L-arginine, and resulting in eNOS *uncoupling* and depletion of endothelial NO. This process is accompanied by increase in free radicals and endothelial inflammation characteristics of dysfunctional endothelium.[50]

5.24 LEPTIN AND THE HEART

A report titled "Modulation of the cardiovascular system by leptin" appeared in a *mini review* titled "Leptin and adiponectin: the Yin and Yang of the adipocyte," published in the journal *Biochimie*. The authors contend that excessive or deficient metabolic effects of leptin make an important contribution to metabolic syndrome, yet many paradoxical observations often preclude a clear definition of the role of leptin. The review presented recent evidence of the direct and indirect regulation of the cardiovascular system by leptin, focusing on cardiac structural and functional, as well as vascular, effects.

In obese patients with left ventricular hypertrophy, a risk factor for congestive heart failure, the higher the leptin levels, the greater the left ventricular mass. High leptin levels are also associated with hypertension.

In fact, writing in the *International Journal of Hypertension*, researchers propose that leptin may play both a direct and an indirect role in cardiovascular and renal regulation by regulating BP and volume. However, with chronic elevated leptin, such as in obesity, it may contribute to the development of hypertension.[51,52]

A review of clinical studies on leptin and cardiovascular function in *Current Hypertension Reports* informs us that in recent years there has been an intense focus on the action of leptin in the cardiovascular system. Plasma leptin concentration has been found to be markedly elevated in obesity and metabolic syndrome, both of which are associated with increased incidence of cardiovascular events. Elevated leptin has been linked to endothelial dysfunction and atherosclerotic CVD.

Additionally, recent evidence suggests that leptin released from perivascular adipose tissue may also have deleterious effects on the underlying vasculature, including the coronary circulation. Perivascular adipose tissue surrounds coronary arteries and may be involved in local stimulation of atherosclerotic plaque formation. The report reviews pertinent literature on leptin-mediated endothelial dysfunction, and leptin as a significant perivascular adipose-derived factor.[53,54]

5.25 LEPTIN RAISES FIBRINOGEN, THUS PROMOTING HAZARD OF BLOOD CLOTTING

In obese subjects, leptin levels are the most significant predictor of higher levels of fibrinogen, one of the most important risk factors for heart disease. Increased body fat and impaired insulin sensitivity are associated with increased concentrations of cardiovascular risk factors. Leptin seems to be involved in this elevation, and emerges as a predictor of circulating fibrinogen concentrations. In a study published in the journal *Clinical Biochemistry*, the authors found that obese participants had higher levels of leptin and fibrinogen, and lower concentrations of HDL-C and lower measures of Quantitative Insulin Sensitivity Check Index (QUICKI). A positive correlation was observed between body fat and fibrinogen, and between plasma leptin concentrations and fibrinogen, and CRP. After adjustment for body fat, leptin emerged as a significant predictor of fibrinogen.[33]

5.26 AGE-RELATED LEPTIN RESISTANCE AND ERECTILE DYSFUNCTION

The *Academic Journal of Second Military Medical University*, published in Shanghai, PRC, featured a clinical study aimed at determining the relationship between serum leptin levels and ED. Outpatients complaining of ED were graded by the International Index of Erectile Function (IIEF-5) questionnaire. The mean IIEF-5 score of the patients was (12.3 ± 2.4), and no one was higher than 21. The patients were divided into three groups—mild, moderate and severe—according to their scores. The fasting serum leptin levels were determined by enzyme-linked immunosorbent assay (ELISA). Body weights and height were measured for BMI.

It was found that serum leptin levels were significantly correlated with BMI in both groups, regardless of patient age and history. The mean leptin level in patients was significantly higher than that in the healthy controls after adjusting for BMI. The leptin levels of the "mild" group were significantly lower than those in the other two groups. The authors concluded that human serum leptin level may be valuable in diagnosis of ED.[55]

ED is often associated with metabolic disorders. Leptin and adiponectin are adipose tissue-derived hormones involved in the regulation of metabolic homeostasis, and are considered important players in the relationship between obesity and CVDs. In a study reported in the *Journal of Sexual Medicine*, leptin, as well as other adipose tissue hormones, were examined with respect to their link to metabolic parameters in men with arteriogenic ED, compared to a control group of men with non-arteriogenic ED.

Diagnosis of ED was based on the IIEF score. The cause of ED was determined with the penile echo-color Doppler technique at baseline and after intracavernous injection of prostaglandin E1. Leptin and the other hormones were measured by ELISA.

In arteriogenic (aka vasculogenic) ED patients, increased levels of insulin, glycated hemoglobin, homeostasis model assessment of insulin resistance (HOMA-IR) index, BMI, leptin, leptin-to-adiponectin ratio (L/A), and decreased levels of total, free, and bioavailable testosterone were observed compared with non-arteriogenic ED patients. Leptin, and L/A-ratio correlated negatively with testosterone/estradiol ratio and positively with BMI, insulin, HOMA-IR. The L/A ratio correlated negatively with HDL, and positively with triglycerides (Sidebar 5.1).[56]

Sidebar 5.1

Adiponectin is an adipocyte-specific protein that is thought to play a role in the development of IR and atherosclerosis. It is usually found in high concentrations, but it is lower in obese subjects than in lean subjects.

Adiponectin levels are also reduced in IR and type-2 diabetes, and visceral adiposity has been shown to be an independent predictor of low adiponectin levels. Adiponectin appears to be an insulin enhancer, and it may be anti-inflammatory, and anti-atherogenic.

Source: *Lihn, A. S., Pedersen S. B., Richelsen, B. (2005). Adiponectin: action, regulation and association to insulin sensitivity. Obes Rev, 6(1):13–21.*

5.27 AGE-RELATED RISE IN INSULIN RESISTANCE

In previous chapters, the *coincidence* (literally) between age-related trends was shown with respect to factors increasing the risk of cardiovascular calamities, and then some. It was also shown that the age-related trend for diabetes was perhaps the most significant with respect to predicting ED. The post-90-year dip in Figure 5.7, age-related change in IR, is curious. There is a sudden dramatic decline in IR in the very elderly. This is not "survivorship bias" because these are not prevalence data. No explanation comes readily to mind.[57] Clearly, IR is not a feature of very long-term survival.

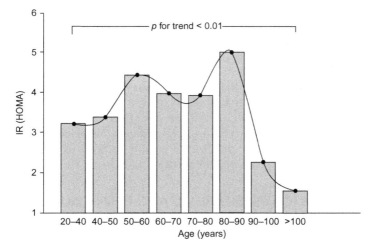

FIGURE 5.7 Age-related differences in IR in healthy people. Source: *Paolisso, G., Barbieri, M., Rizzo, M. R., Carella, C., Rotondi, M., Bonafé, M., et al. (2001). Low insulin resistance and preserved b-cell function contribute to human longevity but are not associated with TH–INS genes.* Exper Gerontol, *37, 149–156. Reproduced with kind permission from* Experimental Gerontology.

Although medicine may view it differently depending on whom you ask, the clinical studies outcome data strongly suggest that "insulin resistance" and "type-2 diabetes" are not interchangeable terms. Clinical studies point to varying degrees of IR—below a certain threshold—that represent neither diabetes nor prospectively determined pre-diabetes. It seems that, consistent with previously cited data, there is also a progressive age-related prevalence of increase in serum IR in the population—up to a point.

The *Journal of Clinical Endocrinology and Metabolism* reported a study titled "Insulin resistance as a predictor of age-related diseases." The title is pretty much self-explanatory: Baseline measurements of IR and related variables were made between 1988 and 1995 in apparently healthy, non-obese individuals (BMI less than 30 kg/m^2) who, between 4 and 11 years later, were evaluated for the presence of these "age-related" diseases: hypertension, CHD, stroke, cancer, and type-2 diabetes.

The effect of IR on the development of clinical events was evaluated by dividing the study group IR measures into tertiles at baseline and comparing the "clinical events" in these three groups. IR was an independent predictor of all clinical events. An age-related clinical event occurred in approximately one out of three healthy individuals in the upper tertile of IR at baseline, followed for an average of six years.[58]

The matter of prevalence of IR was likewise addressed in an article appropriately titled "Prevalence of insulin resistance and the metabolic syndrome with alternative definitions of impaired fasting glucose," appearing in the journal *Atherosclerosis*. The stated purpose of the study was to evaluate the effect of the new definition of impaired fasting glucose (IFG = fasting glucose concentration 100−125 mg/dL among people without diabetes) on the ability to identify IR, as well as the prevalence of metabolic syndrome in apparently healthy individuals.

Data were obtained from the Stanford General Clinical Research Center database (men and women aged 19 to 79 years who had had an insulin suppression test to measure IR), and from the Third National Health and Nutrition Examination Survey (1988−1994), participants older than 20 years. The new definition of IFG expanded the population with IR almost 4-fold, and could expand the population with metabolic syndrome by about 20%.[59]

5.28 METABOLIC SYNDROME IMPAIRS ENDOTHELIAL FUNCTION

A clinical study titled "Endothelial dysfunction in metabolic syndrome may predict cardiovascular risk" was presented at a conference at the Annual Meeting of the *American Society for Clinical Pathology* (ASCP) in Baltimore, Maryland, on October 2, 2008. Brachial artery vasodilation was assessed in all participants by high-resolution ultrasound. The correlates of endothelial dysfunction were based on fasting blood sugar (FBS), systolic BP, levels of HDL and LDL, and age.

HDL was found to be the most significant of all the predictors, as it was inversely associated with endothelial dysfunction. All other variables were directly related to endothelial dysfunction.[60]

The *New England Journal of Medicine* (NEJM) reported a clinical study that aimed to determine the relation of impaired corpora cavernosa relaxation to impotence in diabetic patients. The investigators examined samples of corpus cavernosa tissue obtained at the time of implantation of a penile prosthesis in men with ED who were also diabetic, and in nondiabetic men.

Contraction of the isolated sample of cavernosa smooth muscle *in vitro* was induced by norepinephrine. Relaxation was assessed with electrical stimulation of autonomic nervous system fibers, and also with the administration of three agents, acetylcholine (ACh), and the endothelium-independent muscle relaxants papaverine and sodium nitroprusside.

Autonomic-mediated relaxation via electrical stimulation was less pronounced in the smooth muscle from diabetic men than in those from nondiabetic men. Degree of impairment rose with duration of diabetes. Endothelium-dependent relaxation was also impaired, as shown by lower degree of muscle relaxation after administration of ACh in the tissue from diabetic men than in that from nondiabetic men. The adverse effects of diabetes persisted after controlling for smoking and hypertension. Endothelium-independent relaxation after the administration of nitroprusside and papaverine was similar in tissue samples from the diabetic and nondiabetic men.

The investigators concluded that diabetic men with ED have impairment in both the autonomic nervous system and the endothelium-dependent mechanisms that mediate the relaxation of the smooth muscle of the corpora cavernosa.[2] The significance of these findings is that they point to diabetes-related impairment of endothelial function because smooth muscle relaxation depends in large part on activation of the NO/cGMP pathway.

5.29 METABOLIC SYNDROME AND ERECTILE DYSFUNCTION

The previous section of this chapter gave examples from published clinical research that tell us that metabolic syndrome features elevated BP, CAD, elevated serum lipids, elevated serum glucose, and IR, all cardiovascular and heart catastrophe risks.

Now, here are examples from published clinical research that show the link between metabolic syndrome, along with the abovementioned risks, and ED. The journal *Diabetes Care* reports that 30 million or more American men experience varying degrees of ED. It is estimated to affect perhaps as many as 100 million men worldwide,[61] and it affects an estimated 52% of men between the ages of 40 and 70 years.[62]

In a clinical study appearing in *Diabetes Care*, patients with metabolic syndrome were age-matched and matched for BMI with men in a control group. Compared to the control group, patients with metabolic syndrome had greater prevalence of ED, reduced endothelial function score, and higher circulating concentrations of inflammation marker (CRP). Prevalence of CRP level increased as the number of components of metabolic syndrome increased. Patients' scores on the IIEF were positively associated with the endothelial function score, and negatively associated with CRP level.

Endothelial function was assessed with the L-arginine test where BP in an arm or a limb and platelet aggregation measures were summed before and after intravenous L-arginine administration (3 g).[63]

The results of the study show that increase in CRP level predicts impairment of endothelial function and prevalence of ED as the number of components of the metabolic syndrome increase. These links between IIEF and endothelial function scores support the belief in the presence of common vascular pathways underlying both conditions: Perhaps defective NO activity linked to reduced NO availability could provide a unifying explanation for this association.[1]

The *Journal of Sexual Medicine* reported a study titled "Incidence of metabolic syndrome and IR in a population with organic erectile dysfunction." The study found that metabolic syndrome and FBS predicted increasing severity of ED as evaluated by the Sexual Health Inventory for Men (SHIM) score. Further, men with ED were shown to have a high incidence of metabolic syndrome and IR. The authors contend that early detection of metabolic disease in patients with ED may be a "gateway" to reducing endothelial dysfunction in younger men with increased cardiovascular risk, but who sought only treatment of ED.[64]

A clinical study in the *Asian Journal of Andrology* concluded that ED is present in a high percentage of patients with metabolic syndrome. Among multiple risk factors for ED, metabolic syndrome was most closely linked to ED. The next highest risk group was patients with hypertension and hypercholestrolemia, and obesity as indicated by a BMI greater than 30.[65]

According to a study published in the *Journal of Sexual Medicine*, men with ED have a high incidence of metabolic syndrome and IR. The aim of the clinical study was to determine whether both metabolic syndrome and IR are predictors of CVD based on the fact that men of advanced age tend to have increased prevalence of CVD and ED.

Prior studies revealed that 56% of an ED population had asymptomatic myocardial ischemia, 75% of men with CAD had symptoms of ED, and 91% of ED patients had cardiovascular risks. Patients were evaluated for multiple cardiovascular risk factors and graded on severity of ED. The severity of ED was evaluated by the SHIM questionnaire. The prevalence of metabolic syndrome was determined by National Cholesterol Education Program (NCEP/ATP III) criteria. IR was measured by the QUICKI (QUICKI = 1/log insulin + log glycemia in mg/dL).

The total cholesterol/HDL ratio was moderately and negatively correlated with QUICKI, and similarly for the triglyceride/HDL ratio. Metabolic syndrome was present in 43% of the ED population as opposed to 24% in a matched patient population. Approximately 79% of the total sample had IR, and 73% of the nondiabetic portion had IR, compared to 26% in a general population study. Metabolic syndrome, IR, and FBS greater than 110 mg/dL correlated positively and moderately with increasing SHIM severity of ED scores. The authors concluded that men with ED and metabolic syndrome have a high incidence of IR.[64]

The aim of a clinical study titled "Prevalence of metabolic syndrome and its association with erectile dysfunction among urologic patients: metabolic backgrounds of erectile dysfunction," published in *Urology*, was to identify the prevalence of metabolic syndrome and its association with ED among urologic patients.

Waist circumference, triglycerides, and HDL-C levels were measured in patients aged 40 to 70 years who were admitted to the urology clinics of four different institutions: Group 1 consisted of patients with a waist circumference greater than 102 cm. In Group 2, waist circumference was less than 102 cm. The erectile status of the two groups was compared. Metabolic syndrome was found to be significantly linked to ED. With normal serum HDL and TG levels, greater waist circumference raises relative risk of ED to 1.94; in patients with increased waist circumference and elevated levels of HDL or TG, the relative risk of ED rose to 3-fold. The relative risk of ED in the presence of increased waist circumference plus elevated HDL and triglycerides more than tripled to 3.4.

Note on relative risk: In epidemiology, relative risk is the risk of developing a disease relative to exposure. Relative risk is the ratio of the probability of the event occurring in the exposed group vs. that in a non-exposed group:

The authors concluded that metabolic syndrome was strongly associated with ED, and that fasting blood glucose levels, hypertension, and waist circumference are the best predictors of ED risk factors. However, ED risk is more likely with abdominal obesity, together with altered total triglycerides and HDL cholesterol levels. This suggests an unknown lipid metabolic factor behind the risk of ED.[66]

The title of a review appearing in the journal *Urological Science* is "Bidirectional relationship between metabolic syndrome and erectile dysfunction." The author reports that ED occurs sooner in life in patients with metabolic syndrome than in those with organic ED absent metabolic syndrome. Also, the prevalence of ED increases and the IIEF score decreases as the number of abnormal symptoms of metabolic syndrome increases. The cause of ED in patients with metabolic syndrome is likely attributable to endothelial dysfunction.[67]

Finally, not only has heart disease been designated as a "marker" for ED, but now, so has metabolic syndrome: According to *Current Diabetic Reports*, ED is more commonly seen in men with various components of metabolic syndrome, and it should therefore be considered as a risk marker of metabolic syndrome and its associated conditions. The authors urge that patients with ED should be thoroughly evaluated for coexisting vascular disease because ED is a major unifying etiology for many of the aspects of metabolic syndrome, especially diabetes and CVD.[68]

Likewise, a review in *Current Opinion in Urology* concludes that ED and CVD share risk factors such as hypertension, diabetes, dyslipidemia, and obesity—all implicated in causing endothelial dysfunction—with metabolic syndrome.[69]

5.30 TESTOSTERONE DEFICIENCY

Metabolic syndrome is linked to hypertension, atherosclerosis, CHD, obesity, IR, and ED. What all these also have in common is deficiency in testosterone (also, "T"), yet few who conduct research on metabolic syndrome consider that possibility, though it is well documented.

Clinical studies on metabolic syndrome tend to limit their focus on only two variables at a time: metabolic syndrome is linked to hypertension, metabolic syndrome is linked to atherosclerosis, metabolic syndrome is linked to heart disease, metabolic syndrome is linked to type-2 diabetes, metabolic syndrome is linked to ED . . . etc.

Then, there is hypertension: hypertension is linked to atherosclerosis, hypertension is linked to heart disease, hypertension is linked to diabetes type-2 . . . hypertension is linked to ED, and so on. Much of the research is bivariate analysis of what is clearly multivariate data. Even in studies that entail many variables, the research design rarely results in a component of unaccounted data variability in multivariate research called "interaction," the component that often "spills the beans." What is it that might not be accounted for in connection with metabolic syndrome?

Here is another paired comparison: According to a report in *Therapeutic Advances in Endocrinology and Metabolism*, testosterone deficiency in men is closely linked to metabolic syndrome. The authors report that epidemiological studies have reported a strong link between low testosterone levels and obesity, IR, and an unfavorable lipids profile: there is a high prevalence of *hypogonadism* among men with metabolic syndrome and type-2 diabetes. Observational and experimental data suggest that testosterone replacement results in improved IR, lower weight, a more favorable lipids profile, and improved erectile function.[70] Every one of the risks is linked to progressive age-related, decreasing bioavailable serum testosterone:

This study published in the *International Journal of Andrology* aimed to investigate the relationship between ED and total, bioavailable, and free testosterone levels in men with type-2 diabetes, plus the associations of various cardiovascular risk factors involved in the development of ED in type-2 diabetic men. It was found that bioavailable and free testosterone levels were significantly lower in men with ED than in those without ED. Sex hormone-binding globulin levels were also reduced, but there was no significant difference in total testosterone levels between men with, and those men without, ED.

The severity of ED as assessed by IIEF scores was significantly associated with total testosterone levels, bioavailable testosterone, and free testosterone levels. ED was more frequently observed in men with hypertension and a higher waist circumference.

This study showed that ED is associated with low bioavailable and free testosterone levels, age, visceral adiposity, and hypertension in type-2 diabetic men. There was also a higher prevalence of ED among smokers, but there were no significant associations between ED and alcohol consumption or with BMI greater than 30.[71]

5.31 NORMAL TESTOSTERONE LEVELS

There are two forms of testosterone in blood serum:

- *Free testosterone* is the "active" form, and it is not chemically bound to any proteins, thus it is readily available to bind to androgen receptor sites on cells.
- Most of *bound testosterone* in the body is chemically attached to a protein called "sex hormone binding globulin" (SHBG). Some is bound to albumin, another to protein.

Testosterone levels are usually assessed by a test that measures total bound and free testosterone: Levels of testosterone vary with age and other factors, but generally, a normal test level of combined bound and free testosterone can range from 300 to 1,100 ng/dL (nanograms per deciliter). The level of *free testosterone* is considered more indicative of hormone activity: Levels of free testosterone can range between 0.3 to 5.0% of the total testosterone count, with about 2% considered an optimal level.

5.32 AGE-RELATED TESTOSTERONE DECLINE

The Baltimore Longitudinal Study of Aging (BLSA), as reported in the *Journal of Clinical Endocrinology and Metabolism*, tells us about age-related decline in total and/or free testosterone levels in men. Testosterone and SHBG were assessed by radioimmunoassay (RIA) in the BLSA. Significant independent, age-related effects were found with for both testosterone and free testosterone index. Incidence of hypogonadal testosterone levels increased to about 20% in men over 60, to 30% for those over 70, and to 50% for those over 80 years old (Figure 5.8).

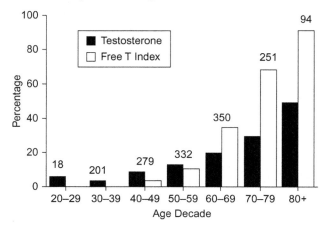

FIGURE 5.8   Prevalence of hypogonadism: Percentage of men in each decade with testosterone values in the hypogonadal range (Total T less than 11.3 nmol/L (325 ng/dL). Source: *Harman, S. M., Metter, E. J., Tobin, J. D., Pearson, J., Blackman, M. R., (2001). Longitudinal effects of aging on serum total and free testosterone levels in healthy men. Baltimore Longitudinal Study of Aging. J Clin Endocrinol Metab, 86, 724-731. Reproduced with permission of* The Journal of Endocrinology and Metabolism.

In the Massachusetts Male Aging Study, 4% of the men 40 to 70 years old had serum total testosterone concentrations below 150 ng/dL (5.2 nmol/L). The lower limit of normal male range in healthy young men was considered to be 275 to 300 ng/dL (9.6–10.5 nmol/L). Thirty percent of men over the age of 60, and 50% over the age of 70, had total testosterone concentration below the lower limit of normal range for healthy young men (325 ng/dL, 11.3 nmol/L). The prevalence of testosterone deficiency was higher for free testosterone values.

Most of the older men with low testosterone levels had a medical disease, whereas fewer than 3% of healthy, older men had low testosterone levels. It was concluded that ill health, rather than aging, was the major factor in testosterone deficiency in older men.[72,73]

Yet, the extent to which decline in testosterone is solely the result of aging, as opposed to other factors, remains unclear. Is ill health the cause of testosterone deficiency, or is testosterone deficiency the cause of ill health? One could stack high two piles of medical journal articles each of which supports the answer to either one of these questions. In addition, see also the work by Harman et al. (2001) for longitudinal effects of aging on serum total and free testosterone levels in healthy men,[72] and the work by Tenover (2003) for declining testicular function in aging men.[74]

5.33 PREVALENCE OF LOW TESTOSTERONE AND CORONARY ARTERY DISEASE

Although high androgen levels have been linked to increased risk of CAD, recent data suggest that low androgen levels are also associated with adverse cardiovascular risk factors, including an atherogenic lipid profile, obesity, and IR.

A study appearing in the *International Journal of Impotence Research* sought to evaluate the relationship between plasma sex hormone levels and degree of CAD in patients undergoing coronary angiography, compared to that in matched controls. The male patients group, 58 years old on average (range 43–72 years), was referred for diagnostic coronary angiography because of symptoms suggestive of CAD, but without acute coronary syndromes or prior diagnosis of low plasma testosterone.

Figure 5.9 shows that there is a linear relationship between plasma level of testosterone and a "coronary artery score," the likelihood and/or degree of CAD. The score is highest when testosterone levels are lowest. The patients were found to have significantly lower levels of bioavailable testosterone than controls and higher levels of gonadotrophin. An inverse relationship between the degree of CAD and plasma testosterone levels was found. The authors concluded that low plasma testosterone may be involved in the increased risk of CAD in men.[75]

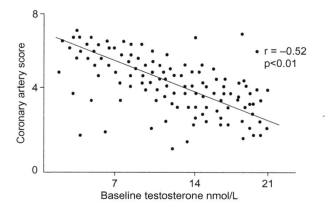

FIGURE 5.9   Relationship between testosterone (nmol/L) and coronary artery score in patients affected by CAD. Source: *Rosano, G. M., Sheiban, I., Massaro, R., Pagnotta, P., Marazzi, G., Vitale, C., et al. (2007). Low testosterone levels are associated with coronary artery disease in male patients with angina. Int J Impot Res, 19(2):176−82. Reproduced with kind permission from the* International Journal of Impotence Research.

5.34 PREVALENCE OF LOW TESTOSTERONE AND METABOLIC SYNDROME

In a report titled "Hypogonadism and metabolic syndrome: implications for testosterone therapy," published in *The Journal of Urology*, data are provided for the age-related prevalence of low serum testosterone, and link that progressive decline to metabolic syndrome. The authors assert that, "Metabolic syndrome, characterized by central obesity, insulin resistance, dyslipidemia and hypertension, is highly prevalent in the United States. When left untreated, it significantly increases the risk of diabetes mellitus and CVD. It has been suggested that hypogonadism may be an additional component of metabolic syndrome."[76]

There being no evidence to the contrary, it could well also be that metabolic syndrome may be a component of hypogonadism. It is possible that hypogonadism is viewed as elemental, whereas metabolic syndrome is viewed as complex, but is that really so?

A review published in the journal *Current Opinion in Endocrinology, Diabetes & Obesity* finds that low testosterone can be a risk factor for the development of diabetes and the metabolic syndrome through such mechanisms as changes in body composition, altered glucose transport, and reduced antioxidant effect. Conversely, diabetes and metabolic syndrome can be risk factors for inducing low testosterone through distinct mechanisms such as increased body weight, decreased SHBG levels, suppression of gonadotrophin release, or Leydig cell testosterone production, cytokine-mediated inhibition of testicular steroid production, and increased aromatase activity contributing to relative estrogen excess.[77]

There have been relatively few data describing the link between SHBG, testosterone, and the risk of developing diabetes. However, one study published in the *American Journal of Epidemiology* concluded that, "Low levels of SHBG and testosterone predict the development of noninsulin-dependent diabetes mellitus in men." The study, relying on stored fasting serum samples from participants enrolled in the Multiple Risk Factor Intervention Trial (MRFIT) at 22 different centers throughout the United States from December 1973 through February 1976, were used to perform a "nested case-control study" (Sidebar 5.2).

Sidebar 5.2

Nested Case Control: In a "nested case control study," only a subset of control subjects from the cohort are compared to the incident cases: a number of controls are selected for each case from that case's matched risk set. By matching factors such as age, and selecting controls from relevant risk sets, the nested case control model is generally more efficient than a case-cohort design with the same number of selected controls.

Higher levels of fasting insulin and lower levels of total and free testosterone and SHBG were significantly associated with diabetes. Low SHBG and testosterone may constitute part of the pre-diabetic state in men along with previously reported variables, such as higher glucose and insulin levels and obesity.[78]

A study titled "Inverse association of testosterone and the metabolic syndrome in men is consistent across race and ethnic groups," published in the *Journal of Clinical Endocrinology & Metabolism*, reports that a strong inverse association was observed between hormone levels and metabolic syndrome. The likelihood of metabolic syndrome increased with decrease in hormone levels. The association between sex hormones and metabolic syndrome was statistically significant across racial/ethnic groups.

The strength of the association of sex hormones with individual components of metabolic syndrome varied: stronger associations were observed with waist circumference and high blood cholesterol levels, and more modest associations with diabetes and elevated blood sugar, and the strong dose-dependent relationship between sex hormone levels and odds of metabolic syndrome in men is consistent also across racial/ethnic groups.[79]

5.35 PREVALENCE OF HYPOGONADISM AND LINKS TO HYPERTENSION, ATHEROSCLEROSIS, AND DIABETES

The Hypogonadism in Males (HIM) study estimated the prevalence of hypogonadism defined as total testosterone (TT) less than 300 ng/dL in men 45 years old or older visiting primary care practices in the United States.

Free testosterone (FT) and bioavailable testosterone (BAT) were assessed, and common symptoms of hypogonadism, co-existing medical conditions, demographics, and reason for visit were recorded.

Prevalence rate of hypogonadism was 38.7%. Similar trends were observed for FT and BAT. Figure 5.10 shows "comorbidity," or medical disorders that coexist with low testosterone. Compared to men with normal testosterone, there is about 15% greater prevalence in hypertension, about 12% higher prevalence of elevated cholesterol, and somewhat more than 10% higher prevalence of diabetes.[80]

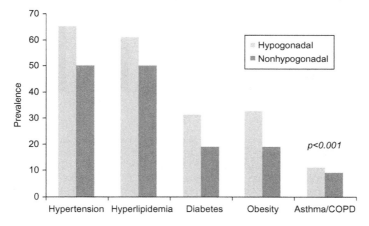

FIGURE 5.10 Comorbidities in hypogonadal men: The HIM Study. First bar: hypogonadal. Second bar: non-hypogonadal. Source: *Mulligan, T., Fric, M. F., Zuraw, Q. C., Stemhagen, A., McWhirter, C., (2006). Prevalence of hypogonadism in males aged at least 45 years: the HIM study.* Int J Clin Pract, *60, 762–769. Reproduced with kind permission from the* International Journal of Clinical Practice.

5.36 LOW TESTOSTERONE LEVEL IS LINKED TO TYPE-2 DIABETES

The journal *European Urology Supplements* reports that hypogonadism and ED, common disorders seen in urology clinics, strongly suggest that both disorders have important associations with metabolic syndrome, type-2 diabetes, and CVD.

Figure 5.11 shows the progressive age-related increase in prevalence of low levels of bioavailable testosterone in men with type-2 diabetes and metabolic syndrome. Hypogonadism and ED have close links to metabolic syndrome, type-2 diabetes, and CVD. Symptoms of hypogonadism are not specific and can be especially confounding in men with chronic diseases such as diabetes. However, testosterone replacement improves well-being and symptoms of hypogonadism, and it may also have vascular and metabolic benefits.

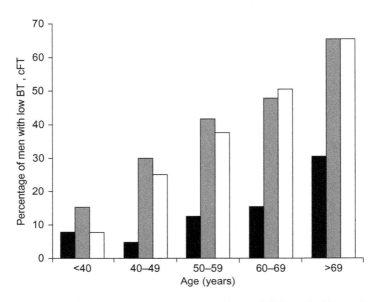

FIGURE 5.11   Prevalence of hypogonadism in men with type-2 diabetes. Incidence of positive symptom score in men with low testosterone by decades of age. (B) Bioavailable testosterone (BT) and calculated free testosterone (cFT). ■, BT < 2.5 nmol/L; ▨, BT < 4 nmol/L; □, cFT < 0.255 nmol/L. Source: *Hugh Jones, T. (2007). Testosterone Associations with Erectile Dysfunction, Diabetes, and the Metabolic Syndrome.* Eur Urol, *Suppl 6, 847−857. Reproduced with permission from* Diabetes Care.

Testosterone replacement may even help erectile function in men who do not respond to phosphodiesterase (PDE) type-5: Men who fail to respond to PDE type-5 inhibitors are more likely to have low testosterone levels; testosterone replacement improves the erectile response and at the same time it can also improve glycemic control, IR, and waist circumference in hypogonadal men with type-2 diabetes.[81]

A strong connection has been shown to exist between low plasma testosterone, type-2 diabetes, and IR. Men with low plasma testosterone are at higher risk for type-2 diabetes. A review appearing in the *Journal of Andrology* aimed to establish the relationships between testosterone, metabolic syndrome, type-2 diabetes, and IR in the context of the implication of these conditions for the development of CVD and ED.

The authors concluded that low testosterone precedes elevated fasting insulin, glucose, and HbA1C values, and may even predict the onset of diabetes. Furthermore, it is noted that treatment of prostate cancer patients with surgical or medical castration exacerbates IR and glycemic control, verifying the link between testosterone deficiency and onset of type-2 diabetes and IR. Testosterone supplementation therapy improved insulin sensitivity, fasting glucose, and HbA1c levels.

The authors suggested that low testosterone levels are associated with IR, type-2 diabetes, metabolic syndrome, and with increased deposition of visceral fat which serves as an endocrine organ, producing inflammatory cytokines and thus promoting endothelial dysfunction and vascular disease.[82]

5.37 LOW TESTOSTERONE COMPROMISES ENDOTHELIAL FUNCTION

A study titled "Low testosterone level is an independent determinant of endothelial dysfunction in men," appearing in the journal *Hypertension Research*, aimed to determine whether low plasma testosterone level is related to endothelial dysfunction in men with coronary risk factors. Outpatients were enrolled who underwent measurement of flow-mediated vasodilation (FMD) of the brachial artery using ultrasonography. The sample average age was 47 ± 15 years.

Total and free testosterone and dehydroepiandrosterone-sulfate (DHEA-S) correlated positively with percent FMD, while estradiol and cortisol did not. Percent FMD in the highest quartile of free testosterone was 1.7-fold higher than that in the lowest quartile. Total and free testosterone were related to percent FMD independent of age, BMI, hypertension, hyperlipidemia, diabetes mellitus, and smoking. They were independent of age, BMI, systolic BP, total cholesterol, HDL-C, fasting plasma glucose, smoking, and nitroglycerin-induced dilation. DHEA-S was not significantly related to percent FMD.

In conclusion, a low plasma testosterone level was associated with endothelial dysfunction in men independent of other risk factors, suggesting a protective effect of endogenous testosterone on the endothelium.[83]

5.38 DIABETES AND TESTOSTERONE DEFICIENCY

The *British Journal of Urology International* reported a study where diabetic and nondiabetic men were paired according to age, BMI, and waist-to-hip ratio (WHR) to evaluate the relationship between type-2 diabetes and serum levels of free testosterone and total testosterone.

In diabetic men, 46% were deficient in serum FT, and 34% in TT. In nondiabetic controls, 24% were deficient in FT, and 23% in TT. Subnormal FT levels were strongly correlated with diabetes, but not with elevated BMI. Subnormal TT levels were more strongly associated with elevated BMI and weight–height ratio (WHR) than with diabetes. The authors concluded that type-2 diabetes is associated with subnormal serum FT levels, whereas serum TT levels are influenced more by obesity and central adiposity.[84]

In a meta-analysis published in *Diabetes Care*, the authors report that obese men with metabolic syndrome and type-2 diabetes also may have low TT and FT, and low SHBG. By the same token, low testosterone and/or SHBG also predict the development of metabolic syndrome and type-2 diabetes.

Abdominal adiposity, believed to have endocrine activity in men with low testosterone, metabolic syndrome, and/or type-2 diabetes, may promote pro-inflammatory factors. These inflammatory markers contribute to vascular endothelial dysfunction with adverse consequences, including risk of CVD and ED.[85]

There are numerous factors that are linked to ED. These principally include hypertension, hyperlipidemia, IR, metabolic syndrome, and hypogonadism. Two related etiological factors have emerged thus far: oxidative stress and endothelium dysfunction. Endothelium dysfunction seems to be at the core of these factors, and oxidative stress may be the principal villain that damages it. It then remains to be determined whether the cardiovascular calamities supply free radicals that damage the endothelium, or whether that damage results from free radical activity.

The next chapter describes the action of yet another factor that has the capacity to cause endothelium damage and, therefore, ED: Asymmetric dimethylarginine (ADMA), a component of blood plasma continuously formed by metabolic protein turnover in all cells of the body. ADMA inhibits NO synthesis, and it is implicated in the pathogenesis and in the progression of CVDs, specifically atherosclerosis.

REFERENCES

1. Esposito K, Giugliano F, Martedì E, Feola G, Marfella R, D'Armiento M, et al. High proportions of erectile dysfunction in men with the metabolic syndrome. *Diab Care* 2005;**28**:1201−3.

2. Saenz de Tejada I, Goldstein I, Azadzoi K, Krane RJ, Cohen RA. Impaired neurogenic and endothelium-mediated relaxation of penile smooth muscle from diabetic men with impotence. *N Engl J Med* 1989;**320**:1025−30.

3. Van Dam RM, Rimm EB, Willett WC, Stampfer MJ, Hu FB. Dietary patterns and risk for type 2 diabetes mellitus in U.S. men. *Ann Intern Med* 2002;**136**:201−9.

4. Ford ES, Giles WH, Dietz WH. Prevalence of the metabolic syndrome among US adults: findings from the third national health and nutrition examination survey. *JAMA* 2002;**287**:356−9.

5. Catravas JD. The 8th international conference on the vascular endothelium: translating discoveries into public health practice − part I. *Vasc Pharm* 2006;**45**:302−7.

6. <http://www2.cscc.unc.edu/aric/>; 2013 [accessed 11.11.13]. Sponsored by National Heart, Lung and Blood Institute (NHLBI).

7. Lutsey PM, Steffen LM, Stevens J. Dietary intake and the development of the metabolic syndrome—The atherosclerosis risk in communities study. *Circul* 2008;**117**:754−61.

8. <http://www.diabetes.org/diabetes-basics/diabetes-myths/>; 2013 [accessed 11.11.13].

9. Esposito K, Marfella R, Ciotola M, Di Palo C, Giugliano F, Giugliano G, et al. Effect of a mediterranean-style diet on endothelial dysfunction and markers of vascular inflammation in the metabolic syndrome: a randomized trial. *JAMA* 2004;**292**:1440−6.

10. Villegas R, Salim A, Flynn A, Perry IJ. Prudent diet and the risk of insulin resistance. *Nutr Metab Cardiovasc Dis* 2004;**14**:334−43.

11. Enas EA, Senthilkumar A, Chennikkara H, Bjurlin MA. Prudent diet and preventive nutrition from pediatrics to geriatrics: current knowledge and practical recommendations. *Indian Heart J* 2003;**55**:310−38.

12. Mule G, Nardi E, Cottone S, Cusimano P, Volpe V, Piazza VG, et al. Influence of metabolic syndrome on hypertension-related target organ damage. *J Int Med* 2005;**257**:503−13.

13. Redon J, Cifkova R, Laurent S, Nilsson P, Narkiewicz K, Erdine S, et al. The metabolic syndrome in hypertension: european society of hypertension position statement. *J Hypertens* 2008;**26**:1891−900.

14. Schillaci G, Pirro M, Vaudo G, Gemelli F, Marchesi S, Porcellati C, et al. Prognostic value of the metabolic syndrome in essential hypertension. *J Am Coll Cardiol* 2004;**43**:1817−22.

15. NIMH online database: <http://www.ncbi.nlm.nih.gov/pubmed>; 2013 [accessed 11.11.13].

16. Er LK, Chen YL, Pei D, Lau SC, Kuo SW, Hsu CH. Increased incidence of metabolic syndrome in older men with high normotension. *Aging Male* 2012;**15**:227−32.

17. Perez-Caballero AI, Alcala-Diaz JF, Perez-Martinez P, Garcia-Rios A, Delgado-Casado N, Marin C, et al. Lipid metabolism after an oral fat test meal is affected by age-associated features of metabolic syndrome, but not by age. *Atherosclerosis* 2012; **S0021−9150**:738−41. [Epub ahead of print].

18. Hulthe J, Bokemark L, Wikstrand J, Fagerberg B. The metabolic syndrome, LDL particle size, and atherosclerosis: the Atherosclerosis and Insulin Resistance (AIR) study. *Arterioscler Thromb Vasc Biol* 2000;**20**:2140−7.

19. McNeill AM, Rosamond WD, Girman CJ, Golden SH, Schmidt MI, East HE, et al. The metabolic syndrome and 11-year risk of incident cardiovascular disease in the atherosclerosis risk in communities study. *Diabetes Care* 2005;**28**:385−90.

20. Milionis HJ, Kalantzi KJ, Papathanasiou AJ, Kosovitsas AA, Doumas MT, Goudevenos JA. Metabolic syndrome and risk of acute coronary syndromes in patients younger than 45 years of age. *Coron Artery Dis* 2007;**18**:247−52.

21. Anderson JL, Horne BD, Jones HU, Reyna SP, Carlquist JF, Bair TL, et al. Which features of the metabolic syndrome predict the prevalence and clinical outcomes of angiographic coronary artery disease? Intermountain Heart Collaborative (IHC) study. *Cardiol* 2004;**101**:185−93.

22. Kaufman FR. Diabesity: the obesity-diabetes epidemic that threatens America—*and what we must do to stop it*. New York: Bantam Books; 2006.

23. Stumvoll M, Tataranni PA, Stefan N, Vozarova B, Bogardus C. Glucose allostasis. *Diabetes* 2003;**52**:903−9.

24. Wilken TG, Metcalf BS. Glucose allostasis: emperor's new clothes? *Diabetologia* 2009;**52**:776−8.

25. Kahn SE, Prigeon RL, McCulloch DK, Boyko EJ, Bergman RN, Schwartz MW, et al. Quantification of the relationship between insulin sensitivity and β-cell function in human subjects: evidence for a hyperbolic function. *Diabetes* 1993;**42**:1663−72.

26. Retnakaran R, Shen S, Hanley AJ, Vuksan V, Hamilton JK, Zinman B. Hyperbolic relationship between insulin secretion and sensitivity on oral glucose tolerance test. *Obesity* 2008;**16**:1901−7.

27. Golay A, Ybarra J. Link between obesity and type 2 diabetes. In: Meier CA, Kiess W, editors. *Adipose tissue as an endocrine organ*, 19. Best Practice & Research Clinical, Endocrinology & Metabolism; 2005. p. 649−63.

28. Lim EL, Hollingsworth KG, Aribisala BS, Chen MJ, Mathers JC, Taylor R. Reversal of type 2 diabetes: normalisation of beta cell function in association with decreased pancreas and liver triacylglycerol. *Diabetologia* 2011;**54**:2506−14.

29. McKnight JR, Satterfield MC, Jobgen WS, Smith SB, Spencer TE, Meininger CJ, et al. Beneficial effects of L-arginine on reducing obesity: potential mechanisms and important implications for human health. *Amino Acids* 2010;**39**:349−57.

30. Yanase T, Fan W-Q, Kyoya K, Mina L, Takayanagi R, Kato S, et al. Androgens and metabolic syndrome: lessons from androgen receptor knock out (ARKO) mice. *J Steroid Biochem Mol Biol* 2008;**109**:254−7.

31. Isidori AM, Caprio M, Strollo F, Moretti C, Frajese G, Isidori A, et al. Leptin and androgens in male obesity: evidence for leptin contribution to reduced androgen levels. *J Clin Endocrinol Metab* 1999;**84**:3673−80.

32. Perego L, Pizzocri P, Corradi D, Maisano F, Paganelli M, Fiorina P, et al. Circulating leptin correlates with left ventricular mass in morbid (grade III) obesity before and after weight loss induced by bariatric surgery: a potential role for leptin in mediating human left ventricular hypertrophy. *J Clin Endocrinol Metab* 2005;**90**:4087−93.

33. Gomez-Ambrosi J, Salvador J, Paramo JA, Orbe J, de Irala J, Diez-Caballero A, et al. Involvement of leptin in the association between percentage of body fat and cardiovascular risk factors. *Clin Biochem* 2002;**35**:315−20.

34. Dunn JP, Kessler RM, Feurer ID, Volkow ND, Patterson BW, Ansari MS, et al. Relationship of dopamine type 2 receptor binding potential with fasting neuroendocrine hormones and insulin sensitivity in human obesity. *Diab Care* 2012;**35**:1105−11.

35. Billes SK, Simonds SE, Cowley MA. Leptin reduces food intake via a dopamine D2 receptor-dependent mechanism. *Mol Metab* 2012;**1**:86−93.

36. Bohnen NI, Muller MLTM, Kuwabara H, Cham R, Constantine GM, Studenski SA. Age-associated striatal dopaminergic denervation and falls in community-dwelling subjects. *J Rehab Res Dev* 2009;**46**:1045−52.

37. Volkow ND, Ding YS, Fowler JS, Wang GJ, Logan J, Gatley SJ, et al. Dopamine transporters decrease with age. *J Nucl Med* 1996;**37**:554−9.

38. Hommel JD, Trinko R, Sears RM, Georgescu D, Liu ZW, Gao XB, et al. Leptin receptor signaling in midbrain dopamine neurons regulates feeding. *Neuron* 2006;**51**:801−10.

39. Li H, Matheny M, Nicolson M, Tümer N, Scarpace PJ. Leptin gene expression increases with age independent of increasing adiposity in rats. *Diabetes* 1997;**46**:2035−9.

40. Carraro G, Albertin G, Aragona F, Forneris M, Casale V, Spinazzi R, et al. Age-dependent decrease in the ghrelin gene expression in the human adrenal cortex: a real-time PCR study. *Int J Mol Med* 2006;**17**:319−21.

41. Serra-Prat M, Palomera E, Clave P, Puig-Domingo M. Effect of age and frailty on ghrelin and cholecystokinin responses to a meal test. *Am J Clin Nutr* 2009;**89**:1410−7.

42. Spiegel K, Esra Tasali E, Penev MD P, Van Cauter E. Brief communication: sleep curtailment in healthy young men is associated with decreased leptin levels, elevated ghrelin levels, and increased hunger and appetite. *Ann Intern Med* 2004;**141**:846−50.

43. Shamsuzzaman ASM, Winnicki M, Wolk R, Svatikova A, Phillips BG, Davison DE, et al. Independent association between plasma leptin and C-reactive protein in healthy humans. *Circulation* 2004;**109**:2181−5.

44. Venugopal SK, Devaraj S, Yuhanna I, Shaul P, Jialal I. Demonstration that C-reactive protein decreases eNOS expression and bioactivity in human aortic endothelial cells. *Circul* 2002;**106**:1439−41.

45. Kimura K, Tsuda K, Baba A, Kawabe T, Boh-oka S, Ibata M, et al. Involvement of nitric oxide in endothelium-dependent arterial relaxation by leptin. *Biochem Biophys Res Commun* 2002;**273**:745−9.

46. de Haro Moraes C, Figueiredo VN, de Faria AP, Barbaro NR, Sabbatini AR, Quinaglia T, et al. High-circulating leptin levels are associated with increased blood pressure in uncontrolled resistant hypertension. J Hum Hypertens, *Jul 19* 2012.

47. Gordo, WM, Faria, AC, Barbaro, NR, Moraes, CH, Sabattini, AR, Silva, TQ, et al. Are controlled and uncontrolled resistant hypertensive patients really similar? *The Journal of Clinical Hypertension* (Abstract Supplement): ASH Annual Scientific Meeting, Vol. 14, New York, NY, 2012 May 19−22.

48. Knudson JD, Dincer UD, Zhang C, Swafford Jr. AN, Koshida R, Picchi A, et al. Leptin receptors are expressed in coronary arteries, and hyperleptinemia causes significant coronary endothelial dysfunction. *Am J Physiol Heart Circ Physiol* 2005;**289**:H48−56.

49. Lee SY, Chung WH, Lee MY, Kim SC, Kim HW, Myung SC. Leptin enhances nitric oxide-dependent relaxation of the clitoral corpus cavernosum. *Korean J Urol* 2011;**52**:136−41.

50. Korda M, Kubant R, Patton S, Malinski T. Leptin-induced endothelial dysfunction in obesity. *Am J Physiol Heart Circ Physiol* 2008;**295**:H1514−21.

51. Kshatriya S, Liu K, Salah A, Szombathy T, Freeman RH, Reams GP, et al. Obesity hypertension: the regulatory role of leptin. *Int J Hypertens* 2011;270624.

52. Mukherjee R, Villarreal D, Reams GP, Freeman RH, Tchoukina I, Spear RM. Leptin as a common link to obesity and hypertension. *Drugs Today (Barc)* 2005;**41**:687−95.

53. Knudson JD, Payne GA, Borbouse L, Tune JD. Leptin and mechanisms of endothelial dysfunction and cardiovascular disease. *Curr Hypertens Rep* 2008;**10**:434−9.

54. Payne GA, Borbouse L, Kumar S, Neeb Z, Alloosh M, Sturek M, et al. Epicardial perivascular adipose-derived leptin exacerbates coronary endothelial dysfunction in metabolic syndrome via a protein kinase C-β pathway. *Arterioscler Thromb Vasc Biol* 2010;**30**:1711−7.

55. Wang J-k, Xu D-f, Liu Y-s, Gao Y, Che J-p, Cui X-g, et al. Determination of serum leptin level in erectile dysfunction patients and its clinical significance. *Academic Journal of Second Military Medical University* 2010;**12**:1346−8.

56. Dozio E, Barassi A, Dogliotti G, Malavazos AE, Colpi GM, D'Eril GV, et al. Adipokines, hormonal parameters, and cardiovascular risk factors: similarities and differences between patients with erectile dysfunction of arteriogenic and nonarteriogenic origin. *J Sex Med* 2012;**9**:2370−7.

57. Barbieri M, Gambardella A, Paolisso G, Varricchio M. Metabolic aspects of the extreme longevity. *Exper Gerontol* 2008;**43**:74−8.

58. Facchini FS, Hua N, Abbasi F, Gerald M, Reaven GM. Insulin resistance as a predictor of age-related diseases. *J Clin Endocrin Metab* 2001;**86**:3574−8.

59. Ford ES, Abbasi F, Reaven GM. Prevalence of insulin resistance and the metabolic syndrome with alternative definitions of impaired fasting glucose. *Atherosclerosis* 2005;**181**:143−8.

60. Zaheer, S Endothelial Dysfunction in Metabolic Syndrome May Predict Cardiovascular Risk. Medscape Conference: American Society for Clinical Pathology (ASCP) 2008 (Oct. 20) Annual Meeting, Baltimore, Maryland. 2008.

61. Lue T. Erectile dysfunction. *N Engl J Med* 2000;**42**:1802−13.

62. Laumann EO, Paik A, Rosen RC. Sexual dysfunction in the United States: prevalence and predictors. *JAMA* 1999;**281**:537−44.

63. Giugliano D, Marfella R, Verrazzo G, Acampora R, Nappo F, Ziccardi P, et al. L-arginine for testing endothelium-dependent vascular functions in humans. *Am J Physiol* 1997;**273**:E606−12.

64. Bansal TC, Guay AT, Jacobson J, Woods BO, Nesto RW. Incidence of metabolic syndrome and insulin resistance in a population with organic erectile dysfunction. *J Sex Med* 2005;**2**:96−103.

65. Gndüz MI, Gümüs BH, Sekuri C. Relationship between metabolic syndrome and erectile dysfunction. *Asian J Androl* 2004;**6**:355−8.

66. Bal K, Oder M, Sahin AS, Karataş CT, Demir O, Can E, et al. Prevalence of metabolic syndrome and its association with erectile dysfunction among urologic patients: metabolic backgrounds of erectile dysfunction. *Urol* 2007;**69**:356−60.

67. Liu W-J. Bidirectional relationship between metabolic syndrome and erectile dysfunction. *Urol Sci* 2011;**22**:58−62.

68. Matfin G, Jawa A, Fonseca VA. Erectile dysfunction: interrelationship with the metabolic syndrome. *Curr Diab Rep* 2005;**5**:64−9.

69. Müller A, Mulhall JP. Cardiovascular disease, metabolic syndrome and erectile dysfunction. *Curr Opin Urol* 2006;**16**:435−43.

70. Muraleedharan V, Hugh Jones T. Review: testosterone and the metabolic syndrome. *Ther Adv Endocrin Metab* 2010;**1**:207−23.

71. Kapoor D, Clarke S, Channer KS, Jones TH. Erectile dysfunction is associated with low bioactive testosterone levels and visceral adiposity in men with type 2 diabetes. *Int J Androl* 2007;**30**:500−7.

72. Harman SM, Metter EJ, Tobin JD, Pearson J, Blackman MR. Longitudinal effects of aging on serum total and free testosterone levels in healthy men. Baltimore Longitudinal Study of Aging. *J Clin Endocrinol Metab* 2001;**86**:724−31.

73. Rodriguez A, Muller DC, Metter EJ, Maggio M, Harman SM, Blackman MR, et al. Aging, androgens, and the metabolic syndrome in a longitudinal study of aging. *J Clin Endocrinol Metab* 2007;**92**:3568−72.

74. Tenover JS. Declining testicular function in aging men. *Int Journal Imp Res* 2003;**15**(Suppl 4):S3−8.

75. Rosano GM, Sheiban I, Massaro R, Pagnotta P, Marazzi G, Vitale C, et al. Low testosterone levels are associated with coronary artery disease in male patients with angina. *Int J Impot Res* 2007;**19**(2):176−82.

76. Makhsida N, Shah J, Yan G, Fisch H, Shabsigh R. Hypogonadism and metabolic syndrome: implications for testosterone therapy. *J Urol* 2005;**174**:827−34.

77. Kalyani RR, Dobs AS. Androgen deficiency, diabetes, and the metabolic syndrome in men. *Curr Opinion Endocrin Diab Obes* 2007;**14**:226−34.

78. Haffner SM, Shaten J, Stem MP, Smith GD, Kuller L. Low levels of sex hormone binding globulin and testosterone predict the development of non insulin dependent diabetes mellitus in men. *Am J Epidemiol* 1996;**143**:889−97.

79. Kupelian V, Hayes FJ, Link CL, Rosen R, McKinlay JB. Inverse association of testosterone and the metabolic syndrome in men is consistent across race and ethnic groups. *J Clin Endocrinol Metab* 2008;**93**:3403−10.

80. Mulligan T, Fric MF, Zuraw QC, Stemhagen A, McWhirter C. Prevalence of hypogonadism in males aged at least 45 years: the HIM study. *Int J Clin Pract* 2006;**60**:762−9.

81. Hugh Jones T. Testosterone associations with erectile dysfunction, diabetes, and the metabolic syndrome. *Eur Urol* 2007;**Suppl 6**:847−57.

82. Traish A, Saad F, Guay A. The dark side of testosterone deficiency: II. Type 2 diabetes and insulin resistance. *J Androl* 2009;**30**:23−32.

83. Akishita M, Hashimoto M, Ohike Y, Ogawa S, Iijima K, Eto M, et al. Low testosterone level is an independent determinant of endothelial dysfunction in men. *Hypertens Res* 2007;**30**:1029−34.

84. Rhode EL, Ribeiro EP, Teloken C, Carlos AV, Souto CAV. Diabetes mellitus is associated with subnormal serum levels of free testosterone in men. *BJU Int.* 2005;**96**:867−70.

85. Wang C, Jackson G, Jones TH, Matsumoto AM, Nehra A, Perelman MA, et al. Low testosterone associated with obesity and the metabolic syndrome contributes to sexual dysfunction and cardiovascular disease risk in men with type 2 diabetes. *Diab Care* 2011;**34**:1669−75.

Chapter 6

Asymmetric Dimethylarginine Impairs Nitric Oxide Bioavailability and Jeopardizes Cardio-Sexual Function

6.1 INTRODUCTION

Asymmetric dimethylarginine (ADMA) is formed by the continuous turnover of protein in body cells, and it makes its way into blood plasma. It appears to be the body's natural nitric oxide (NO) antagonist, keeping NO synthesis in check by its adverse impact on endothelial nitric oxide synthase (eNOS). When it dominates the scene, ADMA reduces NO bioavailability to the detriment of a number of major cardiovascular functions. Clearly—and curiously—ADMA has not received the attention it deserves considering its profound impact on cardiovascular and erectile function.

ADMA as a marker of cardiovascular and erectile hazard is "the new boy on the block." "Superboy," in fact—so says Dr. J. P. Cooke, Division of Cardiovascular Medicine, Stanford University School of Medicine, who refers to ADMA as the *"Über Marker."*[1] The importance of ADMA as an endogenous inhibitor of NOS is now well established,[2] and in fact, one authority states that it may become a goal for pharmacotherapeutic intervention. Among other treatments, the administration of L-arginine has been shown to improve endothelium-dependent vascular function in subjects with elevated ADMA levels.[3]

Endothelial cells produce ADMA, and it acts as a hormone controlling eNOS activity from within the same cell where it is formed. Released into the extracellular space and into blood plasma, ADMA levels are in turn regulated by the enzyme (dimethylarginine dimethylaminohydrolase, DDAH) that metabolizes it. The DDAH/ADMA balance is critical in regulating NO-dependent vascular homeostasis.[4]

Researchers writing in the journal *Hormones and Metabolism Research* confirmed that elevated ADMA levels cause eNOS uncoupling, leading to

Robert Fried: Erectile Dysfunction as a Cardiovascular Impairment.
DOI: http://dx.doi.org/10.1016/B978-0-12-420046-3.00006-8

decreased NO bioavailability and increased production of hydrogen peroxide. They contend that the administration of L-arginine to patients with high ADMA levels improves NO synthesis by antagonizing the harmful effect of ADMA on eNOS function.[5]

Numerous publications report that supplementation with L-arginine restores vascular function and improves the clinical symptoms in various diseases associated with vascular dysfunction linked to elevated concentrations of ADMA.[6] However, infusing ADMA into healthy individuals results in concentrations equivalent to those observed in pathological conditions, decreasing heart rate and cardiac output, and increasing systemic vascular resistance, pulmonary vascular resistance, and causing significant sodium retention and blood pressure increase.[7,8]

6.2 ASYMMETRIC DIMETHYLARGININE CONCENTRATION IS A MAJOR CLUE TO ENDOTHELIAL VIABILITY

The *European Heart Journal* reported a study that aimed to determine the role of serum ADMA in endothelial NO bioavailability and on vascular superoxide radical (O_2^-) production in patients with advanced atherosclerosis. Samples of saphenous veins (SV) and internal mammary arteries (IMA) were collected from patients undergoing coronary bypass surgery. Preoperative measures of serum ADMA were obtained, and the response of saphenous vein segments to acetylcholine (ACh) and bradykinin (Bk) were evaluated (*ex vivo*). Vascular O_2^- was measured in paired SV and IMA by chemiluminescence.

High serum ADMA levels were associated with decreased vasorelaxation of those veins challenged by ACh and Bk. Likewise, high serum ADMA was associated with higher total O_2^- production in both SVs and IMAs and greater vascular O_2^-. Serum ADMA was independently associated with vascular O_2^- in both SVs and IMAs. ADMA was also independently associated with vasorelaxation in response to both ACh and Bk challenge.

The authors conclude that theirs is the first study to demonstrate an association between ADMA and important measures of vascular function such as NO bioavailability directly in human vessels. ADMA is associated with eNOS uncoupling in the human vascular endothelium of patients with coronary artery disease (CAD).[9]

It has consistently been shown that ADMA concentrations rise in the presence of oxidized low-density lipoprotein cholesterol (LDL-C).[10] Writing in the journal *Clinical Chemistry & Laboratory Medicine*, Dr. R. H. Böger, a pioneer in research on the biochemistry and physiology of endothelium function and dysfunction, and L-arginine as a repair mechanism, reports that when rabbits are placed on a diet enriched with 1% cholesterol, their ADMA levels rise within 4 weeks.

Elevated plasma concentrations of ADMA are also present in hypertensive patients, and in other patient groups at high risk of developing cardiovascular disease (CVD). Clinical evidence of endothelium impairment linked to ADMA elevation is observed in measures of endothelium-dependent vasodilation, and in an increased tendency to abnormal platelet aggregation. Dr. Böger contends that there is sufficient evidence now to classify ADMA as a "novel cardiovascular risk factor."[11]

It should be noted also that elevated ADMA plasma levels have been observed in healthy persons with elevated cholesterol and other cardiovascular risk factors long before symptoms emerge.[1,12] These reports all point to ADMA as a marker of the initial stages of atherosclerosis—a direct threat to the endothelium.

Cholesterol-fed animals have shown elevated ADMA levels, which correlates with the extent of carotid artery intima thickening that follows progression of atherosclerosis. As documented in a previous section, carotid artery intima-media thickness has been shown to be related to the progression of atherosclerosis in humans as well.[13]

Elevated ADMA levels are linked to reduced NO production as evidenced by reduced urinary excretion of the NO metabolites, nitrite and nitrate, and observed as impaired endothelium-dependent vasodilation.[12,14] However, it directly impairs nitric oxide/cyclic guanosine monophosphate (NO/cGMP) activity in a way different from the action of other cardiovascular hazards such as hypertension and atherosclerosis; therefore, some sources hold that the detrimental effects of ADMA must be independent of other risk factors and may even add to their effects.

The journal *Circulation* reported a study on plasma ADMA levels in healthy people who had no overt signs of any coronary or peripheral artery disease (PAD). The investigators found a significant relationship between ADMA concentration and age, mean arterial blood pressure, and glucose tolerance; and, a significant relationship between ADMA and carotid artery intima-media wall thickness. The authors likewise concluded that ADMA is a "marker for cardiovascular disease."[15] Numerous clinical studies now show that ADMA plays a key role in the onset and progression of cardiovascular diseases—especially atherosclerosis—and in erectile dysfunction (ED) as well.

Evidence also strongly suggests that ADMA concentration in serum or plasma may provide a more profound risk-warning for a given individual based on ADMA®-ELISA, a simple and rapid, yet specific, sensitive, and fully validated method now available for diagnosis in virtually any laboratory throughout the world.[16-18]

It may be said now that the credibility of the contention cited earlier in this monograph, that "*Aging-associated endothelial dysfunction in humans is reversed by L-arginine*,"[19] is bolstered by evidence that L-arginine supplementation has been shown to reduce ADMA concentration. In an article titled "Does ADMA cause endothelial dysfunction?" appearing in the journal

Atherosclerosis, Thrombosis and Vascular Biology, the author contends that plasma levels of ADMA are elevated in patients with atherosclerosis and in those with risk factors for atherosclerosis. Furthermore, plasma ADMA levels are correlated with the severity of endothelial dysfunction and atherosclerosis, and by inhibiting the production of NO, ADMA may impair blood flow, accelerate atherogenesis, and interfere with angiogenesis. "ADMA may be a novel risk factor for vascular disease.".[20,21]

6.3 ASYMMETRIC DIMETHYLARGININE AND LIPID DISORDERS

One of the earliest clinical findings on ADMA was its elevation in plasma in individuals with hypercholesterolemia who are clinically asymptomatic.[12] In a study reported in *Circulation*, ADMA levels were found to be elevated two-fold as compared to age-matched control participants with normal levels, despite the absence of overt CVD. This observation led to the conclusion that increase in ADMA levels occurs quite early in the time course of atherosclerosis, and may even contribute to its progression.

A pivotal clinical study reported in the *Egyptian Heart Journal* examining patients with various forms of CVD and heart disease clearly shows a positive linear relationship between serum cholesterol and ADMA concentrations.[22] Figure 6.1 shows that as serum cholesterol rises, so does plasma ADMA in all studied patients groups.

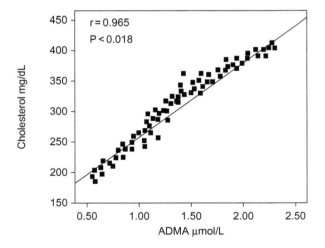

FIGURE 6.1 Correlation between plasma ADMA and serum cholesterol in all studied patient groups. Source: *Tayeh, O., Fahmi, A., Islam, M., Saied, M. (2011). Asymmetric dimethylarginine as a prognostic marker for cardiovascular complications in hypertensive patients.* The Egypt Heart J, *63, 117–124. Reproduced with kind permission from Dr. O. Tayeh and* The Egypt Heart Journal.

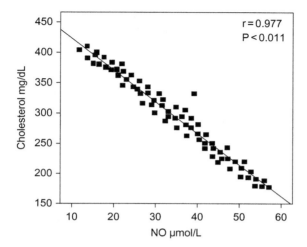

FIGURE 6.2   Correlation coefficient between serum NO and serum cholesterol in all studied patient groups. Source: *Tayeh, O., Fahmi, A., Islam, M., Saied, M. (2011). Asymmetric dimethylarginine as a prognostic marker for cardiovascular complications in hypertensive patients. The Egypt Heart J, 63, 117–124. Reproduced with kind permission from Dr. O. Tayeh and The Egypt Heart Journal.*

Figure 6.2 shows that serum cholesterol varies inversely with NO concentration. What these data tell us is that when cholesterol rises, NO declines and ADMA rises. This does not tell us what causes what.

ADMA concentration seems to track progressive atherogenesis: A study appearing in *Atherosclerosis* aimed to investigate plasma levels of ADMA in populations at high risk for atherosclerosis, and to evaluate the effect of cholesterol lowering therapy. Men with untreated hypercholesterolemia (HC group), men with well-controlled, insulin-dependent diabetes mellitus (DM group), and healthy men controls were given pravastatin (40 mg/day for eight weeks) or a matching placebo.

ADMA levels were significantly higher in the diabetes and the elevated cholesterol groups compared to controls. The L-arginine/ADMA ratios were significantly lower in both of these groups. Pravastatin resulted in significant reductions in total cholesterol and LDL-C levels, whereas no changes were observed in the levels of ADMA or the L-arginine/ADMA ratios.

Significantly elevated ADMA levels and reduced L-arginine/ADMA ratios were found in individuals with type-1 diabetes, as well as in hypercholesterolemia. However, treatment with pravastatin had no effect on the levels of ADMA in hypercholesterolemic men.[23]

According to a study in the journal *Circulation*, ADMA was found to be elevated in young participants with hypercholesterolemia. Researchers aimed to determine whether that elevation correlated with impaired endothelium-dependent, NO-mediated vasodilation, and urinary nitrate excretion, and whether it could be reversed with exogenous L-arginine.

Plasma levels of L-arginine and ADMA were measured by high-performance liquid chromatography (HPLC) in hypercholesterolemic and normocholesterolemic participants. In hypercholesterolemic participants, endothelium-dependent forearm vasodilation was assessed before and after an intravenous infusion of L-arginine, or placebo, and compared with that in normocholesterolemic control subjects.

ADMA levels were found to be significantly elevated in the hypercholesterolemic group compared with the normocholesterolemic group. L-Arginine levels were similar, resulting in a significantly decreased L-arginine/ADMA ratio in hypercholesterolemic participants. In these participants also, intravenous infusion of L-arginine significantly increased the L-arginine/ADMA ratio and normalized endothelium-dependent vasodilation and urinary nitrate excretion. ADMA levels were inversely correlated with endothelium-mediated NO/cGMP vasodilation and urinary nitrate excretion rates.

It was concluded that with ADMA elevation, NO/cGMP vasodilation showed reduced urinary nitrate excretion. This is reversed by administration of L-arginine.[12]

In the above-cited study ADMA concentrations decreased and NO/cGMP endothelial activity improved with L-arginine infusion. Elevated triglycerides also lead to increased ADMA levels, and may induce endothelial dysfunction by the same mechanism. It should be noted however, that endothelial dysfunction in hypercholesterolemia can also be reversed by oral supplementation of L-arginine.[24] In fact, that has been shown to be the case: Supplementation with L-arginine restores vascular function and improves the clinical symptoms of various diseases associated with vascular dysfunction.[6]

Parenthetically, it may seem curious that supplementing L-arginine does not necessarily result in the same outcome in endothelial impairment disorders with different etiologies: The *Journal of the American College of Cardiology* reports on the effects of intravenous L-arginine infusion on brachial artery FMD and glyceryl trinitrate (GTN)-mediated dilation in young participant subjects (18 to 40 years old) without clinical atherosclerosis, healthy participants, smokers, hypercholesterolemic patients, and controls. Baseline FMD was significantly impaired in hypercholesterolemic patients, smokers, and diabetic subjects, compared to control participants.

After L-arginine infusions, FMD improved significantly in hypercholesterolemic subjects and in smokers, but there was no change in FMD in diabetic subjects and, of course, in controls.[25] A sufficient number of studies have shown FMD improvement in diabetes with oral supplementation or infusion of L-arginine, so as to make it hard to know what to make of this outcome.

6.4 THE HOMOCYSTEINE—ASYMMETRIC DIMETHYLARGININE LINK IN ENDOTHELIAL DYSFUNCTION

Endothelial dysfunction is common in individuals with hyperhomocysteine-mia, and it is held that this condition is responsible for impaired bioavailability of NO, perhaps due to accumulation of ADMA, and therefore, increased oxidative stress. Investigators reported in *Circulation* the outcome of experimentally induced homocysteinemia on ADMA levels in patients with different cardiovascular disorders.

Having previously shown that homocysteine inhibits NO production in cultured endothelial cells by causing the accumulation of ADMA, the aim of the study was to determine if the same mechanism operates in people as well. There were patients included in the trials who had documented PAD (age, 64 ± 3 years), there were age-matched individuals at risk for atherosclerosis (age, 65 ± 1 year), and young control subjects (age, 31 ± 1 year) without evidence of, or risk factors for, atherosclerosis.

Endothelial function was measured by brachial artery FMD before and 4 hours after an orally administered methionine-loading test to induce homocysteinenemia. Blood was drawn at both time points to measure homocysteine and ADMA concentrations.

Plasma homocysteine rose in all participants after the methionine-loading test. Plasma ADMA levels rose in all participants. FMD was significantly reduced in all participants. Furthermore, positive correlations between plasma homocysteine and ADMA concentrations, as well as ADMA and FMD, obtained. It was concluded that experimental hyperhomocysteinemia leads to accumulation of the endogenous NO synthase inhibitor, ADMA, accompanied by varying degrees of endothelial dysfunction according to the preexisting state of cardiovascular health.[26]

Another clinical study published in *Circulation* also asked the question: Could homocysteine increase ADMA levels? Endothelial or nonvascular cells exposed to DL-homocysteine, or to its precursor L-methionine, raised ADMA concentration in the cell culture medium as a dose- and time-dependent function. This effect was associated with the reduced activity of DDAH. Furthermore, homocysteine-induced accumulation of ADMA was associated with reduced NO synthesis by endothelial cells. An antioxidant preserved DDAH activity and reduced ADMA accumulation. It was concluded that homocysteine inhibits DDAH enzyme activity, causing ADMA to accumulate and inhibit NO synthesis. This may explain why homocysteine impairs the endothelium NO/cGMP pathway.[27]

A report in *Cardiovascular Research* sought to determine whether oral supplementation with B vitamins or L-arginine normalizes endothelium-dependent, flow-dependent vasodilation (FDD) in patients with peripheral arterial occlusive disease (PAOD) and hyperhomocysteinemia. Patients with these conditions received combined B vitamins (folate, 10 mg; vitamin B-12, 200 μg; vitamin B-6, 20 mg/day), L-arginine (24 g/day), or placebo, for 8 weeks. Radial artery FDD was determined by high-resolution ultrasound.

It was found that vitamin B supplementation significantly lowered plasma homocysteine concentration, however, that had no significant effect on FDD. In contrast, whereas L-arginine treatment did not affect homocysteine levels, it significantly improved FDD, possibly by countering the impact of elevated ADMA concentration and reducing the oxidative stress.[28]

6.5 ASYMMETRIC DIMETHYLARGININE AND HYPERTENSION

The authors of a two-part study first reported impaired NO/cGMP responses and eNOS activity in subcutaneous vessels dissected from patients with essential hypertension. In the follow-up study appearing in the *American Journal of Physiology—Regulatory, Integrative & Comparative Physiology*, they tested the hypothesis that the patients in the study have increased circulating levels of the eNOS inhibitor, ADMA, or the lipid peroxidation product of linoleic acid (HODE), which is a marker of reactive oxygen species (ROSs).

Patients had significantly elevated plasma levels of ADMA, but similar levels of L-arginine accompanied by significantly increased rates of renal ADMA excretion and decreased rates of renal ADMA clearance. They had significantly increased plasma levels of HODE and renal HODE excretion. For the combined group of normal and hypertensive subjects, the individual values for plasma levels of ADMA and HODE were both significantly and inversely correlated with microvascular NO, and positively correlated with mean blood pressure. In conclusion, elevated levels of ADMA and oxidative stress in a group of hypertensive patients could contribute to the associated microvascular endothelial dysfunction and elevated blood pressure.[29]

In the report cited above, concerning patients with various forms of CVD and heart disease, it is clear that there is also a positive linear relationship between elevated blood pressure and ADMA concentrations: as one rises, for whatever cause, so does the other.[22]

As the following figures show, there was also a negative linear relationship between blood pressure and serum concentration of NO. These two findings are entirely consistent insofar as it has been shown in previously cited studies that blood pressure and FMD indices of atherosclerosis and endothelial impairment follow NO bioavailability—up or down. It is clear so far that blood pressure varies with serum NO concentration, as does ADMA (Figures 6.3 and 6.4).

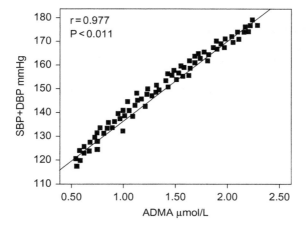

FIGURE 6.3   Correlation between plasma ADMA and blood pressure in all studied patient groups. Source: *Tayeh, O., Fahmi, A., Islam, M., Saied, M. (2011). Asymmetric dimethylarginine as a prognostic marker for cardiovascular complications in hypertensive patients. The Egypt Heart J, 63, 117–124. Reproduced with kind permission from Dr. O. Tayeh and* The Egypt Heart Journal.

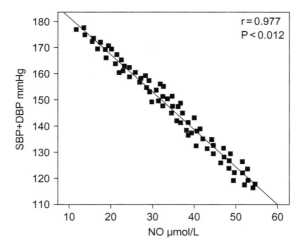

FIGURE 6.4   Correlation between serum NO and blood pressure in all studied patient groups. Source: *Tayeh, O., Fahmi, A., Islam, M., Saied, M. (2011). Asymmetric dimethylarginine as a prognostic marker for cardiovascular complications in hypertensive patients. The Egypt Heart J, 63, 117–124. Reproduced with kind permission from Dr. O. Tayeh and* The Egypt Heart Journal.

6.6 HOW INDEPENDENT IS THE ENDOTHELIUM-INDEPENDENT VASODILATION PATHWAY?

Medical control of essential hypertension is commonly accomplished by classes of meds that inhibit the action of endogenous autonomic nervous system vasoconstrictors, as well as hormones that cause the body to retain sodium. These are said to be endothelium-independent: The ACE-inhibition and beta blocker interventions act on endothelium-independent vasodilation physiological mechanism thought to be—in fact, treated as though it were—entirely separate from the endothelium-dependent (NO/cGMP) vasodilation physiology. There is tentative reason to doubt the validity of this arbitrary separation, reason based on overlapping factors—ADMA, for one.

A study published in *Hypertension* sought to test the hypothesis that angiotensin II increases microvascular ROSs and ADMA, and switches endothelial function from vasodilator to vasoconstrictor pathways. ACh-induced endothelium-dependent responses of mesenteric resistance arterioles were assessed by myography and vascular NO and ROSs by fluorescent probes in groups of male rats infused for 14 days with angiotensin II (200 ng/kg/min), or given a sham infusion. Additional groups of angiotensin or sham-infused rats were given antioxidant oral tempol (2 mmol . l−1; tempol is a free radical scavenger that mimics the action of SOD).

Angiotensin II infusion increased mean blood pressure and raised malondialdehyde, a plasma marker of oxidative stress, and decreased maximal endothelium-dependent relaxation responses and NO activity. Yet, it enhanced endothelium-dependent contraction responses and ROS production. Angiotensin II increased ADMA in vessels, but not plasma. Tempol prevented any significant changes with angiotensin II.

The authors concluded that angiotensin redirected endothelial responses from relaxation to contraction, reduced vascular NO, and increased ADMA. These effects were dependent on ROSs, and could therefore be targeted with effective antioxidant therapy.[30] A related study also found that angiotensin II increased cellular ADMA and ROSs.[31]

Yet another related study in the *American Journal of Physiology—Regulatory, Integrative & Comparative Physiology* aimed to test the hypothesis that angiotensin II-induced hypertension in the absence of renal injury is associated with increased oxidative stress and plasma and renal cortex ADMA levels (in rats).

Arterial pressure was increased after 3 and 6 weeks of angiotensin II. Renal injury was evident only after 6 weeks of exposure. No changes in renal cortex ADMA were observed. Three weeks of angiotensin II hypertension in the absence of renal injury was not found to increase ADMA. However, when the severity and duration of the treatment were increased, plasma ADMA increased. The authors concluded that the elevated blood pressure alone (for up to 3 weeks) in the absence of renal injury does not play an important role in the regulation of ADMA. However, the presence of renal injury and sustained hypertension for 6 weeks increases ADMA levels and contributes to NO deficiency and CVD.[32]

Another study was designed to examine whether ACE activity is also involved in the mechanism of ADMA elevation in type-2 diabetes. A study was performed to determine if ACE inhibition with perindopril (4 mg/day; an ACE-inhibitor commonly used to treat hypertension) for 4 weeks decreases serum ADMA concentration and plasma von Willebrand factor (vWF; a blood component that helps prevent bleeding) level, a marker of endothelial injury, in patients with type-2 diabetes.

None of the patients was treated with insulin or oral hypoglycemic drugs, and none had major diabetic complications. At the start, serum ADMA and plasma vWF were significantly higher in the type-2 diabetes patients when compared with control subjects without diabetes. Perindopril did not affect blood pressure or glucose metabolism, but did significantly decrease serum ADMA and plasma vWF.

These results suggest that endothelial injury associated with ADMA elevation may be present even in patients with non-complicated type-2 diabetes, and that increased activity of ACE may be involved in such endothelial dysfunction.[33] ADMA is a potential proinflammatory factor, and may be involved in the inflammatory reaction induced by angiotensin II.

6.7 ENDOTHELIUM-DERIVED CONTRACTING FACTOR

This monograph has focused so far principally on endothelium-derived *relaxing* factors, specifically, NO. However, in the presence of adverse conditions including hypertension, endothelial cells can also become a source of endothelium-derived *contracting* factors (EDCFs) in response to a number of agents and physical stimuli, including endothelins and angiotensin II, and particularly cyclooxygenase (COX)-derived ROSs (Sidebar 6.1).

Sidebar 6.1

COX is prostaglandin-endoperoxide synthase (PTGS), an enzyme responsible for forming prostaglandins, prostacyclin, and thromboxane. Pharmacological inhibition of COX can provide relief from the symptoms of inflammation and pain. Nonsteroidal anti-inflammatory drugs such as aspirin and ibuprofen exert their effects through inhibition of COX.

The latter were at first identified as responsible for impaired endothelium-dependent vasodilation in patients with essential hypertension. However, COX-dependent EDCF production is characteristic of the aging process, and essential hypertension seems to only anticipate the phenomenon.

It will be shown in a later chapter that (trans-)resveratrol is antioxidant and anti-inflammatory and beneficial to cardio-sexual function because it inhibits COX-2. It is worth noting that both in aging and in hypertension, EDCF production is associated with a decreased NO availability, suggesting that these substances could actually raise oxygen free radicals. Accordingly, in hypertension, antioxidants have been shown to increase the vasodilation to ACh by restoring NO availability.

In patients with essential hypertension, ADMA levels have been shown to be about two-fold higher than those in normotensive controls, and urinary excretion of nitrate is significantly decreased in these patients.[34] Studies have shown that intra-arterial infusion of ADMA causes local vasoconstriction in the corresponding circulation,[8] and in doses that lead to a certain circulating level (about 2 μmol/l), total peripheral resistance is significantly increased.[2] The conclusion drawn from these studies is that ADMA impairs normal blood pressure regulation. It is a small leap from that conclusion to the next one, that early-on, pre-symptomatic atherosclerosis may be responsible for elevating blood pressure.

It has been said that elevated serum cholesterol *leads to* elevated blood pressure. In fact, elevated ADMA found in the presence of elevated serum cholesterol may be the mediating culprit.

Researchers reported in the *Journal of the American College of Cardiology* that urinary nitrate excretion is reduced concomitantly with elevated ADMA plasma levels in patients with essential hypertension, suggesting that NO/cGMP production is impaired in these patients. They found that plasma ADMA, though within normal limits, was higher in hypertensive than in normotensive patients, and inversely related with ACh-stimulated forearm blood flow (FBF).[35] Their study provides the first demonstration that in essential hypertension, relatively higher ADMA plasma levels impair endothelium-dependent vasodilation.

That is not the only process that inhibits endothelium-dependent vasodilation by inhibiting NO bioavailability in hypertension. It has also been shown that it is thwarted by impaired L-arginine transport system y + L, with resulting reduced availability of L-arginine limiting NO synthesis in blood cells (detailed earlier).[36] A detailed description of the link between ADMA and the y + transport system is beyond the scope of this monograph, but it can be learned in Teerlink (2005)[37]

Because it has been shown that impaired endothelium-dependent, NO-mediated vasodilation is a key feature of essential hypertension and may precede the increase in blood pressure, investigators reported in the journal *Circulation* that they investigated whether transport of the NO precursor L-arginine is related to decreased endothelial function. Similar to their hypertensive counterparts, normotensive participants in their study at high risk for the development of hypertension showed evidence of impaired L-arginine transport, which may represent the link between a defective L-arginine/NO pathway and the onset of essential hypertension. The observed transport defect is not apparently due to elevated ADMA.[38]

Perticone *et al.* (2005), cited above, concluded that, "There is a subtle increase in plasma ADMA in essential hypertensives. Such an increase seems functionally relevant because relatively higher plasma ADMA levels underlie endothelial dysfunction. Plasma L-arginine concentration in hypertensive patients parallels plasma ADMA concentration, and relatively higher L-arginine levels are associated with compromised endothelial function, a phenomenon in keeping with the hypothesis that in essential hypertension the transport of this amino acid is altered."[35]

Yet, the cause of high plasma ADMA concentration in essential hypertension is unknown. Among others, a *shear stress theory* is proposed. This study published in the journal *Hypertension* investigated the effect of shear stress on ADMA release in two types of cells: transformed human umbilical vein endothelial cells (HUVECs cell line ECV-304), and HUVECs. The release of ADMA was increased by shear stress at 15 dyne/cm^2, but was not affected by shear stress at 25 dyne/cm^2, a seeming paradox.

Low to moderate shear stress enhances ADMA release, but at higher magnitudes it facilitates the degradation of ADMA, also raising eNOS activity, and thus returning ADMA release levels to baseline.[39] Alternatively, high ADMA may result from reduced catabolic rate secondary to oxidative stress, a well-known feature of human hypertension.[2]

The study cited above,[35] published in the *Journal of the American Medical Association*, aimed to determine the relationship between plasma levels of ADMA and endothelium-dependent vasodilation in participants with essential hypertension who had received no prior treatment, and in healthy normotensive participants. Endothelial function was measured during arterial infusion of ACh alone, and during co-infusion of L-arginine and sodium nitroprusside at increasing doses. Concentrations of ADMA and L-arginine in plasma were measured by high-performance liquid chromatography (HPLC). Hypertensive participants had significantly higher ADMA and L-arginine plasma concentrations than normotensive healthy controls. The ACh-stimulated FBF was significantly reduced in hypertensive participants in comparison to normotensive controls (Figure 6.5).

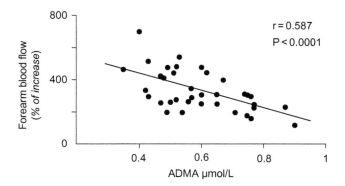

FIGURE 6.5   Correlation in hypertensive patients between ADMA levels and peak increase in ACh-stimulated forearm blood flow. Source: *Perticone, F., Sciacqua, A., Maio, R., Maria Perticone, M., Renke Maas, R., Boger, R.H., et al. (2005). Asymmetric dimethylarginine, L-Arginine, and endothelial dysfunction in essential hypertension.* J Am Coll Cardiol, 46, 518–523. *Reproduced with kind permission from the* Journal of the American College of Cardiology.

Arterial co-infusion of L-arginine induced a significant enhancement in ACh-stimulated vasodilation in hypertensive participants in whom ADMA was strongly and inversely associated with the peak increase in FBF. The investigators concluded that the L-arginine and ADMA are inversely related to endothelial function in essential hypertension.[35] An unexpected companion finding in this study was that circulating L-arginine is directly related to plasma ADMA and, like plasma ADMA, is inversely related to endothelial function. Until this clinical study, relatively little was known of the relationship between ADMA and essential hypertension. There had been only a few reports of plasma ADMA levels that found it to be higher in hypertensive patients than in normotensive healthy persons.

A report in the *American Journal of Cardiology*, for instance, examined hypertensive patients with microvascular angina before and after receiving oral L-arginine (2 g, two times daily, for 4 weeks). It was found that L-arginine significantly reduced angina, improved systolic blood pressure at rest, and improved quality of life. Maximal FBF, plasma L-arginine, the ADMA/L-arginine ratio, and cGMP increased significantly after treatment. The authors concluded that, "in medically treated hypertensive patients with microvascular angina, oral L-arginine may represent a useful therapeutic option."[40]

6.8 ASYMMETRIC DIMETHYLARGININE IN TYPE-2 DIABETES

Clinical studies have shown that ADMA concentrations are significantly elevated in type-2 diabetes.[41] Even a single high-fat meal can dramatically raise ADMA levels in type-2 diabetic patients.[42] The mechanisms underlying elevated ADMA plasma levels in diabetes mellitus are thought to be oxidative stress induced by dysregulation of DDAH, the enzyme that inactivates ADMA, linked to insulin resistance (IR).

In a trial involving both rats fed a high-glucose diet to induce type-2 diabetes and human endothelial cells (HMEC-1), a glucose-induced impairment of DDAH causes ADMA accumulation, and may contribute to endothelial vasodilator dysfunction in type-2 diabetes.[43] A review in *Frontiers of Biosciences* also reports that a high-fat diet and carbohydrate-rich diet raise serum ADMA levels, thus inhibiting NO synthesis. As noted above, ADMA levels are elevated in patients with hypertension, poor control of hyperglycemia, and dyslipidemia. One of the earliest signs of vascular dysfunction and IR, which are present in hypertension and type-2 diabetes mellitus, is an elevation in serum ADMA levels.

Reducing plasma ADMA by oral supplementation of L-arginine restores endothelial function by augmenting endothelial NO generation. Strict control of hyperglycemia decreases serum ADMA levels. These and other studies suggest that serum ADMA levels could predict the development of hypertension and type-2 diabetes in those who are at high risk to develop these diseases.[44]

A study titled "Impaired nitric oxide synthase pathway in diabetes mellitus. Role of asymmetric dimethylarginine and dimethylarginine dimethylaminohydrolase," published in *Circulation*, found that in rats a glucose-induced impairment of DDAH causes ADMA accumulation and may contribute to endothelial vasodilator dysfunction in type-2 diabetes.[43]

A clinical trial published in the *Journal of the American Medical Association* (JAMA) reported a significant relationship between IR and plasma concentrations of ADMA. Pharmacological intervention with rosiglitazone (an antidiabetic drug; for results found with metformin, also an antidiabetic drug, see[45]) enhanced insulin sensitivity and reduced ADMA levels. Increases in plasma ADMA concentrations may contribute to the endothelial dysfunction observed in insulin-resistant patients.[46]

6.9 ASYMMETRIC DIMETHYLARGININE AND ERECTILE DYSFUNCTION

The case was made and documented in previous chapters that in most instances, ED is linked to endothelial impairment. For whatever reason—hypertension, atherosclerosis, IR, oxidative stress, metabolic syndrome, or any combination of these ills (not to mention aging)—the NO/cGMP pathway has been shown to be impaired. Now, we can add another "ill" to the list, elevated ADMA (alternatively, elevated DDAH).

In many persons with ED, the L-arginine—NO/cGMP pathway is disturbed, and now, it turns out, by ADMA as well. Referring once more to the study cited earlier,[22] the prospects as we age, are not promising (Figures 6.6 and 6.7). With age, serum NO progressively declines. As it declines, serum ADMA rises. ADMA is highest when NO is lowest. Is the effect causal? Who can say? But the relationship is clearly linear—dose-related, in common terms.

ADMA levels have been shown to be linked at baseline with some cardiovascular risk factors, including the endothelial inflammation factor, E-selectin, and lipoprotein (a), a risk factor for atherosclerosis in patients with ED.[47] Penile vascular insufficiency ahead of overt symptoms predicts elevated ADMA: An interesting study published in the journal *European Urology* aimed to determine whether the extent of ultrasonographically documented penile vascular disease is associated with higher ADMA levels. ED patients aged 56 ± 9 years without manifest atherosclerotic disease, and participants with normal erectile function matched for age and traditional risk factors, were engaged in the study. Penile dynamic color Doppler parameters of arterial insufficiency, such as peak systolic velocity, veno-occlusive dysfunction, such as end diastolic velocity, and measured systemic inflammatory markers/mediators, were examined.

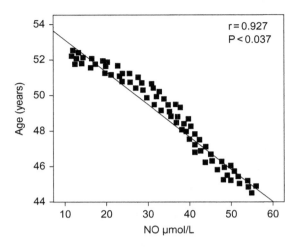

FIGURE 6.6 Correlation between serum NO and age in all studied patient groups. Source: *Tayeh, O., Fahmi, A., Islam, M., Saied, M. (2011). Asymmetric dimethylarginine as a prognostic marker for cardiovascular complications in hypertensive patients.* The Egypt Heart J, 63, 117–124. *Reproduced with kind permission from Dr. O. Tayeh and* The Egypt Heart Journal.

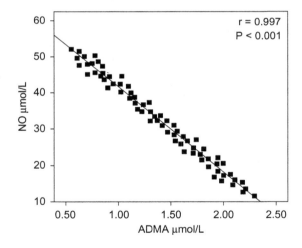

FIGURE 6.7 Correlation coefficient between serum NO and ADMA in all studied patient groups. Source: *Tayeh, O., Fahmi, A., Islam, M., Saied, M. (2011). Asymmetric dimethylarginine as a prognostic marker for cardiovascular complications in hypertensive patients.* The Egypt Heart J, 63, 117–124. *Reproduced with kind permission from Dr. O. Tayeh and* The Egypt Heart Journal.

Compared to men without ED, ED patients had significantly higher ADMA levels. ADMA was significantly increased in patients with severe arterial insufficiency compared to participants with borderline insufficiency, and men with normal penile arterial function. Analysis adjusting for age, mean pressure, other risk factors, high-sensitivity C-reactive protein (CRP), testosterone, and treatment, showed independent inverse association between ADMA level and peak systolic velocity. The combination of higher ADMA level with arterial insufficiency showed greater impact on 10-year risk of a cardiovascular calamity compared to either parameter alone.

ADMA level is independently associated with ultrasonographically documented poor penile arterial inflow. This finding underlines the important role of ADMA as a marker of penile arterial damage, and implies a contribution of this factor to the generalized vascular disease associated with ED.[48]

6.10 THE ASYMMETRIC DIMETHYLARGININE–L-ARGININE CONNECTION

The *International Journal of Andrology* reported a study aimed at investigating the relationship between ADMA, symmetric dimethylarginine (SDMA) and L-arginine concentrations, and ED. They compared plasma levels of ADMA, SDMA and L-arginine in healthy men with ED of arteriogenic and non-arteriogenic origin. Diagnosis of ED was based on the International Index of Erectile Function (IIEF) score, and its etiology was classified with penile echo-color-Doppler in basal condition and after intracavernous injection of prostaglandin E1.

The ADMA and SDMA concentrations were significantly higher in men with arteriogenic ED compared with those with ED of non-arteriogenic origin, and the concentrations in both subgroups were significantly higher than in controls. There was a negative correlation between ADMA and IIEF score only in arteriogenic ED subgroup. L-Arginine did not differ significantly either between the two ED subgroups or between each of the two ED subgroups and controls.

The L-arginine/ADMA and the L-arginine/SDMA ratios in arteriogenic ED subgroups were significantly lower than both in controls and in non-arteriogenic ED patients. The two ratios in non-arteriogenic ED patients did not differ from those in the controls.

The investigators concluded that ADMA and SDMA concentrations are significantly higher, and L-arginine/ADMA ratio lower, in patients who have arteriogenic ED compared with both patients with non-arteriogenic ED and controls. The negative correlation between ADMA and severity of ED is present only in patients with arteriogenic ED.[49]

Endothelial dysfunction: This review in *Vascular Medicine* addresses the pathophysiology of ED with a special focus on new insights into NO-mediated pathways, oxidative stress, and parallels to endothelial dysfunction. It concludes that NO appears to be the key mediator promoting endothelium-derived vasodilation and penile erection. It is thought that elevated plasma concentrations of ADMA, an endogenous eNO inhibitor, provide an additional mechanism for various forms of ED associated with cardiovascular risk factors and disease.[50]

The heart connection: The purpose of the study published in the journal *European Urology* was to investigate the role ADMA in ED with and without underlying CAD. Plasma ADMA levels were determined in a group of men with ED who were then assigned to one of two groups: a group with underlying CAD (ED-CAD), and a group without clinical evidence of underlying CAD (ED-No-CAD). Diagnosis of ED was based on the IIEF-5.

Plasma ADMA concentrations in the ED-CAD group were significantly elevated vs. the ED-no-CAD group. In analysis adjusting for hypertension, hypercholesterolemia, low HDL cholesterol, and diabetes or fasting glucose greater than or equal to 6.1 mmol/L, ADMA remained a strong and independent predictor for presence of CAD. The authors report that this study provides further strong evidence for the close interrelation of CAD and ED. Determination of ADMA concentration may help to pinpoint underlying CVD in men with ED.[51]

A study titled "Coronary endothelial dysfunction is associated with erectile dysfunction and elevated asymmetric dimethylarginine in patients with early atherosclerosis," appearing in the *European Heart Journal*, aimed to determine whether coronary endothelial dysfunction (CED) is associated with ED in men with early coronary atherosclerosis. In addition, the study examined the role of the eNOS inhibitor ADMA as an independent marker for cardiovascular disease.

ADMA levels were determined in men without obstructive CAD who underwent coronary endothelial function testing. They were all asked to complete the IIEF-5 to assess erectile function. They were then separated in accordance with the presence or absence of CED.

It was found that men with CED had significant impairment of erectile function and significantly higher ADMA levels compared with men with normal endothelial function. Erectile function positively correlated with coronary endothelial function, independent of age, body mass index (BMI), high-density lipoprotein (HDL), CRP, homeostasis model assessment of IR index, and smoking status. CED is independently associated with ED and plasma ADMA concentration in men with early coronary atherosclerosis.[52]

A study published in the journal *International Urology and Nephrology* examined the relation between plasma ADMA concentration, severity of ED, and CAD. Investigators measured plasma levels of ADMA in a group of patients, then assigned them to one of three groups: group 1, patients with ED and without CAD; group 2, patients with stable CAD; group 3, control group of patients without CAD or ED. Erectile function was evaluated with the conventional ED domain of the international index of erectile function (IIEF-EFD).

Group 1 had significantly higher concentrations of plasma ADMA than groups 2 and 3. There was negative correlation between ADMA and

IIEF-EFD score in all groups. ADMA was found to be an independent predictor of severe ED.

First, it was found that ADMA concentrations are significantly higher in patients who have ED when compared to patients with CAD, absent ED, and controls. Second, elevated levels of circulating ADMA are an independent risk factor for severe ED, and ADMA may be a link between CAD and ED.[53]

Clinical studies show that patients with ED and concomitant CAD or diabetes mellitus not only have elevated ADMA concentration, but also decreased L-arginine levels. The ratio of L-arginine to ADMA is highly unfavorable in these patients. In contrast, patients with trauma-related ED show no elevation of ADMA. Response to L-arginine treatment is usually better when pretreatment urinary excretion rates of NO metabolites are found to be low.

Elevated plasma levels of ADMA have been shown to result in uncoupling eNOS. L-arginine can thwart that process. The next chapter examines additional properties of L-arginine that benefit endothelium function, cardiovascular health and, down the road, erectile function.

REFERENCES

1. Cooke JP. Asymmetrical dimethylarginine—The "Über Marker"? *Circulation* 2004;**109**: 1813−8.
2. Achan V, Broadhead M, Malaki M, Whitley G, Leiper J, MacAllister R, et al. Asymmetric dimethylarginine causes hypertension and cardiac dysfunction in humans and is actively metabolized by dimethylarginine dimethylaminohydrolase. *Arterioscler. Thromb. Vasc. Biol* 2003;**23**:1455−9.
3. Böger RH. The emerging role of asymmetric dimethylarginine as a novel cardiovascular risk factor. *Cardiovasc Res* 2003;**59**:824−33.
4. Fiedler L. The DDAH/ADMA pathway is a critical regulator of NO signalling in vascular homeostasis. *Cell Adh Migr* 2008;**2**:149−50.
5. Toutouzas K, Riga M, Stefanadi E, Stefanadis C. Asymmetric dimethylarginine (ADMA) and other endogenous nitric oxide synthase (NOS) inhibitors as an important cause of vascular insulin resistance. *Horm Metab Res* 2008;**40**:655−9.
6. Böger R, Ron ES. L-arginine improves vascular function by overcoming the deleterious effects of ADMA, a novel cardiovascular risk factor. *Altern Med Rev* 2005;**10**:14−23.
7. Kielstein JT, Tsikas D, Fliser D. Effects of asymmetric dimethylarginine (ADMA) infusion in humans. *Eur J Clin Pharm* 2006;**62**(Issue, 1 Supplement):39−44.
8. Calver A, Collier J, Leone A, Moncada S, Vallance P. Effect of local intra-arterial asymmetric dimethylarginine (ADMA) on the forearm arteriolar bed of healthy volunteers. *J. Hum. Hypertens* 1993;**7**:193−4.
9. Antoniades C, Shirodaria C, Leeson P, Antonopoulos A, Warrick N, Van-Assche T, et al. Association of plasma asymmetrical dimethylarginine (ADMA) with elevated vascular superoxide production and endothelial nitric oxide synthase uncoupling: implications for endothelial function in human atherosclerosis. *Eur Heart J* 2009;**30**:1142−50.

10. Böger RH, Sydow K, Borlak J, Thum T, Lenzen H, Schubert B, et al. LDL cholesterol upregulates synthesis of asymmetric dimethylarginine (ADMA) in human endothelial cells. Involvement of S-adenosylmethionine-dependent methyltransferases. *Circ. Res* 2000;**87**: 99–105.

11. Böger RH. Association of asymmetric dimethylarginine and endothelial dysfunction. *Clin. Chem. Lab. Med* 2003;**41**:1467–72.

12. Böger RH, Bode-Böger SM, Szuba A, Tangphao O, Tsao PS, Chan JR, et al. Asymmetric dimethylarginine: a novel risk factor for endothelial dysfunction. Its role in hypercholesterolemia. *Circul* 1998;**98**:1842–7.

13. Zoccali C, Benedetto FA, Maas R, Mallamaci F, Tripepi G, Malatino L, et al. Asymmetric dimethylarginine (ADMA), C-reactive protein, and carotid intima media-thickness in end-stage renal disease. *J. Am. Soc. Nephrol* 2002;**13**:490–6.

14. Böger RH, Bode-Böger SM, Thiele W, Junker W, Alexander K, Frölich JC. Biochemical evidence for impaired nitric oxide synthesis in patients with peripheral arterial occlusive disease. *Circul* 1997;**95**:2068–74.

15. Miyazaki H, Matsuoka H, Cooke JP. Endogenous nitric oxide synthase inhibitor. A novel marker of atherosclerosis. *Circul* 1999;**99**:1141–6.

16. <http://www.dld-diagnostika.de>; 2013 [accessed 12.11.13].

17. <http://www.germediq.de>; 2013 [accessed 12.11.13].

18. <http://www.allaboutadma.com>; 2013 [accessed 12.11.13].

19. Chauhan A, More RS, Mullins PA, Taylor G, Petch C, Schofield PM. Aging-associated endothelial dysfunction in humans is reversed by L-arginine. *J Am Coll Cardiol* 1996;**28**:1796–804.

20. Cooke JP. Does ADMA cause endothelial dysfunction? *Arterioscler Thromb Vasc Biol* 2000;**20**:2032–7.

21. Böger RH, Tsikas D, Bode-Böger SM, Phivthong-ngam L, Schwedhelm E, Frölich JC. Hypercholesterolemia impairs basal nitric oxide synthase turnover rate: a study investigating the conversion of L-[guanidino-$^{15}N_2$]-arginine to ^{15}N-labeled nitrate by gas chromatography–mass spectrometry. *Nitric Oxide* 2004;**11**(Issue 1):1–8.

22. Tayeh O, Fahmi A, Islam M, Saied M. Asymmetric dimethylarginine as a prognostic marker for cardiovascular complications in hypertensive patients. *The Egypt Heart J* 2011;**63**:117–24.

23. Eid HM, Eritsland J, Larsen J, Arnesen H, Seljflot I. Increased levels of asymmetric dimethylarginine in populations at risk for atherosclerotic disease. Effects of pravastatin. *Atherosclerosis* 2003;**166**:279–84.

24. Lundman P, Eriksson MJ, Stühlinger M, Cooke P, Hamsten A, Tornvall P. Mild-to-moderate hypertriglyceridemia in young men is associated with endothelial dysfunction and increased plasma concentrations of asymmetric dimethylarginine. *J. Am. Coll. Cardiol* 2001;**38**:111–6.

25. Thorne S, Mullen MJ, Clarkson P, Donald AE, Deanfield JE. Early endothelial dysfunction in adults at risk from atherosclerosis: different responses to l-arginine. *J Am Coll Cardiol* 1998;**32**:110–6.

26. Stühlinger MC, Oka RK, Graf EE, Schmölzer I, Upson BM, Kapoor O, et al. Endothelial dysfunction induced by hyperhomocyst(e)inemia: role of asymmetric dimethylarginine. *Circulation* 2003;**108**(26):933–8.

27. Stühlinger MC, Tsao PS, Her JH, Kimoto M, Balint RF, Cooke JP. Homocysteine impairs the nitric oxide synthase pathway: role of asymmetric dimethylarginine. *Circulation* 2001;**104**:2569–75.

28. Sydow K, Schwedhelm E, Arakawa N, Bode-Böger SM, Tsikas D, Hornig B, et al. ADMA and oxidative stress are responsible for endothelial dysfunction in hyperhomocyst(e)inemia: effects of L-arginine and B vitamins. *Cardiovasc Res* 2003;**57**:244–52.

29. Wang D, Strandgaard S, Iversen J, Wilcox CS. Asymmetric dimethylarginine, oxidative stress, and vascular nitric oxide synthase in essential hypertension. *Am J Physiol Regul Integr Comp Physiol* 2009;**296**:R195–200.

30. Wang D, Luo Z, Wang X, Jose PA, Falck JR, Welch WJ, et al. Impaired endothelial function and microvascular ADMA in angiotensin II infused rates: effects of tempol. *Hypertension* 2010;**56**:950–5.

31. Luo Z, Teerlink T, Griendling K, Aslam S, Welch WJ, Wilcox CS. Angiotensin II and NADPH oxidase increase ADMA in vascular smooth muscle cells. *Hypertension* 2010;**56**:498–504.

32. Sasser JM, Moningka NC, Cunningham Jr. MW, Croker B, Baylis C. Asymmetric dimethylarginine in angiotensin II-induced hypertension. *Am J Physiol Regul Integr Comp Physiol* 2010;**298**:R740–6.

33. Ito A, Egashira K, Narishige T, Muramatsu K, Takeshita A. Angiotensin-converting enzyme activity is involved in the mechanism of increased endogenous nitric oxide synthase inhibitor in patients with type 2 diabetes mellitus. *Circul. J* 2002;**66**:811–5.

34. Surdacki A, Nowicki M, Sandmann J, Tsikas D, Böger RH, Bode-Böger SM, et al. Reduced urinary excretion of nitric oxide metabolites and increased plasma levels of asymmetrical dimethylarginine in men with essential hypertension. *J. Cardiovasc. Pharmacol* 1999;**33**:652–8.

35. Perticone F, Sciacqua A, Maio R, Maria Perticone M, Renke Maas R, Boger RH, et al. Asymmetric dimethylarginine, L-Arginine, and endothelial dysfunction in essential hypertension. *J Am Coll Cardiol* 2005;**46**:518–23.

36. Moss MB, Brunini TMC, DeMoura RS, Malagris LEN, Roberts NB, Ellory JC, et al. Diminished L-arginine bioavailability in hypertension. *Clin Sci (Lond)* 2004;**107**:391–7.

37. Teerlink T. ADMA metabolism and clearance. *Vascular Medicine* 2005;**10**:S73–81.

38. Schlaich MP, Parnell MM, Ahlers BA, Finch S, Marshall T, Zhang WZ, et al. Impaired L-arginine transport and endothelial function in hypertensive and genetically predisposed normotensive subjects. *Circulation* 2004;**110**:3680–6.

39. Osanai T, Saitoh M, Sasaki S, Tomita H, Matsunaga T, Okumura K. Effect of shear stress on asymmetric dimethylarginine release from vascular endothelial cells. *Hypertension* 2003;**42**:985–90.

40. Palloshi A, Fragasso G, Piatti P, Monti LD, Setola E, Valsecchi G, et al. Effect of oral L-arginine on blood pressure and symptoms and endothelial function in patients with systemic hypertension, positive exercise tests, and normal coronary arteries. *Am J Cardiol* 2004;**93**:933–5.

41. Abbasi F, Asagmi T, Cooke JP, Lamendola C, McLaughlin T, Reaven GM, et al. Plasma concentrations of asymmetric dimethylarginine are increased in patients with type 2 diabetes mellitus. *Am. J. Cardiol* 2001;**88**:1201–3.

42. Fard A, Tuck CH, Donis JA, Sciacca R, Di Tullio MR, Wu HD, et al. Acute elevations of plasma asymmetric dimethylarginine and impaired endothelial function in response to a high-fat meal in patients with type 2 diabetes. *Arterioscler. Thromb. Vasc. Biol* 2000;**20**:2039–44.

43. Lin KY, Ito A, Asagami T, Tsao PS, Adimoolam S, Kimoto M, et al. Impaired nitric oxide synthase pathway in diabetes mellitus. Role of asymmetric dimethylarginine and dimethylarginine dimethylaminohydrolase. *Circulation* 2002;**106**:987–92.

44. Das UN, Repossi G, Dain A, Eynard AR. L-arginine, NO and asymmetrical dimethylarginine in hypertension and type 2 diabetes. *Front Biosci* 2011;**16**:13–20.

45. Asagami T, Abbasi F, Stuelinger M, Lamendola C, McLaughlin T, Cooke JP. Metformin treatment lowers asymmetric dimethylarginine concentrations in patients with type 2 diabetes. *Metab* 2002;**51**:843–6.

46. Stühlinger M, Abbasi F, Chu JW, Lamendola C, McLaughlin TL, Cooke JP, et al. Relationship between insulin resistance and an endogenous nitric oxide synthase inhibitor. *J. Am. Med. Assoc* 2002;**287**:1420–6.

47. Wierzbicki AS, Solomon H, Lumb PJ, Lyttle K, Lambert-Hammill M, Jackson G. Asymmetric dimethyl arginine levels correlate with cardiovascular risk factors in patients with erectile dysfunction. *Atherosclerosis* 2006;**185**:421–5.

48. Ioakeimidis N, Vlachopoulos C, Rokkas K, Aggelis A, Terentes-Printzios D, Samentzas A, et al. Relationship of asymmetric dimethylarginine with penile Doppler ultrasound parameters in men with vasculogenic erectile dysfunction. *Eur Urol* 2011;**59**:948–55.

49. Paroni R, Barassi A, Ciociola F, Dozio E, Finati E, Fermo I, et al. Asymmetric dimethylarginine (ADMA), symmetric dimethylarginine (SDMA) and L-arginine in patients with arteriogenic and non-arteriogenic erectile dysfunction. *Int J Androl* 2012;**35**:660–7.

50. Maas R, Schwedhelm E, Albsmeier J, Böger RH. The pathophysiology of erectile dysfunction related to endothelial dysfunction and mediators of vascular function. *Vasc. Med* 2002;**7**:213–25.

51. Maas R, Wenske S, Zabel M, Ventura R, Schwedhelm E, Steenpass AH, et al. Elevation of asymmetrical dimethylarginine (ADMA) and coronary artery disease in men with erectile dysfunction. *Eur.Urol* 2005;**48**:1004–11.

52. Elesber AA, Solomon H, Lennon RJ, Mathew V, Prasad A, Pumper G, et al. Coronary endothelial dysfunction is associated with erectile dysfunction and elevated asymmetric dimethylarginine in patients with early atherosclerosis. *Eur.Heart J* 2006;**27**:824–31.

53. Aktoz T, Aktoz M, Tatlı E, Kaplan M, Turan FN, Barutcu A, et al. Assessment of the relationship between asymmetric dimethylarginine and severity of erectile dysfunction and coronary artery disease. *Int Urol Nephrol* 2010;**42**:873–9.

Chapter 7

Arginine Supplementation in Cardiovascular Disorders

7.1 INTRODUCTION

The amino acid L-arginine is the substrate from which the endothelium forms nitric oxide (NO). We know now that NO formation is essential to many body functions, but that in extreme cases NO can have adverse side effects— even toxicity. Nevertheless, many cardio-sexual calamities are linked to NO insufficiency.

Endothelium-derived NO is formed from L-arginine by the NO synthase enzyme (eNOS) with the cofactor tetrahydrobiopterin (BH4). It is a neuro-transmitter in the brain; it is essential to our immune system and to the endo-thelium of our blood vessels, where it mediates vasodilation. The typical Western diet supplies about 5 g of L-arginine each day, and most common disease conditions do not significantly reduce plasma levels, though there are exceptions.

7.2 PHARMACODYNAMICS OF L-ARGININE

What happens to L-arginine when it is either infused intravenously or ingested as an oral supplement—the pharmacodynamics of the process? In the study cited below, this is shown for one-time administration of L-arginine by intravenous infusion and by oral supplementation in healthy men. Intravenous infusion of L-arginine or oral supplementation induces peripheral vasodilation via the nitric oxide/cyclic guanosine monophosphate (NO/cGMP) pathway:

This study published in the *British Journal of Clinical Pharmacology* aimed to determine the physiological time-course of L-arginine after a single intravenous infusion of 30 g, or after a single oral application of 6 g, com-pared to placebo, in healthy men (Figure 7.1).

The vasodilation was assessed non-invasively by blood pressure (BP) moni-toring and impedance cardiography. Urinary nitrate and cGMP excretion rates were measured as non-invasive indicators of endogenous NO production.

Robert Fried: Erectile Dysfunction as a Cardiovascular Impairment.
DOI: http://dx.doi.org/10.1016/B978-0-12-420046-3.00007-X

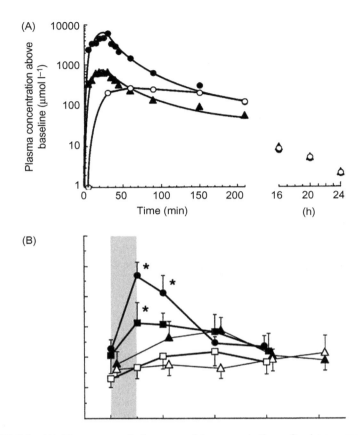

FIGURE 7.1 (A) Time course of plasma L-arginine concentrations after intravenous or oral administration of L-arginine in healthy human subjects. (B) Urinary nitrate excretion rates of nitrate (a) and cyclic GMP (b) in healthy human subjects before and after the infusion of 30 g (•) or 6 g (■) L-arginine or placebo (□), or oral ingestion of 6 g L-arginine (▲) or placebo. Source: *Bode-Böger, S. M., Böger, R. H., Galland, A., Tsikas, D., Frölich, J.C. (1998). L-arginine-induced vaso-dilation in healthy humans: pharmacokinetic-pharmacodynamic relationship.* Br J Clin Pharmacol, *46, 489–497. Reproduced with kind permission from the* British Journal of Clinical Pharmacology.

Plasma L-arginine levels rose after intravenous infusion of 30 g and 6 g L-arginine, respectively, and after oral ingestion of 6 g L-arginine. BP and total peripheral resistance declined after intravenous infusion of 30 g L-arginine, but were not significantly changed after oral or intravenous administration of 6 g L-arginine.

L-arginine (30 g) also significantly increased urinary nitrate and cGMP excretion rates. After infusion of 6 g L-arginine, urinary nitrate excretion also significantly increased, as did cGMP, although to a lesser and more variable extent than it did after 30 g of L-arginine.

The onset and the duration of the vasodilator effect of L-arginine and its effects on endogenous NO production closely corresponded to the plasma concentration half-life of L-arginine, as indicated by an equilibration half-life of 6 ± 2 (3.7−8.4) min between plasma concentration. The vascular relaxation effects of L-arginine are closely correlated with plasma concentrations.[1]

After intravenous infusion, urinary nitrate and cGMP excretion rates peaked in the first urine sample after the end of the infusion (30 min), whereas after oral L-arginine administration, the peak in urinary nitrate excretion was delayed until 120 min after intake. It seemingly took four times as long for oral supplement to peak, but there is a considerable difference not only in dosage but in absorption from oral supplement as there is a good portion that is taken up by macrophages in the digestive system. Thus, these data are revealing, but not to be taken as a basis for comparison of intravenous and oral supplement bioavailability.

The rising interest in the role of L-arginine supplementation as a new effective strategy of improving endothelial function in patients with hypertension led to this study published in the *European Review for Medical and Pharmacological Science*. It aimed to evaluate the effect of a 28-day oral supplementation of L-arginine on plasma levels of asymmetric dimethylarginine (ADMA), L-citrulline, L-arginine, and total antioxidant status (TAS), in male and female patients with mild arterial hypertension.

Ambulatory blood pressure monitoring (ABPM) was used for assigning patients to either the healthy control group or the hypertensive treatment group. Randomly assigned treatment consisted of either L-arginine (2 g tid or 4 g tid) or placebo for 28 days. TAS, plasma level of ADMA, L-citrulline, and L-arginine were measured on five consecutive visits.

In patients with mild hypertension treated with L-arginine, significant increase in TAS and plasma level of arginine and citrulline were observed. Additionally, after 28 days of L-arginine supplementation, plasma ADMA concentrations rose significantly above pretreatment concentrations.

L-Arginine supplementation increases plasma arginine, citrulline, and TAS in patients with mild arterial hypertension. It confirms that augmented concentrations of L-arginine stimulate NO biosynthesis, leading to reduction of oxidative stress.[2]

7.3 NITRIC OXIDE-DEPENDENT AND -INDEPENDENT VASCULAR ACTION OF L-ARGININE

Clinical research has shown that either intravenous administration or dietary supplementation of relatively large doses of L-arginine enhance NO formation in individuals with impaired endothelial function. In a number of clinical trials, long-term administration of L-arginine has been shown to improve the symptoms of cardiovascular disease (CVD). However, we are cautioned that there are also published reports where L-arginine did not prove

TABLE 7.1 Comparison of the Actions of NO-dependent (NO/cGMP) and NO-Independent Vascular Actions of L-arginine

NO-Dependent Vascular Actions	NO-Independent Vascular Actions
↑ Smooth muscle cell relaxation	↑ Polarization of EC membranes
↑ EC proliferation and angiogenesis	↑ Extracellular and intracellular pH
↓ Endothelin-1 release	↑ Release of insulin, GH, glucagon, and prolactin
↓ Leukocyte adhesion	↑ Synthesis of urea, creatine, PRO, and PA
↓ Platelet aggregation	↑ Plasmin generation and fibrinogenolysis
↓ Superoxide production	↓ Leukocyte adhesion to non-EC matrix
↓ Expression of cell adhesion molecules	↓ Blood viscosity
↓ Expression of monocyte chemotactic peptides	↓ Angiotensin-converting enzyme activity
↓ Proliferation of smooth muscle cells	↓ O_2^- release & lipid peroxidation
↓ EC apoptosis	↓ Formation of TXB_2, fibrin, platelet-fibrin

Abbreviations: EC, endothelial cells; GH, growth hormone, PA, polyamines; PRO, proline; TXB_2, thromboxane B_2. The symbols ↑ and ↓ denote increase and decrease, respectively.
Source: Wu, G., & Meininger, C. J. (2000). Cardiovascular benefits of L-arginine. *J Nutr*, 130: 2626–2629. Reproduced with permission from the *Journal of Nutrition*. Source: Wu, G., & Meininger, C. J. (2000). Cardiovascular benefits of L-arginine. J Nutr, 130: 2626–2629.

therapeutic, and in one study it is reported that there was a higher mortality rate among participants receiving L-arginine than among those receiving placebo. This clinical trial is detailed later in this chapter. Table 7.1 compares vascular actions of L-arginine.

It was also shown in the last chapter that metabolic byproducts of protein turnover yield various serum concentrations of ADMA, a competitive inhibitor of L-arginine formation by NOS. ADMA can affect the therapeutic response to L-arginine supplementation: L-Arginine is relatively ineffective in the presence of low ADMA levels, whereas in those with high ADMA levels, L-arginine restores the L-arginine/ADMA ratio to normal levels and normalizes endothelial function.

Dr. R. H. Böeger tells us in a publication titled "The pharmacodynamics of L-arginine," published in the *Journal of Nutrition*, that, "the effects of L-arginine supplementation on human physiology appear to be multi-causal and dose-related. Doses of 3–8 g/day appear to be safe and not to cause acute pharmacologic effects ...."[3]

Many different conditions can lead to inadequate NO bioavailability, including eNOS uncoupling. In the following clinical studies it is generally assumed that supplementation of L-arginine proves insufficiency of NO as the basis of the disorder, but there is reason to believe that L-arginine insufficiency is not likely the case.

The first five chapters spelled out how hypertension, atherosclerosis, heart disease, obesity and insulin resistance, oxidative stress, and metabolic syndrome seem all to be largely progressive age-related undesirable cardio-sexual changes that are also said to be causally linked to ED. The common factor is thought likewise to be age-related progressive endothelium dysfunction. This parcel of health and well-being calamities certainly paints a bleak picture of the so-called "Golden Years." Now the good news: The headline "L-ARGININE REVERSES AGING" appeared in virtually every newspaper one day in 1995. L-Arginine reverses aging?

In 1996, the *Journal of the American College of Cardiology* published a study titled "Aging-associated endothelial dysfunction in humans is reversed by L-arginine." This was one of the earlier studies of the hypothesis that aging impairs endothelium-dependent function and that the process may be reversible by administration of L-arginine.

The endothelium-independent vasodilators papaverine and glyceryl trinitrate, and the endothelium-dependent vasodilator acetylcholine (ACh), were infused into the left coronary artery of patients with atypical chest pain, negative exercise test results, whose age ranged between 27 and 73 years, with completely normal findings on coronary angiography, and no coronary risk factors. Coronary blood flow was measured with an intracoronary Doppler catheter. The papaverine and acetylcholine infusions were repeated in some of the patients after intracoronary infusion of L-arginine.

The study reports an initially significant inverse relationship between aging and the peak coronary blood flow response evoked by ACh. However, there was no correlation to papaverine and glyceryl trinitrate. The peak coronary blood flow response evoked by acetylcholine correlated significantly with aging before L-arginine infusion, and this inverse relationship disappeared after infusion. The results suggest that aging selectively impairs endothelium-dependent coronary vascular function, and that this impairment can be restored by intracoronary infusion of L-arginine.[4]

This means basically that in this clinical procedure where L-arginine was administered directly into the coronary artery, an "aged" endothelium responded to L-arginine as would a "younger" one ... don't try this at home.

Flow-mediated dilation (FMD) of the brachial artery is impaired in elderly individuals with CVD and reduced vascular NO bioavailability. However, there are benefits to L-arginine supplementation even where one is aged and is not suffering from CVD. As noted previously, the endothelium ages along with us as we age, and age-related progressive decline in endothelial function, even absent CVD, impairs erectile function.

A study published in *Vascular Medicine* aimed to determine whether oral L-arginine can improve impaired FMD in healthy "very old people." Healthy participants (age 73.8 ± 2.7 years) took L-arginine (8 g p.o. two times daily) or placebo for 14 days each, separated by a washout period of 14 days.

L-Arginine significantly improved FMD, whereas placebo had no effect. After L-arginine administration, plasma levels of L-arginine increased significantly, but placebo had no effect. As NO synthesis can be antagonized by its endogenous inhibitor, ADMA, it was determined that ADMA plasma concentrations were elevated at baseline. ADMA remained unchanged during treatment, but L-arginine supplementation normalized the L-arginine/ADMA ratio. It was concluded that in healthy "very old age," endothelial function is impaired and may be improved by oral L-arginine supplementation, probably due to normalization of the L-arginine/ADMA ratio (clarification of the L-arginine/ADMA ratio is given in the next chapter).[5]

7.4 THE 1992 *SCIENCE* "MOLECULE OF THE YEAR"

In 1992, the prestigious journal *Science*, took the very unusual step of declaring the gas, NO, the *Molecule of the Year*. The entire field of biological and medical science was at first in utter disbelief that a gas (NO) played a crucial biological regulatory role in the body. But today, if one were to search for the term "nitric oxide" on the US National Library of Medicine website (PubMed), the search would generate more than 130,000 medical science journal title entries linking it to virtually every known health condition, including asthma, CVD and heart disease, diabetes, kidney failure, toxemia of pregnancy, cancer, sickle cell disease, impotence, and many more.

Science first discovered in the 1980s that NO plays a crucial biological role in the body. It has since been repeatedly shown to:

- Lower BP.[6]
- Reduce the workload of the heart.[7]
- Be antioxidant.[8]
- Reduce insulin resistance in type-2 diabetes.[9]
- Strengthen penile erection.[10]

It is now widely recognized that maintaining normal BP, heart action, healthy serum cholesterol levels, and healthy blood sugar are absolutely essential to attaining and maintaining vigorous sexual performance. Failure to do so can dramatically accelerate the otherwise normal age-related progressive decline in cardiovascular function that leads to erectile dysfunction (ED).

7.5 L-ARGININE—A BUILDING BLOCK OF PROTEIN

Every cell in every tissue in the body contains proteins constituting about 20% of total body weight. Amino acids are the basic building blocks of proteins. There are twenty of them in the proteins our bodies need to function. Some also contain trace metals such as iron, chromium, and zinc. We get amino acids from protein-containing foods such as meat, fish, dairy products, nuts, and beans. When we consume foods that contain protein, digestive enzymes break the proteins into their component amino acids. These building blocks are reabsorbed and rebuilt as needed for growth, maintenance, and control of body processes.

There are two types of amino acids: Eight amino acids are considered to be *essential* because the body cannot make them in sufficient amounts and they must be supplied by food. The twelve *non-essential* amino acids can be made in adequate amounts by the body. The body cannot store amino acids, and so it is constantly breaking down and remaking proteins. This protein turnover or recycling process must be continuously resupplied by proteins in the diet, and there may often be either too much or not enough of some amino acids available to the body.

L-Arginine is an essential amino acid. Dietary sources of L-arginine include: animal sources such as dairy products, beef, pork, lamb, poultry, wild game, fish and shell fish; plant sources such as wheat germ and flour, buckwheat, granola, oatmeal, nuts (especially walnuts and almonds), seeds (pumpkin, sesame, sunflower), and chickpeas. Just one cup of chickpeas (164 g), for instance, holds 1,365 mg of L-arginine. Parenthetically, that same cup of chickpeas also holds 282 mcg of folate, a B vitamin—71% of the RDA. Folate also strongly supports healthy endothelium function, and deficiency is implicated in endothelium dysfunction, the principal cause of ED.

7.6 L-ARGININE BLOOD CELL TRANSPORT

The principal amino acid transport system in the body is the "y+L" system. Reporting in the journal *Experimental Biology*, researchers found that in human umbilical vein endothelial cells (HUVECs), the L-arginine transport via the y+L system plays a role in NO synthesis, and it is also rate-limiting and, parenthetically, may be affected by the impairment of endothelial function in type-2 diabetes.[11,12] System y+L exhibits two distinctive properties: (a) it can bind and transport *cationic* and neutral amino acids, and (b) its specificity varies depending on the ionic composition of the medium.[13]

7.7 IS DEFECTIVE TRANSPORT A POSSIBLE CAUSE OF CARDIO-SEXUAL IMPAIRMENT?

Defective L-arginine blood platelet transport was reported in a study published in the journal *Blood Cells and Metabolic Diseases* that examined the effects of obesity and metabolic syndrome on the L-arginine−NO/cGMP pathway in platelets from a population of adolescents. Adolescent patients and healthy volunteers participated in this study. Transport of L-arginine, NOS activity and cGMP content in platelets were analyzed, as well as platelet function, plasma levels of L-arginine, and metabolic and clinical markers.

L-arginine transport in platelets by system y+L was diminished in obese subjects and metabolic syndrome patients compared to that in controls. The y+L transport system activity was inversely related to insulin levels and the Homeostasis Model Assessment of Insulin Resistance Index (HOMA IR). No differences in eNOS activity and cGMP content were found among the groups. Moreover, plasma levels of L-arginine were not affected by obesity or metabolic syndrome.

This study provides the first evidence that obesity and metabolic syndrome are linked to a dysfunction of L-arginine influx that is inversely related to insulin resistance. These findings could be an early marker of future cardiovascular complications in adulthood.[14]

7.8 ENDOTHELIAL NITRIC OXIDE FROM L-ARGININE—NITRIC OXIDE SYNTHASE

There are three known forms of the enzyme nitric oxide synthase (NOS): two are *constitutive* (cNOS) and the third is *inducible* (iNOS). The cNOS form has two functions, one is brain-constitutive, and the other is eNOS. The third form is the inducible iNOS that supplies the immune system cells with NO. The form of NOS of principal concern in connection with cardio-sexual function is the eNOS that generates NO in blood vessels, including the cavernosae of the penis, and regulates vascular function.

All of the cardio-sexual and related disorders benefit from L-arginine supplementation. For instance, a clinical report on the treatment of heart failure published in *Circulation* examined patients with moderate to severe heart failure that typically have impaired blood circulation both at rest and during exercise. Patients were given six weeks of oral L-arginine hydrochloride (5.6 to 12.6 g/day) and six weeks of matched placebo capsules in random sequence. Compared with placebo, supplemental oral L-arginine significantly increased forearm blood flow (FBF) during forearm exercise. Furthermore, their functional status was significantly better on L-arginine compared with placebo, as indicated by increased distances during a 6-minute walk test and lower scores on the Living with Heart Failure questionnaire. Oral L-arginine also improved arterial compliance and reduced circulating levels of endothelin.[7]

7.9 L-ARGININE AND NITRIC OXIDE BIOAVAILABILITY IN AGING

The age-related decline in NO bioavailability noted in Figure 7.2, and also reported earlier in this monograph, explains both progressive age-related endothelial impairment and declining erectile function.

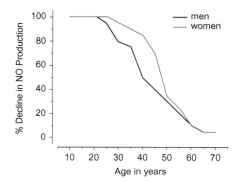

Compilation of data from multiple published reports in humans
(a) Gerhard et al Hypertension 1996
(b) Celermajer et al JACC 1994
(c) Taddei et al Hypertension 2001
(d) Egashira et al Circulation 1993

FIGURE 7.2   NO production declines with age.
Based on compilation of data from multiple published reports. Sources: *Based on compilation of data from multiple published reports: (a) Gerhard, M., Roddy, M. A., Creager, S. J., Creager, M. A. (1996). Aging progressively impairs endothelium-dependent vasodilation in forearm resistance vessels of humans.* Hypertension, *27, 849−853; (b) Celermajer, D. S., Sorensen, K. E., Bull, C., Robinson, J., Deanfield, J. E. (1994). Endothelium-dependent dilation in the systemic arteries of asymptomatic subjects relates to coronary risk factors and their interaction.* J Am Coll Cardiol. *24, 1468−1474; (c) Taddei, S., Virdis, A., Ghiadoni, L., Salvetti, G., Bernini, G., Magagna, A. (2001). Age-related reduction of NO availability and oxidative stress in humans.* Hypertension, *38, 274−279; (d) Egashira, K., Inou, T, Hirooka, Y., Kai, H., Sugimachi, M., Suzuki, S. (1993). Effects of age on endothelium-dependent vasodilation of resistance coronary artery by acetylcholine in humans.* Circul, *88, 77−81. Reprinted with kind permission from these sources:* Circulation, Hypertension, and Journal of the American College of Cardiology.

What it does not tell us is why this is occurring. What does "aging endothelium" actually mean? A number of possibilities come to mind: damaged endothelium, uncoupled or otherwise inhibited eNOS … and there are other possibilities as well.

The website "Integrative Practitioner" features a comprehensive and illuminating article titled "The role of nitric oxide insufficiency in aging & disease" by Dr. Nathan S. Bryan. Dr. Bryan is Assistant Professor of Molecular Medicine at The University of Texas Graduate School of Biomedical Sciences, Houston, TX. The article makes a number of cogent points relevant to the context of cardiovascular function and dysfunction in aging: Age is the most significant predictor of NO insufficiency because we gradually lose the ability to produce NO over the lifespan.[15]

When we are young and healthy, endothelial production of NO/cGMP derived from L-arginine is efficient and sufficient to produce NO when and where needed in the body, but with time, it progressively declines. By the time we are about 40 years old we are producing about half of the output at age 10, and by age 60, 15%. One possible explanation is that endothelium function is downgraded by oxidative stress and, as noted previously, endogenous antioxidant activity declines with age also.

Dr. Bryan concurs with previously presented evidence of a link between progressive age-related structural and functional changes that are a hallmark of hypertension and other cardiovascular risk factors. The article cites the clinical reports that show there to be:

- A gradual decline in endothelial function due to aging, with greater than 50% loss in endothelial function in the oldest age group tested as measured by FBF assays.[16]
- A loss of 75% of endothelium-derived NO in 70- to 80-year-old patients compared to young, healthy 20-year-olds.[17]
- Evidence of the upregulation of arginase in aged blood vessels and the corresponding changes of eNOS activity.[18]

Dr. Bryan concludes that EDD in resistance vessels declines progressively with increasing age as well as in healthy adults with no cardiovascular risk factors. Most of these studies report that impaired EDD is clearly evident by the fourth decade, though endothelium-independent vasodilation does not change significantly with aging. This demonstrates that the responsiveness to NO does not change only the ability to generate it. Dr. Bryan further states that, "These observations enable us to conclude that reduced availability of endothelium-derived NO occurs as we age. Being able to diagnose and intervene early on is the key to optimal health and disease prevention."[19]

This conclusion essentially summarizes the collective evidence corralled in the first five chapters of this monograph that the cardio-sexual calamities that life may have in store for us may in fact be more likely determined by hereditary dictates of the individual time course of the normal, natural, progressive age-related changes in our body, than by diet and lifestyle factors ... excepting extreme cases. This is far from saying that the latter are trivial: One can think of them perhaps as accelerants. That is not a minor detail.

In keeping with "Being able to diagnose and intervene early on is the key to optimal health and disease prevention," Dr. Nathan Bryan developed the Neogenis Test Strips™[20] that can be used by an individual to determine adequacy of NO formation.

The aim of a study titled "Aging progressively impairs endothelium-dependent vasodilation (EDV) in forearm resistance vessels of humans" that appeared in *Hypertension* was to determine whether increasing age results in altered EDV (NO/cGMP) in the forearm resistance vessels of healthy people. FBF was measured in healthy participants 19 to 69 years old by venous occlusion plethysmography. Brachial artery infusions of methacholine chloride were used to assess EDV (NO/cGMP) and infusions of sodium nitroprusside to assess endothelium-independent vasodilation (i.e. NO/cGMP-independent).

Analyses were intended to relate vascular reactivity to selected variables, including age, lipids, and BP.

EDV (NO/cGMP) was progressively impaired with increasing age. The decline in EDV was already evident by the fourth decade of life. However, endothelium-independent vasodilation did not change with age. Age, total cholesterol, and low-density lipoprotein cholesterol (LDL-C) also predicted degree of EDV. Age was the most significant predictor of endothelium-dependent vasodilator responses: In healthy people, EDV (NO/cGMP) declines steadily with increasing age.[15]

7.10 CARDIO-SEXUAL BENEFITS OF ORAL L-ARGININE

The "Cardiovascular benefits of L-arginine" are reported in the *Journal of Nutrition*. They include the following, selected from the original list in accordance with relevance to cardio-sexual function:

- Arginine availability and vascular effects: the study was funded by a grant from The *American Heart Association* (AHA) and the *Juvenile Diabetes Foundation* (JDF). It summarized a number of the findings about L-arginine and cardiovascular function that will be detailed below.
- Diet is the ultimate source of L-arginine. Dietary intake by the average American adult has been estimated to be about 5.4 g/day. Because of the relatively high arginase activity in the small intestinal mucosa, approximately 40% of dietary L-arginine is absorbed by the body. For other reasons, including the action of the transport system for amino acids, the digestibility of protein-bound L-arginine is about 90%. Also, macrophages take up quite a bit of it in the stomach.
- Alternate availability: Although extracellular L-arginine is the major source for endothelial NO synthesis, intracellular protein degradation, the arginine—citrulline cycle, may provide L-arginine for supporting short-term NO production by endothelial cells when extracellular L-arginine is limited.
- Although extensive cardiovascular research has focused historically on dietary fat and cholesterol due to the recognition of their role in atherosclerosis, studies over the last 10 years have shown the promise of using L-arginine to reverse endothelial dysfunction associated with major cardiovascular risk factors.[16]

7.11 THE ARGININE PARADOX

The enzyme eNOS is ordinarily saturated with intracellular L-arginine, and endothelial NO synthesis should not respond to alterations in extracellular L-arginine concentrations. However, despite saturation, increasing extracellular L-arginine concentrations in a dose-dependent manner increased NO production in cultured endothelial cells, and elevating plasma L-arginine levels has been shown to enhance vascular NO production *in vivo.*[21] In other words, the *arginine paradox* is the dependence of cellular NO production on exogenous L-arginine concentration despite the saturation of eNOS with intracellular L-arginine. A number of theories have been proposed to explain this paradox.

The journal of the Japanese Pharmacological Society, *Nihon Yakurigaku Zasshi*, published several theories attempting to explain the phenomenon, but concluded that none of them explains fully how exogenous L-arginine causes NO-mediated biological effects, despite the fact that eNOS is theoretically saturated with L-arginine.[22]

Dr. Brendan Lee, professor of molecular and human genetics at Howard Hughes Medical Institute, Baylor College of Medicine, avers that the key to unraveling the arginine paradox is the enzyme argininosuccinate lyase, critical to forming arginine. His down-to-earth explanation: "Think of it as though you were baking a cake. You have tons of eggs in the bakery but you can bake only so many cakes each day. You should be saturated in terms of your requirement for eggs. For some reason, though, when a truck brings in an extra 10 cases of eggs, you make more cakes.

"To carry our bakery story further, this enzyme not only delivers the eggs to the bakery, it also transfers the eggs in the bakery into the blender for use in baking the cakes."[23] While this describes the "paradox" it doesn't really resolve it.

Dr. Lee said that argininosuccinate lyase has two separate functions: First, to make arginine, and the second, to hold together a complex of proteins that transfers arginine inside the cell, or into the "oven," that makes NO. "What our work suggests is that this enzyme is the central way of regulating all of nitric oxide production in the body" said, Dr. Lee.[24]

There is a condition where this enzyme is lacking—Argininosuccinate Lyase Deficiency: The early form entails an unpleasant mix of neonatal symptoms, including elevated levels of ammonia, vomiting, lethargy, hypothermia, and poor feeding. In the absence of treatment, lethargy, seizures, and coma worsen, resulting in death. There is a later development form as well. Its symptoms include ADHD, developmental disability, seizures, learning disability, and hypertension.[25]

In the clinical studies attributing endothelium-dependent NO formation to L-arginine supplementation, how can we determine that it is actually the L-arginine-to-NO process that is responsible?

7.12 MEASURING NITRIC OXIDE FORMATION DIRECTLY AFTER L-ARGININE SUPPLEMENTATION

NO is detectable in the exhaled air of normal individuals, and may be used to monitor the formation of NO in the respiratory tract. The *American Journal of Respiratory and Critical Care Medicine* reports comparing three different techniques in asthma patients, asymptomatic smokers and non-smoking healthy adults. Patients performed a slow vital capacity maneuver

(a) through the mouth directly into an NO chemiluminescence analyzer (peak oral NO), (b) through the mouth into a collection bag (mean oral NO), and (c) through the nose into a collection bag (mean nasal NO).

Mean oral NO levels were significantly lower than peak oral NO levels, but still higher in patients with asthma in comparison with nonsmoking healthy control subjects and asymptomatic smokers. There was no significant difference in mean nasal NO levels between the three groups. There was a significant correlation between peak oral NO and mean oral NO levels. Determination of exhaled oral NO levels is qualitatively independent of the technique used, but nasal exhalation may affect NO determination in conditions associated with airway inflammation.[26]

As reported in the journal *Clinical Science (London)*, oral L-arginine causes significant increases in the concentration of NO in exhaled air at doses of 0.1 and 0.2 mg/kg, which was maximal 2 h after administration. This was associated with an increase in the concentration of L-arginine and nitrate in plasma. An increase in the amount of substrate for NOS can increase the formation of endogenous NO.[27]

Examples of the determination of NO level in blood by different means can be found in:

- Electrochemical detection of NO in biological fluids. In *Methods Enzymol.*[28]
- Determination of NO in serum and plasma of human blood. Method of high-pressure liquid chromatography (HPLC). In *Biomeditsinskaia Khimiia.*[29]
- Measurement of NO levels in the red cell: Validation of tri-iodide-based chemiluminescence with acid-sulfanilamide pretreatment. In *Journal of Biological Chemistry.*[30]

In another study appearing in the journal *Chest*, in a randomized, placebo-controlled, crossover trial, healthy men received a 30-min infusion of 0.5 g/kg L-arginine hydrochloride. BP and heart rate were continuously monitored in addition to intermittent measures of mixed expired NO concentration, and plasma L-arginine and L-citrulline levels.

Infused L-arginine caused a significant fall in BP and raised expired NO concentration (FeNO), an increase in the rate of pulmonary NO excretion, as well as a rise in plasma L-citrulline. There was a significant correlation between the hypotensive response to L-arginine and the increase in expired NO. The hypotensive effect of L-arginine appears to be mediated, at least in part, by NO synthase metabolism of L-arginine and increased endogenous NO production as indicated both by increased plasma L-citrulline and by increased expired NO.[31]

7.13 L-ARGININE AND HYPERTENSION

Hypertension contributes to endothelium damage by promoting atherosclerosis, especially in the coronary arteries. As noted previously, hypertension has been repeatedly linked to ED, and it is said to be a factor in impaired NO/cGMP formation. Because NO is formed from L-arginine, there has been a suggestion that perhaps hypertension is due, at least in part, to L-arginine insufficiency. However, that is unlikely.

A study titled "Diminished L-arginine bioavailability in hypertension," appearing in the journal *Clinical Science (Lond)*, was designed to investigate the L-arginine/NO process in hypertension. Transport of L-arginine into red blood cells and platelets, eNOS activity, and amino acid profiles in plasma were analyzed in hypertensive patients and in an animal model of hypertension. It was found that influx of L-arginine into red blood cells was mediated by the y+ and y+L amino acid transport systems, whereas in platelets, influx was mediated only via system y+L.

Chromatographic analyses revealed higher plasma levels of L-arginine in hypertensive patients compared with control subjects. L-arginine transport via system y+L, but not y+, was significantly reduced in red blood cells from hypertensive patients compared with controls. Basal eNOS activity was decreased in platelets from hypertensive patients compared to controls. Their findings provided the first evidence that hypertension is associated with an inhibition of system y+L L-arginine transport with reduced availability of L-arginine limiting NO synthesis in blood cells.[32]

Another clinical trial reported in the journal *Circulation* titled "Impaired L-arginine transport and endothelial function in hypertensive and genetically predisposed normotensive subjects" measured forearm and peripheral blood mononuclear cell arginine uptake in hypertensive participants and in two groups of healthy volunteers with and without a family history of hypertension. FBF responses to ACh and sodium nitroprusside were measured before and after supplemental intra-arterial infusion of L-arginine. It was found that similar to their hypertensive counterparts, normotensive individuals are at high risk for the development of hypertension as shown by impaired L-arginine transport, which may represent the link between a defective L-arginine/NO pathway and the onset of essential hypertension.[33]

7.14 ORAL L-ARGININE LOWERS BLOOD PRESSURE

A recent article in the *American Heart Journal* reported the "Effect of oral L-arginine supplementation on blood pressure: a meta-analysis of randomized, double-blind, placebo-controlled trials." It included 11 randomized, double-blind, placebo-controlled trials. Oral L-arginine intervention ranged from 4 to 24 g/day. L-Arginine intervention significantly lowered both systolic BP and diastolic BP. Analyses included only trials lasting 4 weeks or longer, and trials where participants did not use antihypertensive medications. The authors concluded that the meta-analysis upholds evidence that oral L-arginine supplementation significantly lowers both systolic and diastolic BP.[34]

In a prospective randomized, double-blind trial titled "Oral L-arginine improved endothelial dysfunction in patients with essential hypertension," appearing in the *International Journal of Cardiology*, patients with essential hypertension received either 6 g L-arginine or placebo. Effect of treatment was assessed for endothelium-dependent FMD of the brachial artery before administration of L-arginine or placebo, and 1½ hours afterwards. At the end of the protocol, nitrate-induced, endothelium-independent vasodilatation was also evaluated.

The two groups, L-arginine and placebo, were similar regarding age, sex, blood lipids, smoking, diabetes, coronary artery disease (CAD), body mass index (BMI), intima-media thickness of the common carotid artery, clinical BP, and baseline brachial artery parameters.

Administration of L-arginine or placebo did not significantly change heart rate, BP, baseline diameter, or blood flow. However, L-arginine resulted in a significant improvement of FMD while placebo did not. The effect of L-arginine on FMD was significantly different from the effect of placebo. The authors concluded that oral administration of L-arginine acutely improves endothelium-dependent brachial artery FBF in patients with essential hypertension.[35]

In another study published in the *American Journal of Cardiology*, hypertensive patients with microvascular angina were studied before and after receiving oral L-Arginine (4 weeks, 2 g, 3 times daily). L-arginine significantly improved angina class, systolic BP at rest, and quality of life. Maximal FBF, plasma L-arginine, and cGMP increased significantly after treatment.[36]

A clinical study from the Marcinkowski University of Medical Science, Poland, found a strong association between L-arginine supplementation and BP reduction: Both systolic and diastolic BP were significantly lower after 4 weeks of oral L-arginine supplementation in the subgroup of patients treated with 12 g of L-arginine daily. A stronger effect was observed during the day.[37]

Here, in brief, are a few more of the many clinical studies that demonstrate the benefits of oral L-arginine supplementation under medical supervision:

- Oral L-arginine, 3 g/h for 10 h, lowered systolic and diastolic BP, and increased plasma citrulline in patients with type-2 diabetes. In the journal *Atherosclerosis*.[38]
- Patients with mild hypertension treated with 28-day oral L-arginine supplement (2 g tid or 4 g tid) showed significant increase in TAS and plasma level of arginine and citrulline. Additionally, plasma ADMA concentrations after 28 days of L-arginine supplementation significantly exceeded initial concentrations. Augmented concentrations of L-arginine stimulate NO biosynthesis, which leads to reduction of oxidative stress. In *European Review of Medical Pharmacology Science*.[2]

- In normotensive subjects, inhibiting the action of eNOS as measured by response to ACh decreased with advancing age. In normotensive individuals, an earlier primary dysfunction of the NO system, and a later production of oxidative stress, causes age-related reduction in EDV. These alterations are similar but anticipated in hypertensive patients compared with normotensive subjects. In *Hypertension*.[16]

7.15 ORAL L-ARGININE IMPROVES DAMAGED AND AGING ENDOTHELIUM

The endothelial lining the cavernosae of the penis are especially vulnerable to free radical damage and atherosclerosis formation and, as previously noted, are liable to those hazards somewhat sooner than endothelium elsewhere in the body. Normal age-related progression of endothelial impairment can be accelerated by "poor" diet. Endothelial dysfunction detected by FMD declines with increasing age in healthy people: Age predicts impaired endothelial function.[39,40]

According to a review titled "L-Arginine as a nutritional prophylaxis against vascular endothelial dysfunction with aging," published in the *Journal of Cardiovascular Pharmacology & Therapeutics*, "With advancing age, peripheral conduit and resistance arteries lose the ability to effectively dilate owing to endothelial dysfunction. This vascular senescence contributes to increased risk of CVD with aging." The review reports on oral L-arginine as a "novel" nutritional strategy to reduce progression of vascular dysfunction with aging and CVD. Emphasis is placed on the ability of L-arginine to modulate the vascular inflammatory and systemic hormonal factors, which may result in a positive effect on vascular endothelial function.[41]

In support of their conclusion that a considerable amount of attention has been placed on the ability of oral L-arginine to benefit vascular endothelial function, the authors cite the following clinical findings:

- In young men with CAD, oral L-arginine improves endothelium-dependent dilatation and reduces monocyte/endothelial cell adhesion. In the journal *Atherosclerosis*.[42]
- Short-term oral L-arginine effectively improves vascular endothelial function when the baseline FMD is low. In the *American Journal of Clinical Nutrition*.[43]
- In healthy very old age, endothelial function is impaired and may be improved by oral L-arginine supplementation, probably due to normalization of the L-arginine/ADMA ratio. In the journal *Vascular Medicine*.[5]
- Dietary supplementation with L-arginine significantly improves endothelium-dependent dilation (EDD) in hypercholesterolemic young adults, and this may impact favorably on the atherogenic process. In the *Journal of Clinical Investigation*.[44]

- The impairment of endothelium-dependent vasodilatation induced by old age is due to an altered NO signaling mechanism in skeletal muscle arterioles. It is not the result of increased arginase activity and limited L-arginine substrate. In the *Journal of Physiology.*[45]
- L-arginine improves endothelial function in hypercholesterolemia and with atherosclerosis. Clinical trials to date support potential clinical applications of L-arginine in the treatment of CAD and peripheral artery disease (PAD). In *Journal of Nutrition.*[46]
- Aging diminishes endothelium-dependent vasodilatation and tetrahydrobiopterin content in skeletal muscle arterioles. In the *Journal of Physiology.*[47]
- Supplementing L-arginine reduces endothelial dysfunction and oxidative stress induced by postprandial hypertriglyceridemia. In *International Journal of Cardiology.*[48]
- Postprandial endothelial impairment is partly abolished by L-arginine administration. In the *Journal of Clinical Pharmacology and Therapeutics.*[49]
- Enteral or parenteral L-arginine reverses endothelial dysfunction associated with major cardiovascular risk factors such as hypercholesterolemia, smoking, hypertension, diabetes, obesity/insulin resistance and aging, and ameliorates many common cardiovascular disorders, including coronary and PAD, and heart failure. In the *Journal of Nutrition.*[50]
- Oral L-arginine supplement improved endothelial function as shown by FMD, and reduced LDL oxidation in stable CAD patients. In *Clinical Nutrition.*[51]

A study published in *Cardiovascular Pharmacology* aimed to determine effects of long-term L-arginine supplementation on arterial compliance, and inflammatory and metabolic parameters in patients with multiple cardiovascular risk factors. Patients were assigned to daily oral L-arginine, or matching placebo, capsules. They were evaluated for lipid profile, glucose, HbA1C, insulin, hs-CRP, rennin, and aldosterone. Arterial elasticity was evaluated with pulse wave contour analysis (HDI CR 2000, Eagan, Minnesota).

Large artery elasticity index (LAEI) did not differ significantly between the groups at baseline, but at the end of the study LAEI was significantly greater in patients treated with L-arginine than in the placebo group. Systemic vascular resistance was significantly lower in patients treated with L-arginine than in the placebo group after 6 months. Small artery elasticity index (SAEI) did not differ significantly between the groups at baseline or at the end of the study. Serum aldosterone decreased significantly in Group 1, but it did not change in the placebo group.

Oral L-arginine supplementation improves large artery elasticity in patients with multiple cardiovascular risk factors. This improvement was associated with a decrease in systolic BP, peripheral vascular resistance, as well as a decrease in aldosterone levels. The results suggest that long-term L-arginine supplementation has beneficial vascular effects in pathologic disease states associated with endothelial dysfunction.[52]

A report in *Clinical Nutrition* informs us that consumption of either L-arginine or vitamin C significantly increased brachial artery FMD. Neither oral L-arginine nor vitamin C affected lipid profiles and circulating levels of inflammatory markers, however, in those patients whose LDL susceptibility to oxidation was determined, lag time significantly increased by 27.1% after consumption of L-arginine for 4 weeks: Oral L-arginine supplement improved endothelial function and reduced LDL oxidation in stable CAD patients.[51]

What all these studies point to is that inadequate NO/cGMP formation is most likely due to deficient transport or eNOS uncoupling often erroneously interpreted as "L-arginine deficiency."

7.16 L-ARGININE, INSULIN RESISTANCE, AND TYPE-2 DIABETES

Type-2 diabetes is a prominent player in cardiovascular calamities and in ED, and it is likewise linked to NO deficiency, but is it causally linked? Numerous studies indicate that elevated blood sugar and oxidative stress are factors responsible for endothelium dysfunction and the subsequent blood vessel impairment. Increased production of free radicals in vascular endothelium causes disturbance in production and/or decreases bioavailability of NO.

It has been suggested that L-arginine supplementation is a reasonable method to increase endothelium NO production and lower free radical formation. There is growing evidence that dietary supplementation of L-arginine (by whatever mechanism) reverses endothelial dysfunction associated with major cardiovascular risk factors, and ameliorates many common cardiovascular disorders.

An *in vitro* study published in the *Journal of Pharmacotherapy and Experimental Therapeutics* aimed to evaluate the effect of supplementation with L-arginine on both functional relaxation and cGMP generation in response to ACh in the induced diabetic rat aorta. Concentration of L-arginine in plasma and aortic tissue were both decreased by diabetes: Acute incubation *in vitro* with L-arginine augmented the impaired relaxation to acetylcholine in diabetic rings. It did not alter relaxation in control rings.[53]

Amino acids are important modulators of glucose metabolism, insulin secretion, and insulin sensitivity. However, little is known about the changes in amino acid metabolism in patients with diabetes. While rat aorta is a legitimate model of human aorta it is, after all, only a model. In a study titled "Selective amino acid deficiency in patients with impaired glucose tolerance and type 2 diabetes," appearing in the journal *Regulatory Peptides*, the circulating amino acid levels were determined in patients with type-2 diabetes, non-diabetic participants with impaired glucose tolerance (IGT), and control subjects.

Sidebar 7.1

Homeostatic Model Assessment (HOMA): A computer-solved model predicts plasma glucose and insulin concentrations arising from varying degrees of beta-cell deficiency and insulin resistance. Comparison of a patient's fasting values with the model's predictions allows a quantitative assessment of the contributions of insulin resistance and deficient *beta*-cell function to the fasting hyperglycemia

Source: Matthews, D. R., Hosker, J. P., Rudenski, A. S., Naylor, B. A., Treacher, D. F., Turner, R. C. (1985). Homeostasis model assessment: insulin resistance and beta-cell function from fasting plasma glucose and insulin concentrations in man. Diabetologia. 28, 412–9.

Patients with type-2 diabetes had significant reductions in the concentrations of various amino acids, including L-arginine. The plasma levels of essential amino acids were positively related to fasting and post-challenge glucose levels, HOMA (Sidebar 7.1) insulin resistance, and fasting glucagon levels.

Total amino acid levels were similar in patients with diabetes, impaired glucose tolerance participants, and controls, but individual levels of several amino acids differ significantly between these groups. These alterations may contribute to the disturbances in insulin secretion and action in diabetic patients.[54]

A clinical study published in the *American Journal of Physiology, Endocrinology & Metabolism* aimed to evaluate the effects of a long-term oral L-arginine therapy on adipose fat mass and muscle free-fat mass distribution, daily glucose levels, insulin sensitivity, endothelial function, oxidative stress, and adipokine release in obese type-2 diabetic patients with insulin resistance who were treated with a combined period of hypocaloric diet and exercise training. The patients, participating in a hypocaloric diet plus an exercise training program for 21 days, were divided into two groups: the first group, also treated with L-arginine (8.3 g/day), and the second group, given a placebo.

Long-term oral L-arginine treatment resulted in an additive effect on glucose metabolism and insulin sensitivity compared with a diet and exercise training program alone. Furthermore, it improved endothelial function, oxidative stress, and adipokine release in obese type-2 diabetic patients with insulin resistance.[55]

A study appearing in the journal *Diabetes Care* reported that in lean patients with type-2 diabetes, L-arginine treatment significantly improved, but did not completely normalize, insulin sensitivity.[9] Here are several more of the many studies that report the benefits of oral L-arginine supplementation:

- Oral two-month supplementation with L-arginine (3×2 g/day) had no effect on fasting glucose and HbA1 levels in diabetic patients with atherosclerotic PAD of lower extremities at Fontaine's stage II. The supplementation of L-arginine, however, led to substantial increase in NO concentration and TAS level in these patients, suggesting its indirect antioxidative effect. In *European Review for Medical & Pharmacological Science.*[56]

- Patients with diabetes mellitus received two daily dosages of 1 g L-arginine free base for three months. Treatment reduced the lipid peroxidation product malondialdehyde, evidence that treatment with L-arginine may counteract lipid peroxidation and thus reduce microangiopathic long-term complications in diabetes mellitus. In *Free Radical Biology & Medicine.*[57]

- Healthy participants and patients with obesity and non-insulin-dependent diabetes mellitus (NIDDM) were given low-dose intravenous supplementation of L-arginine (0.52 mg kg-1 min-1) to note the effect on insulin-mediated vasodilatation and insulin sensitivity measured by venous occlusion plethysmography during the insulin suppression test, evaluating insulin sensitivity. L-Arginine restored the impaired insulin-mediated vasodilatation observed in obesity and noninsulin dependent diabetes. Insulin sensitivity improved significantly in all three groups. Results suggest that defective insulin-mediated vasodilatation in obesity and NIDDM can be normalized by intravenous L-arginine. Furthermore, L-arginine improves insulin sensitivity in obese patients and NIDDM patients as well as in healthy subjects, indicating a possible mechanism that is different from the restoration of insulin-mediated vasodilatation. In the *European Journal of Clinical Investigation.*[58]

7.17 OXIDATIVE STRESS AND METABOLIC SYNDROME

A review appearing in the *Open Biochemistry Journal* summarizes the links between L-arginine–NO pathway and oxidative stress in metabolic syndrome and cardiovascular events at the systemic level, as well as the effects of exercise on this syndrome. Emerging evidence suggests that NO, inflammation and oxidative stress play an important role in this syndrome.[59]

The journal *Amino Acids* reported that dietary L-arginine supplementation reduces adiposity in obese people with type-2 diabetes mellitus, and it holds great promise as a safe and cost-effective nutrient to reduce adiposity, increase muscle mass, and improve the metabolic profile.[60]

7.18 ORAL L-ARGININE COMBATS ATHEROSCLEROSIS

According to a review appearing in the journal *Vascular Medicine*, oral L-arginine has shown considerable benefit in treatment of hypercholesterolemia and atherosclerosis.[61]

The following citations support the conclusion. The titles speak for themselves:

- Oral L-arginine improves endothelium-dependent dilatation and reduces monocyte adhesion to endothelial cells in young men with CAD. In *Atherosclerosis.*[42]

- Oral L-arginine improves EDD in hypercholesterolemic young adults. In the *Journal of Clinical Investigation.*[44]
- L-Arginine improves EDD in hypercholesterolemic humans. In the *Journal of Clinical Investigation.*[62]
- Correlation of endothelial dysfunction in coronary microcirculation of hypercholesterolemic patients by L-arginine. In the *Lancet.*[63]
- Long-term L-arginine supplementation improves small-vessel coronary endothelial function in humans. In *Circulation.*[64]
- Oral L-arginine administration attenuates postprandial endothelial dysfunction in young healthy males. In the *Journal of Clinical Pharmacology & Therapeutics.*[49]
- Dietary L-arginine supplementation normalizes platelet aggregation in hypercholesterolemic humans. In the *Journal of the American College of Cardiology.*[8]

A study in the *Journal of Nutrition*, titled "Oral L-arginine improves hemodynamic responses to stress and reduces plasma homocysteine in hypercholesterolemic men," describes a hemodynamic mechanism for the hypotensive effect of oral L-arginine (12 g/day for 3 wk). It is first to show substantial reductions in homocysteine with oral administration. Compared with placebo, L-arginine changed cardiac output, diastolic BP, pre-ejection period, and plasma homocysteine. The change in plasma L-arginine was inversely related to change in plasma homocysteine, but it did not affect plasma insulin or plasma glucose, C-reactive protein (CRP), or lipids.[65]

The journal *Medical Science Monitor* reported a study designed to assess the effect on NO concentration and TAS in patients with atherosclerotic PAD at Fontaine's classification of PAD stages II and III (Sidebar 7.2).

Sidebar 7.2

Fontaine's classification of PAD consists of several stages of severity:
- Stage II – Intermittent claudication. This stage takes into account the fact that patients usually have a very constant distance at which they have pain:
  - Stage IIa – Intermittent claudication after more than 200 meters of pain free walking.
  - Stage IIb – Intermittent claudication after less than 200 meters of walking.
- Stage III – Rest pain. Rest pain is especially troubling for patients during the night. The reason for this is twofold: First, the legs are usually raised up on to a bed at night, thus diminishing the positive effect gravity may have had during the day when the legs were dependent. Second, during the night the lack of sensory stimuli allow patients to focus on their legs

---

Source: *Weinberg, I. (2010). Fontaine Classification. http://www.angiologist.com/arterial-disease/ fontaine-classification/ (Accessed November 14, 2013).*

The treatment regimen consisted of two different doses of L-arginine over 28 days of supplementation. Patients were assigned to group A receiving L-arginine at 3i2 g/day, or group B, 3i4 g/day. NO concentration was determined with Oxis using a Hyperion Micro Reader. TAS in mmol/L was established in serum with the RANDOX NX2332 test.

Group A showed substantially higher NO levels after 14 and 28 days of therapy. In group B, the NO level increase was substantial after 28 days. Noticeably higher TAS was noted in both groups. Group A showed this only after 28 days of treatment, while group B exhibited substantial increase in TAS after 7, 14, and 28 days of L-arginine supplementation.

It was concluded that oral supplementation of L-arginine for 28 days leads to substantial increases in NO and TAS levels, comparable in both groups of patients, in the blood of patients with atherosclerotic PAD at Fontaine's stages II and III. The TAS concentration rise points to an antioxidative effect of L-arginine oral supplementation.[66]

In this study published in the *Journal of Clinical Investigation*, high-resolution external ultrasound was used to study blood vessel status in participants aged 19 to 40 years with known endothelial dysfunction and elevated LDL-C levels. Each patient was studied before and after 4 wk of L-arginine (7 grams × 3/day) or placebo powder, with 4 wk washout, in a randomized double-blind crossover study. Brachial artery diameter was measured at rest, during increased flow causing EDD, and after sublingual glyceryl trinitrate causing endothelium-independent dilation.

After oral L-arginine, plasma L-arginine levels rose, and EDD improved. In contrast, there was no significant change in response to glyceryl trinitrate causing endothelium-independent dilation. After placebo there were no changes in endothelium-dependent or -independent vascular responses. Lipid levels were unchanged after L-arginine and placebo. Dietary supplementation with L-arginine significantly improves EDD in hypercholesterolemic young adults, and this may impact favorably on the atherogenic process.[44]

7.19 ORAL L-ARGININE AND HEART DISEASE

Heart disease, and in particular CAD, has been declared a distinct threat to erectile function by numerous medical health authorities. Some even went so far as to state that ED may be an early warning sign, if not a symptom, of heart disease. A clinical study on patients with normal coronary angiograms and those with CAD appearing in the journal *Circulation* based the conclusion that, "... there is a relative deficiency of L-arginine in diseased coronary arteries" on blood vessel dilation measures following intracoronary infusion of various substances such as substance P, NG-monomethyl- L-arginine monoacetate salt (L-NMMA), L-arginine, and nitroglycerin.[67]

The journal *Heart* reported a clinical trial where patients with CAD and stable angina were given an L-arginine infusion. Infusion significantly increased the diameter of all the coronary segments and stenoses. L-arginine dilates coronary segments and stenosis, but does not increase the magnitude of the response to atrial pacing in proximal and distal segments and in coronary stenoses and their reference segments. These findings provide evidence that the shear stress responsive mechanism is absent at stenosis, but present in non-stenotic segments of diseased coronary arteries. They also indicate a relative deficiency of L-arginine, except in the shear response mechanism.[68]

The validity of the concept of "L-arginine deficiency" is doubtful, as noted previously. Nevertheless, clinical data stand even if the interpretation of the outcome of the trial may be questionable. It is likely that at least some of the studies that conclude that there may exist an L-arginine deficiency may be reinterpreted in terms of eNOS uncoupling or another more recently determined probable cause, such as a dysfunctional transport system.

A clinical study titled "Oral L-arginine improves endothelium-dependent dilatation and reduces monocyte adhesion (Sidebar 7.3) to endothelial cells in young men with coronary artery disease" appeared in *Atherosclerosis*.

Sidebar 7.3

Monocytes are a type of white blood cell, part of the immune system that serves numerous functions but principally respond to inflammation, moving quickly to sites of infection and differentiating into macrophages and dendritic cells that elicit an immune response. Inflammation-related blood-borne factors cause them to firmly attach to the endothelium. Macrophages trigger scavenger function and take up and internalize oxLDL, thus transforming into foam cells that migrate from the intima into the media in formation of a fibrous atheroma

Sources: *Kinlay, S., Selwyn, A. P., Libby, P., Ganz, P. (1998) Inflammation, the endothelium, and the acute coronary syndromes. J Cardiovasc Pharmacol. 32:S62–S66; and Newby, A. C., & Zaltsman, A. B. (1999) Fibrous cap formation or destruction—the critical importance of vascular smooth muscle cell proliferation, migration and matrix formation. Cardiovasc Res. 41, 345–360.*

The aim of the study was to determine the effect of oral L-arginine on endothelial function in humans with established atherosclerosis. Men aged 41 to 92 years with angiographically proven coronary atherosclerosis took 7 g, three times per day, of L-arginine or placebo for 3 days each, with a washout period of 10 days.

After L-arginine supplementation, plasma levels of arginine rose, and EDD of the brachial artery measured using external vascular ultrasound was improved. No such change was seen in the placebo phase. No changes were noted in endothelium-independent dilatation of the brachial artery measured as the change in diameter in response to sublingual nitroglycerine, BP, heart rate, or fasting lipid levels. After L-arginine and placebo, serum from six of the participants was then added to HUVECs for 24 h, before monocytes obtained by centrifuge were added and cell adhesion assessed by light microscopy. Adhesion was reduced following L-arginine, compared to placebo.

In young men with CAD, oral L-arginine improves EDD and reduces monocyte-endothelial cell adhesion.[42] In addition:

- Long-term oral L-arginine supplementation (3 g TID) for 6 months improves coronary small-vessel endothelial function and symptoms. A role for L-arginine as a therapeutic option for patients with coronary endothelial dysfunction and nonobstructive CAD is proposed. In the journal *Circulation*.[64]
- In patients with myocardial ischemia, L-arginine (3 g/day for 7 days) increased the activity of superoxide dismutase (SOD), total thiols (T-SH), and plasma ascorbate levels, and decreased serum cholesterol, the activity of xanthine oxidase (XO), and malondialdehyde (MDA) levels as noted. In the *Indian Journal of Biochemistry & Biophysics*.[69]

7.20 CAVEAT

In the studies cited in this monograph, supplementation of L-arginine in whatever form was undertaken in clinical settings with medical supervision. Based on these studies, it appears that oral L-arginine consumed in the dosages per time period indicated is safe and well tolerated. However, we are not told whether any participants were eliminated from clinical trials due to adverse reactions to the supplements. Perhaps there were none, but we do not know.

Suggested oral dosages listed by the Mayo Clinic, therapeutic dosage, the maximum dose considered to be safe, is 400−6,000 mg, are given for information purposes only. It is not possible for this author to know what is a safe and effective dosage for any purpose in any given person at any given time.[70]

Mild—and occasionally severe—adverse reactions have been reported. There are also contraindications for supplementation, as is the case, for instance, with active herpes. The reader is advised to also examine these in Chapter 8.

Note—animal or vegetable: The author has no knowledge of the source of L-arginine in any conventional supplement formulations. However, the VasoRect Ultra® product formulated by the author for Real Health Laboratories, Div. PharmaCare, US, is considered vegetarian with a starting material of corn. It is vegetarian if removed from the capsule, which is gelatin (bovine source, not porcine).

Additional consideration of safety of supplementation is given at the end of the next chapter.

REFERENCES

1. Bode-Böger SM, Böger RH, Galland A, Tsikas D, Frölich JC. L-arginine-induced vasodilation in healthy humans: pharmacokinetic-pharmacodynamic relationship. *Br J Clin Pharmacol* 1998;**46**:489−97.

2. Jabłecka A, Ast J, Bogdaski P, Drozdowski M, Pawlak-Lemaska K, Cielewicz AR, et al. Oral L-arginine supplementation in patients with mild arterial hypertension and its effect on plasma level of asymmetric dimethylarginine, L-citruline, L-arginine and antioxidant status. *Eur Rev Med Pharmacol Sci* 2012;**16**:1665−74.

3. Böger RH. The pharmacodynamics of L-arginine. *J. Nutr* 2007;**137**:1650S−5S.

4. Chauhan A, More RS, Mullins PA, Taylor G, Petch C, Schofield PM. Aging-associated endothelial dysfunction in humans is reversed by L-arginine. *J Am Coll Cardiol* 1996;**28**:1796−804.

5. Bode-Boger SM, Muke J, Surdacki A, Brabant G, Boger RH, Frolich JC. Oral L-arginine improves endothelial function in healthy individuals older than 70 years. *Vasc Med* 2003;**8**:77−81.

6. Bode-Böger SM, Böger RH, Creutzig A, Tsikas D, Gutzki FM, Alexander K, et al. L-arginine infusion decreases peripheral arterial resistance and inhibits platelet aggregation in healthy subjects. *Clinical Science* 1994;**87**:303−10.

7. Rector TS, Bank AJ, Mullen KA, Tschumperlin LK, Sih R, Pillai K, et al. Randomized, double-blind, placebo-controlled study of supplemental oral L-arginine in patients with heart failure. *Circul* 1996;**93**:2135−41.

8. Wolf A, Zalpour C, Theilmeier G, Wang B-Y, Ma A, Anderson B, et al. Dietary L-arginine supplementation normalizes platelet aggregation in hypercholesterolemic humans. *J Am Coll Cardiol* 1997;**29**:479−85.

9. Piatti P-M, Monti LD, Valsecchi G, Magni F, Setola E, Marchesi F, et al. Long-term oral L-arginine administration improves peripheral and hepatic insulin sensitivity in type 2 diabetic patients. *Diab Care* 2001;**24**:875−80.

10. Zorgniotti AW, Lizza EF. Effect of large doses of the nitric oxide precursor, L-arginine, on erectile dysfunction. *Int J Impotence Res* 1994;**6**:33−35; discussion 36.

11. Arancibia-Garavilla Y, Toledo F, Casanello P, Sobrevia L. Nitric oxide synthesis requires activity of the cationic and neutral amino acid transport system y+L in human umbilical vein endothelium. *Exp Physiol* 2003;**88**:699−710.

12. Sala R, Rotoli BM, Colla E, Visigalli R, Parolari A, Bussolati O, et al. Two-way arginine transport in human endothelial cells: TNF-alpha stimulation is restricted to system y(+). *Am J Physiol Cell Physiol* 2002;**282**:C134−43.

13. Deves R, Angelo S, Rojas AM. Glucose transport: a functional approach system y+L: the broad scope and cation modulated amino acid transporter. *Exper Physiol* 1998;**83**:211−20.

14. Assumpção CRL, Brunini TMC, Pereira NR, Godoy-Matos AF, Siqueira MAS, Mann GE, et al. Insulin resistance in obesity and metabolic syndrome: is there a connection with platelet L-arginine transport? *Blood Cells, Mol Dis* 2010;**45**:338−42.

15. Gerhard M, Roddy MA, Creager SJ, Creager MA. Aging progressively impairs endothelium-dependent vasodilation in forearm resistance vessels of humans. *Hypertension* 1996;**27**:849−53.

16. Taddei S, Virdis A, Ghiadoni L, Salvetti G, Bernini G, Magagna A, et al. Age-related reduction of NO availability and oxidative stress in humans. *Hypertension* 2001;**38**:274−9.

17. Egashira K, Inou T, Hirooka Y, Kai H, Sugimachi M, Suzuki S, et al. Effects of age on endothelium-dependent vasodilation of resistance coronary artery by acetylcholine in humans. *Circul* 1993;**88**:77−81.

18. Berkowitz DE, White R, Li D, Minhas M, Cernetich A, Kim S, et al. Arginase reciprocally regulates nitric oxide synthase activity and contributes to endothelial dysfunction in aging blood vessels. *Circul* 2003;**108**:2000−6.

19. <http://www.integrativepractitioner.com/article.aspx?id = 19192>; [accessed 14.11.13].

20. <http://www.neogenis.com/neogenis-test-strips/>; 2013 [accessed 14.11.13].

21. Lee J, Ryu H, Ferrante RJ, Morris Jr. SM, Ratan RR. Translational control of inducible nitric oxide synthase expression by arginine can explain the arginine paradox. *Proc Natl Acad Sci USA* 2003;**100**:4843−8.

22. Nakaki T, Hishikawa K. The arginine paradox. *Nihon Yakurigaku Zasshi* 2002;**119**:7−14.

23. Erez A, Nagamani SCS, Shchelochkov OA, Premkumar MH, Campeau PM, Chen Y, et al. Requirement of argininosuccinate lyase for systemic nitric oxide production. *Nature Med* 2011;**17**:1619−26.

24. <http://www.bcm.edu/news/item.cfm?newsID = 4725>; 2013 [accessed 14.11.13].

25. Sreenath Nagamani SC, Erez A, Lee B, (2011). Argininosuccinate Lyase Deficiency. *Medical Genetic Searches*. <http://www.ncbi.nlm.nih.gov/books/NBK51784/>; 2013 [accessed 12.13].

26. Robbins RA, Floreani AA, Von Essen SG, Sisson JH, Hill GE, Rubinstein I, et al. Measurement of exhaled nitric oxide by three different techniques. *Am J Respir Crit Care Med* 1996;**153**:1631−5.

27. Kharitonov SA, Lubec G, Lubec B, Hjelm M, Barnes PJ. L-arginine increases exhaled nitric oxide in normal human subjects. *Clin Sci (Lond)* 1995;**88**:135−9.

28. Allen BW, Liu J, Piantadosi CA. Electrochemical detection of nitric oxide in biological fluids. *Methods Enzymol* 2005;**396**:68−77.

29. Kobylianskiĭ AG, Kuznetsova TV, Soboleva GN, Bondarenko ON, Pogorelova OA, Titov VN, et al. Determination of nitric oxide in serum and plasma of human blood. Method of high pressure liquid chromatography. *Biomed Khim* 2003;**49**:597−603.

30. Wang X, Bryan NS, MacArthur PH, Rodriguez J, Gladwin MT, Feelisch M. Measurement of nitric oxide levels in the red cell: validation of tri-iodide-based chemiluminescence with acid-sulfanilamide pretreatment. *J Biol Chem* 2006;**281**:26994−7002.

31. Mehta S, Stewart DJ, Levy RD. The hypotensive effect of L-arginine is associated with increased expired nitric oxide in humans. *Chest* 1996;**109**:1550−5.

32. Moss MB, Brunini TMC, Soares De Moura R, Novaes Malagris LE, Roberts NB, Ellory JC, et al. Diminished L-arginine bioavailability in hypertension. *Clin Sci (Lond)* 2004;**107**:391−7.

33. Schlaich MP, Parnell MM, Ahlers BA, Finch S, Marshall T, Zhang WZ, et al. Impaired L-arginine transport and endothelial function in hypertensive and genetically predisposed normotensive subjects. *Circulation* 2004;**110**:3680−6.

34. Dong JY, Qin LQ, Zhang Z, Zhao Y, Wang J, Arigoni F, et al. Effect of oral L-arginine supplementation on blood pressure: a meta-analysis of randomized, double-blind, placebo-controlled trials. *Am Heart J* 2011;**162**:959−65.

35. Lekakis JP, Papathanassiou S, Papaioannou TG, Papamichael CM, Zakopoulos N, Kotsis V, et al. Oral L-arginine improves endothelial dysfunction in patients with essential hypertension. *Int J Cardiol* 2002;**86**:317−23.

36. Palloshi A, Fragasso G, Piatti P, Monti LD, Setola E, Valsecchi G, et al. Effect of oral L-arginine on blood pressure and symptoms and endothelial function in patients with systemic hypertension, positive exercise tests, and normal coronary arteries. *Am J Cardiol* 2004;**93**:933−5.

37. Ast J, Jablecka A, Bogdanski P, Smolarek I, Krauss H, Chmara E. Evaluation of the antihypertensive effect of L-arginine supplementation in patients with mild hypertension assessed with ambulatory blood pressure monitoring. *Med Sci Monit* 2010;**16**: CR266−71.

38. Huynh NT, Tayek JA. Oral arginine reduces systemic blood pressure in type 2 diabetes: its potential role in nitric oxide generation. *J Am Coll Nutr* 2002;**21**:422−7.

39. Yavuz BB, Yavuz B, Sener DD, Cankurtaran M, Halil M, Ulger Z, et al. Advanced age is associated with endothelial dysfunction in healthy elderly subjects. *Gerontol* 2008;**54**:153−6.

40. Welsch MA, Dobrosielski DA, Arce-Esquivel AA, Wood RH, Ravussin E, Rowley CA, et al. The association between flow-mediated dilation and physical function in older men. *Med Sci Sports Exerc* 2008;**40**:1237−43.

41. Heffernan KS, Fahs CA, Ranadive SM, Patvardhan EA. L-arginine as a nutritional prophylaxis against vascular endothelial dysfunction with aging. *J Cardiovasc Pharmacol Ther* 2010;**15**:17−23.

42. Adams MA, McCredie R, Jessup W, Robinson J, Sullivan D, Celermajer DS. Oral L-arginine improves endothelium-dependent dilatation and reduces monocyte adhesion to endothelial cells in young men with coronary artery disease. *Atherosclerosis* 1997; **129**:261−9.

43. Bai Y, Sun L, Yang T, Sun K, Chen J, Hui R. Increase in fasting vascular endothelial function after short-term oral L-arginine is effective when baseline flow-mediated dilation is low: a meta-analysis of randomized controlled trials. *Am J Clin Nutr* 2009;**89**: 77−84.

44. Clarkson P, Adams MR, Powe AJ, Donald AE, McCredie R, Robinson J, et al. Oral L-arginine improves endothelium-dependent dilation in hypercholesterolemic young adults. *J Clin Invest* 1996;**97**:1989−94.

45. Delp MD, Behnke BJ, Spier SA, Wu G, Muller-Delp JM. Ageing diminishes endothelium-dependent vasodilatation and tetrahydrobiopterin content in rat skeletal muscle arterioles. *J Physiol* 2008;**586**:1161−8.

46. Gornik HL, Creager MA. Arginine and endothelial and vascular health. *J Nutr* 2004;**134** (10 suppl):2880S−7S (discussion 2895S).

47. Heffernan KS, Vieira VJ, Valentine RJ. Microvascular function and ageing: L-arginine, tetrahydrobiopterin and the search for the fountain of vascular youth. *J Physiol* 2008;**586**:2041−2.

48. Lin CC, Tsai WC, Chen JY, Li YH, Lin LJ, Chen JH. Supplements of L-arginine attenuate the effects of high-fat meal on endothelial function and oxidative stress. *Int J Cardiol* 2008;**127**:337−41.

49. Marchesi S, Lupattelli G, Siepi D, Roscini AR, Vaudo G, Sinzinger H, et al. Oral L-arginine administration attenuates postprandial endothelial dysfunction in young healthy males. *J Clin Pharm Ther* 2001;**26**:343−9.

50. Wu G, Meininger CJ. Cardiovascular benefits of L-arginine. *J Nutr* 2000;**130**:2626−9.

51. Yin WH, Chen JW, Tsai C, Chiang MC, Young MS, Li SJ. L-arginine improves endothelial function and reduces LDL oxidation in patients with stable coronary artery disease. *Clin Nutr* 2005;**24**:988−97.

52. Guttman H, Zimlichman R, Boaz M, Matas Z, Shargorodsky M. Effect of long-term L-arginine supplementation on arterial compliance and metabolic parameters in patients with multiple cardiovascular risk factors: randomized, placebo-controlled study. *J Cardiovasc Pharmacol* 2010. Jun 7. [Epub ahead of print].

53. Pieper GM, Dondlinger LA. Plasma and vascular tissue arginine are decreased in diabetes: acute arginine supplementation restores endothelium-dependent relaxation by augmenting cGMP production. *J Pharmacol Exp Ther* 1997;**283**:684−91.

54. Menge BA, Schrader H, Ritter PR, Ellrichmann M, Uhl W, Schmidt WE, et al. Selective amino acid deficiency in patients with impaired glucose tolerance and type 2 diabetes. *Regul Pept* 2010;**160**:75−80.

55. Lucotti P, Setola E, Monti LD, Galluccio E, Costa S, Sandoli EP, et al. Beneficial effects of a long-term oral L-arginine treatment added to a hypocaloric diet and exercise training program in obese, insulin-resistant type 2 diabetic patients. *Am J Physiol Endocrinol Metab* 2001;**291**:E906−12.

56. Jabłecka A, Bogdański P, Balcer N, Cieślewicz A, Skołuda A, Musialik K. The effect of oral L-arginine supplementation on fasting glucose, HbA1c, nitric oxide and total antioxidant status in diabetic patients with atherosclerotic peripheral arterial disease of lower extremities. *Eur Rev Med Pharmacol Sci* 2012;**16**:342−50.

57. Lubec B, Hayn M, Kitzmüller E, Vierhapper H, Lubec G. L-arginine reduces lipid peroxidation in patients with diabetes mellitus. *Free Radic Biol Med* 1977;**22**:355−7.

58. Wascher TC, Graier WF, Dittrich P, Hussain MA, Bahadori B, Wallner S, et al. Effects of low-dose L-arginine on insulin mediated vasodilatation and insulin sensitivity. *Eur J Clin Investig* 1997;**27**:690−5.

59. Assumpção CR, Brunini TMC, Matsuura C, Resende AC, Mendes-Ribeiro AC. Impact of the L-arginine-nitric oxide pathway and oxidative stress on the pathogenesis of the metabolic syndrome. *Open Biochem J* 2008;**2**:108−15.

60. McKnight JR, Satterfield MC, Jobgen WS, Smith SB, Spencer TE, Meininger CJ, et al. Beneficial effects of L-arginine on reducing obesity: potential mechanisms and important implications for human health. *Amino Acids* 2010;**39**:349−57.

61. Tousoulis D, Antoniades C, Tentolouris C, Goumas G, Stefanadis C, Toutouzas P. L-arginine in cardiovascular disease: dream or reality? *Vasc Med* 2002;**7**:203−11.

62. Creager MA, Gallagher SJ, Girerd XJ, Coleman SM, Dzau VJ, Cooke JP. L-arginine improves endothelium dependent vasodilation in hypercholesterolemic humans. *J Clin Invest* 1992;**90**:1248−53.

63. Drexler H, Zeiher AM, Meinzer K, Just. H. Correlation of endothelial dysfunction in coronary microcirculation of hypercholesterolemic patients by L-arginine. *Lancet* 1991;**338**: 1546−50.

64. Lerman. A, Burnett Jr. JC, Higano ST, McKinley LJ, Holmes Jr. DR. Long-term L-arginine supplementation improves small-vessel coronary endothelial function in humans. *Circul* 1998;**97**:2123−8.

65. West SG, Likos-Krick A, Brown P, Mariotti F. Oral L-arginine improves hemodynamic responses to stress and reduces plasma homocysteine in hypercholesterolemic men. *J Nutr* 2005;**135**:2212−7.

66. Jabłecka A, Checiński P, Krauss H, Micker M, Ast J. The influence of two different doses of L-arginine oral supplementation on nitric oxide (NO) concentration and total antioxidant status (TAS) in atherosclerotic patients. *Med Sci Monit* 2004;**10**:CR29−32.

67. Tousoulis D, Davies GJ, Tentolouris C, Crake T, Katsimaglis G, Stefanadis C, et al. Effects of changing the availability of the substrate for nitric oxide synthase by L-arginine administration on coronary vasomotor tone in angina patients with angiographically narrowed and in patients with normal coronary arteries. *Am J Cardiol* 1998;**82**:1110−3 A6.

68. Tousoulis D, Davies GJ, Tentolouris C, Crake T, Goumas G, Stefanadis C, et al. Effects of L-arginine on flow mediated dilatation induced by atrial pacing in diseased epicardial coronary arteries. *Heart* 2003;**89**:531−4.

69. Tripathi P, Misra MK. Therapeutic role of L-arginine on free radical scavenging system in ischemic heart diseases. *Indian J Biochem Biophys* 2009;**46**:498−502.

70. <http://www.mayoclinic.com/health/L-arginine/NS_patient-arginine/ DSECTION = safety>; 2013 [accessed 14.11.13]. Reprinted with kind permission from www.naturalstandard.com, copyright owner.

Chapter 8

Arginine and Arginine-Combinations in Treatment of Erectile Dysfunction

8.1 INTRODUCTION

Previous chapters documented the mounting clinical evidence that erectile dysfunction (ED) is part and parcel of a mix of progressing cardiovascular health hazards linked to the gradual and seemingly inexorable impairment of the vascular endothelium over the lifespan, due largely to oxidative stress. In fact, as the focus in diagnosis and treatment of ED has shifted from psychology and psychiatry first to urology, and then to medicine, it has become increasingly clear that ED may well in fact be a symptom of early onset "cardiovascular disease."

Researchers writing in the *Journal of the American Association of Echocardiography* went one step further and titled their report, "Erectile dysfunction as a generalized vascular dysfunction." Based on brachial artery flow-mediated dilation (FMD) with shear stress and nitroglycerin as stimuli for assessing endothelium-dependent FMD and non-endothelium-dependent dilation of the brachial artery, they compared patients with ED, but absent atherosclerosis or other major risk factors, to healthy volunteers.

FMD was significantly decreased in the ED group compared with the control group. The relationship between ED and FMD was significant, whereas no relationship was found between ED and non-endothelium-dependent dilation. Aortic strain and distensibility were found to be significantly lower in the ED group than in the control group. The relationship between ED and aortic stiffness was also significant.

The authors concluded that, "Aortic and brachial artery functions are impaired in men with ED without cardiovascular disease or its major risk factors, indicating a more generalized vascular disease."[1] Comparable findings were also reported in an article in the *Journal of the American College of Cardiology* with the self-explanatory title, "Impaired brachial artery endothelium-dependent and -independent vasodilation in men with erectile dysfunction and no other clinical cardiovascular disease."[2]

Robert Fried: Erectile Dysfunction as a Cardiovascular Impairment.
DOI: http://dx.doi.org/10.1016/B978-0-12-420046-3.00008-1

The evidence shows that ED often precedes any other manifestation of cardiovascular and heart disease, together with measurable endothelial impairment for which reason it has been said to be a "harbinger," warning us of impending blood vessel and heart problems. This idea is further bolstered by the evidence that successful treatment for ED is basically treatment of conditions that may ultimately lead to cardiovascular and heart disease, including hypertension, atherosclerosis, coronary artery disease (CAD), insulin resistance, metabolic syndrome and the oxidative damage caused by type-2 diabetes.

That is the main point of treatment of ED with L-arginine: If previous clinical trials have taught us anything, it is that treatment of ED with that supplement should not differ in any way from treatment of any other cardiovascular disorders such as hypertension and atherosclerosis, for instance. Therefore, the proper context for L-arginine supplementation is that if geared to cardiovascular treatment, it should alleviate ED as well, and if geared to treating ED, it should alleviate cardiovascular disorders as well.

Albeit a distinct entity from a conventional clinical point of view, ED is not different from hypertension or atherosclerosis: both are also distinct entities, yet labeled cardiovascular hazards. In fact, many clinical studies now refer to "vascular ED." The underlying problem, endothelial impairment, is the same. In fact, recovery of erectile function has been reported after "aggressive hypertensive therapy,"[3] and a report in *Circulation* informs us that ED is very common as men age and that it is frequently a sign of atherosclerosis.[4]

It is evident that the importance of the first blood vessel signal system, the acetylcholine/nitric oxide/cyclic guanosine monophosphate (ACh/NO/cGMP) pathway, has been mostly overlooked. That means that conventional treatments for hypertension, for instance, did not address improving endothelial function because it aimed at lowering blood pressure via the non-endothelium-dependent blood vessel control system, perhaps unaware of the existence of the other.

With that emerging insight, different questions are now being asked ... and answered, and ancillary treatment strategies are emerging. Many of these center on the role of nitric oxide and the substrate, the amino acid L-arginine and linked enzyme systems, in blood vessel function. This new vista into a parallel endothelium-dependent blood vessel control system requires a new way of gearing up to understanding the link between cardiovascular health hazards and ED.

Historically, it was universally understood that men's sexual "performance" vitality—erectile function—is subject to the vagaries of three separate but interlinked factors: desire (libido), performance (behavior), and social control (real or imagined-psychological). Until the advent of VIAGRA®, performance issues as manifest by ED were attributed to either low or absent desire (libido), or socially induced emotional inhibition (psychological/subconscious), or both.

There is little doubt that social and emotional issues modulate sexual behavior and can gain the upper hand and put a damper on performance.

However, chronic ED is rarely due to those factors, but rather to the inability of the endothelium to respond to arousal and deliver adequate NO/cGMP to set erection in motion and keep it going for as long as circumstances (i.e. arousal-triggered ACh release) dictate.

We also know now that while ED results from the complex interaction of heredity, progressive aging, and diet and lifestyle factors, it usually, though not invariably, boils down to a select few factors such as:

- Disruption of arginase activity.
- Uncoupled nitric oxide synthase (eNOS).
- Oxidative stress or otherwise impaired endothelium function and NO biosynthesis.

8.2 SOME MOLECULAR FACTORS AFFECTING NITRIC OXIDE BIOSYNTHESIS

It is beyond the scope of this chapter to detail here all the molecular factors that play a key role in erectile function or dysfunction, but here is a brief recapitulation of some of the key elements involved in failure. The recapitulation is provided so as to make plausible the implementation of a regimen of supplementation. It would otherwise be illogical to assume *a priori* that the supplementation of L-arginine is a sensible strategy if there were no basis for it in the biochemistry and physiology of penile erection as well as in other cardiovascular disorders. In brief:

Arginase is the enzyme involved in forming L-arginine. It coexists with eNOS in the blood vessels in the genitals of both men and women. eNOS and arginase compete for L-arginine, and excess production of arginase can affect NOS activity and NO-dependent (cGMP) smooth muscle relaxation by depleting the pool of L-arginine that would otherwise be available to NO synthase.[5-7]

Studies show in animal models that arginase production affects NO synthesis *in vitro* in endothelium, and that basal levels of arginase limit endothelial synthesis of substances that play a key role in cardiovascular function.[8] Studies in animal models also show that supplementing the diet with L-arginine decreases the competition between arginase and NOS by providing extra substrate for each enzyme activity.[9]

eNOS is found in vascular endothelium as well as other homologous structures. It generates the key signaling molecule (NO), as detailed in previous chapters. Diminished NO availability also contributes to hypertension, atherosclerosis,[10] and ED. Aged or otherwise damaged endothelium can result in eNOS uncoupling that in turn promotes free radical formation, further damaging the endothelium.[11] This process results in reduction in NO bioavailability.

Tetrahydrobiopterin (BH4): eNOS enzymatic activity depends on the availability of its cofactor tetrahydrobiopterin. When BH4 levels are adequate, eNOS produces NO; when BH4 levels are low, eNOS becomes enzymatically uncoupled and instead generates superoxide, contributing to vascular oxidative stress and endothelial dysfunction. BH4 bioavailability is determined by a balance of enzymatic synthesis and recycling, versus oxidative degradation in dysfunctional endothelium.[12]

Cardiovascular conditions such as hypertension, for instance, are linked to oxidation of BH4: Oxidation of BH4 leads to uncoupling of eNOS in hypertension.[13] Parenthetically, ascorbate increases BH4 levels in cultured endothelial cells by preventing oxidation,[14] and preventing oxidation of BH4 may explain the observed beneficial effect of long-term ascorbate treatment on blood pressure in patients with hypertension.[15]

As previously detailed, endothelium may be exposed to damage by hypertension, atherosclerosis, oxidative stress, and elevated blood sugar, usually chronically due to type-2 diabetes. For instance, researchers reporting in the *International Journal of Impotence Research* found that endothelial function was impaired in ED patients with no apparent cardiovascular disease or diabetes mellitus. Conventional patient assessment, endothelium "challenge," and standard measurement techniques were employed.[16]

8.3 VASCULAR ERECTILE DYSFUNCTION CANNOT BE CAUSED BY INSUFFICIENT L-ARGININE

It might seem logical to suppose that insufficient NO/cGMP formation might be due to L-arginine insufficiency. However, that is unlikely. The "arginine paradox" mentioned previously is that while cellular L-arginine may be adequate, more L-arginine releases more NO. The requirement for more L-arginine to release NO, despite adequate supply, makes it appear as though the problem is insufficiency, when in fact the problem may be eNOS uncoupling.

In the following study, "global" may be taken to mean systemic: A clinical study in the *Journal of the American College of Cardiology*, titled "Diminished global arginine bioavailability and increased arginine catabolism as metabolic profile of increased cardiovascular risk," addressed the issue of L-arginine insufficiency in patients with "significantly obstructive" CAD. The procedure entailed measuring plasma levels of free arginine, ornithine, citrulline, and the endogenous NOS inhibitor, asymmetric dimethylarginine (ADMA), by liquid chromatography coupled with tandem mass spectrometry.

The researchers examined the relationship of global arginine bioavailability ratio (GABR, defined as arginine/[ornithine + citrulline]) vs. arginine and its catabolic metabolites to prevalence of significantly obstructive CAD and incidence of major adverse cardiovascular events, including myocardial infarction, stroke, and death over a 3-year follow-up in more than 1,000 patients undergoing cardiac catheterization.

It was concluded that patients with significantly obstructive CAD had significantly lower GABR and L-arginine levels than those without that condition. After adjusting for Framingham risk score, C-reactive protein (CRP), and renal function, lower GABR (but not arginine levels) and higher citrulline levels remained significantly associated with both the prevalence of significantly obstructive CAD and 3-year risk for the incidence of major adverse cardiovascular events (MACE).

The investigators contend that GABR might serve as a more comprehensive concept of reduced NO synthesis capacity compared with systemic arginine levels. Diminished GABR and high citrulline levels are associated with both development of significantly obstructive atherosclerotic CAD and heightened long-term risk for MACE.[17]

It would have been helpful to ascertain whether there were factors that diminished the bioavailability of eNOS, without which information conclusions might be tentative: Simply measuring body levels of L-arginine gives an incomplete picture.

"Arginases and arginine deficiency syndromes" is the title of a review in *Current Opinion in Clinical Nutrition & Metabolic Care*: Many body functions involve arginine availability regulated by arginase. Some conditions result in elevated arginase activity as well as consequences of arginine deficiency in two broad categories of L-arginine deficiency syndromes, involving either T-cell dysfunction or endothelial dysfunction.[18]

The arginine paradox explains the dependence of cellular NO production on exogenous L-arginine concentration despite the theoretical saturation of eNOS with intracellular L-arginine. The question of actual or functional arginine insufficiency in cardiovascular disorders and ED must for the present remain moot.

8.4 SUPPLEMENTS WITH L-ARGININE IN TREATMENT OF ERECTILE DYSFUNCTION

What does oral supplementation of L-arginine actually do in the conduit blood vessels to the penis and to the cavernosae (Sidebar 8.1)? In general, we can infer that from the action of oral L-arginine supplementation on the brachial artery as shown by FMD. It has been noted that brachial artery changes due to FMD maneuvers are not invariably identical to vascular dilation response in other regions of the body. Nevertheless FMD is presently the *gold standard* for vasodilation studies.

A clinical study of the effects of oral L-arginine on endothelial function, intravascular oxidative stress, and circulating inflammatory markers in patients with stable CAD was reported in the journal *Clinical Nutrition*. Patients with stable CAD were randomly assigned to oral L-arginine (10 g), or vitamin C (500 mg, an antioxidant here used as active control) daily for 4 weeks, with crossover to the alternate therapy after two weeks off therapy. Brachial artery endothelial function and serum concentrations of lipids and inflammatory markers were measured at baseline (the end of each 4-week treatment period), and at the 2-week wash-out period.

Sidebar 8.1

A note of caution in connection with L-arginine supplementation in patients with ST-segment elevation myocardial infarction. It appeared in the *Journal of the American Medical Association* (JAMA) titled "L-arginine therapy in acute myocardial infarction: the vascular interaction with age in myocardial infarction (VINTAGE MI) randomized clinical trial":

"L-Arginine, when added to standard postinfarction therapies, does not improve vascular stiffness measurements or ejection fraction and may be associated with higher postinfarction mortality. L-Arginine should not be recommended following acute myocardial infarction."

No such warning about possible adverse effects of L-arginine supplementation were found in connection with treatment of ED, per se, absent this condition.

Source: *Schulman, S. P., Becker, L. C., Kass, D. A., Champion, H.C., Terrin, M.L., Forman, S., et al. (2006). L-arginine therapy in acute myocardial infarction: the vascular interaction with age in myocardial infarction (VINTAGE MI) randomized clinical trial. JAMA. 295, 58–64.*

As shown in Figure 8.1, L-arginine or vitamin C significantly increased brachial artery FMD. Neither L-arginine nor vitamin C affected lipid profiles and circulating levels of inflammatory markers, however, in the patients with low-density lipoprotein (LDL) susceptibility to oxidation, lag time significantly increased for 4 weeks after L-arginine: Oral L-arginine supplement improved endothelial function.[19]

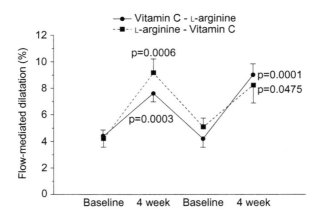

FIGURE 8.1 Mean ± SE of FMD graphed over time by treatment groups showing FMD impaired at baseline. After treatment with L-arginine and vitamin C, FMD increased significantly. Source: *Yin, W-H., Chen, J-W., Tsai, C., Chiang, M-C., Young, M.S., Lin, S-J. (2005). L-arginine improves endothelial function and reduces LDL oxidation in patients with stable coronary artery disease. Clin Nutr, 24, 988–997. Reproduced with kind permission from* Clinical Nutrition.

Disregarding the effects of vitamin C, what Figure 8.1 shows is that daily oral supplementation with a clinical dosage of 10 g of L-arginine raises FMD in otherwise impaired endothelium in the brachial artery. Brachial artery FMD is the "model" for systemic and penile vasodilation response, but can we rely on the model? Will oral L-arginine also increase the vasodilation response in the cavernosae?

Researchers from the Department of Urology, Tulane Health Sciences Center, New Orleans, LA, reported on the relaxant effects of repetitive administration of L-arginine hourly in human corpus cavernosum. Samples of human corpus cavernosum were suspended in an organ chamber for measurements of isometric tension. After precontraction with phenylephrine (10 μM), concentration–response curves were performed for L-arginine at 2-hour intervals (1 to 10 hours). Underlying mechanisms of relaxation were evaluated by inhibitory and stimulatory agents.

L-Arginine induces slow and prolonged relaxation of human corpus cavernosum, presumably by restoring the endogenous amino acid pool for NO synthesis and by other NO signaling involving KCa channels and other pathways. The use of Sildenafil® combined with L-arginine further facilitates erections and benefits men with more severe ED.[20] Indeed it does.

The *British Journal of Urology International* reported a prospective randomized, double-blind, placebo-controlled study of the effect of six weeks of "high-dose" (5 g/day) orally administered L-arginine in men with organic ED. O'Leary's sexual function questionnaire, specially designed to obtain detailed medical and sexual history, and a sexual activity diary, were obtained from each patient.[21] Complete physical examination included assessment of bulbocavernosus reflex and penile hemodynamics. Plasma and urine nitrite and nitrate, designated NOx, were determined at the end of the placebo run-in period, and after 3 and 6 weeks.

Oral administration of L-arginine in high doses resulted in significant subjective improvement in sexual function in men with organic ED who also had decreased NOx marker excretion or production. The hemodynamics of the corpus cavernosum were not affected by oral L-arginine at the dosage used.[22] What is curious in this study is that men who did not show evidence of low NOx did not report subjective improvement in sexual function. It is difficult to know what to make of this, and the correlation between NO levels and subjective assessment of one's sexual function is likewise obscure.

A two-part clinical study titled "Effect of large doses of the nitric oxide precursor, L-arginine on erectile dysfunction" appeared in the *International Journal of Impotence Research*. In the first part, men with ED were given daily oral doses of L-arginine (2,800 mg) vs. placebo for a period of two weeks. Some participants did not complete the trial, but of the remaining, six of fifteen reported improved erection, while none in the placebo group reported improvement.

In the second part of the study, ED patients in a stable marriage were recruited as volunteers. To qualify, they had to be free of hypertension and not on anti-hypertensive medication, non-smokers, and free of severe vascular insufficiency. They were given oral L-arginine (1,500 mg/day) for two weeks. Only six of fifteen improved. However, none in the placebo group showed improvement.[23]

In another study by Klotz, Mathers, Braun *et al.*, appearing in *Urology International*, oral L-arginine (1,500 mg/day) produced mixed results.[24] Considering typical oral dosages of 5 g to 9 g/day given to patients with cardiovascular disorders, who responded favorably, it is a wonder that any participants responded at all to 1.5 g/day.

Given conventional dosage ranges reported in clinical trials, it is difficult to understand why anyone would think that an oral dosage of 1,500 mg/day of L-arginine—what one would find in less than a cup of garbanzo beans—would have any impact on established vascular ED. If dosages of L-arginine supplement are significantly lower in ED treatment studies than those in, let's say, hypertension, then ED is not being treated seriously as a vascular disorder.

Here are some typical examples of L-arginine supplementation dosages in clinical trials in cardiovascular disorder treatment:

- Oral L-arginine 2 g tds (6 g/day) for 1 week: "L-arginine restored impaired endothelium-dependent vasodilation in newly diagnosed borderline hypertensives. The effect may explain the hypotensive effect of L-arginine." In *Journal of the American College of Cardiology.*[25]
- Oral L-arginine hydrochloride 5.6 to 12.6 g/day for six weeks: "Supplemental oral L-arginine had beneficial effects in patients with heart failure." In *Circulation.*[26]
- Oral L-arginine 7 g three times per day for 3 days (21 g/day): "In young men with coronary artery disease, oral L-arginine improves endothelium-dependent dilatation and reduces monocyte/endothelial cell adhesion." In *Atherosclerosis.*[27]
- Oral L-arginine 8.3 g/day for 21 days: "Long-term oral L-arginine treatment resulted in an additive effect compared with a diet and exercise training program alone on glucose metabolism and insulin sensitivity. Furthermore, it improved endothelial function, oxidative stress, and adipokine release in obese type-2 diabetic patients with insulin resistance." In *American Journal of Physiology, Endocrinology, & Metabolism.*[28]
- Oral L-arginine intervention ranging from 4 to 24 g/day: "This meta-analysis provides further evidence that oral L-arginine supplementation significantly lowers both systolic and diastolic BP." In *American Heart Journal.*[29]

This study by Klotz, Mathers, Braun *et al.*[24] which is most often cited as an example of the futility of oral L-arginine treatment in ED, appeared in the journal *Urology International*, titled, "Effectiveness of oral L-arginine in first-line treatment of erectile dysfunction in a controlled crossover study."

Patients averaging 51.6 years old, with *mixed-type impotence* (my italics), diagnosed according to the results of sexual history and urological examination, were enrolled in a randomized, placebo-controlled, crossover comparison of an oral placebo with 1,500 mg oral L-arginine/day. The treatment consisted of two 17-day courses (50 tablets). After a 7-day crossover and washout period, the patients were switched to L-arginine and vice versa. The authors concluded that, "Oral L-arginine 3×500 mg/day is not better than placebo as a first-line treatment for mixed-type impotence."[24]

"No better than placebo" is not surprising when applied to "mixed-type impotence" patients, and considering the conventional range of dosages and range of treatment duration, that proved effective in cardiovascular disorders.

The *mixed type of impotence* is characterized by typically unidentified multiple reasons for the ED. Mostly it may include one or a combination of medical diseases, as well as organic urological, endocrine, and psychological causes. Treatment with L-arginine is indicated only for the vasculogenic-type. In the study, there is no indication of any attempt to un-mix the types, and so the applicability of a treatment specifically targeting known vascular impotence remains questionable in any case.

The conclusion in this study that 1,500 mg/day of L-arginine supplementation is ineffective, given the overwhelming evidence that L-arginine enhances vascular function at typically much higher dosages, and that the treatment is applicable specifically to vascular ED and not to "mixed-types," is consistent with what we know about effective treatment.

There are numerous commercial L-arginine supplement consumer products available at pharmacies countrywide. They supply L-arginine in one of several different forms for oral consumption in capsules that range in dosage from 250 mg to 1,000 mg each. With notable exceptions, there is generally no way to know if the products are pharmaceutical-grade and meet conventional standards set by the United States Food and Drug Administration (USFDA), unless there is a specific citation of violation(s) posted for public access.

8.5 UNITED STATES GOVERNMENT SUPPLEMENTS REGULATIONS AND SAFETY INFORMATION

Comprehensive information about dietary supplements is provided by the National Center for Complementary and Alternative Medicine (NCCAM) of the US Department of Health and Human Services of the National Institutes of Health (NIH): "Using Dietary Supplements wisely."[30] See also the websites for the USFDA,[31] and for general information, the website for the USDA NAL.[32]

8.6 L-ARGININE DIETARY SUPPLEMENTS TO SUPPORT ERECTILE VITALITY

Arginine supplements available to the consumer public without prescription are—and must be—intended solely to support healthy function and not as a medical treatment for any medical condition. For that reason, to effectively support sexual vitality, dosages need not—indeed should not—reach the clinical dosage levels administered under medical supervision as detailed in treatment trials reported in this monograph.

Supplements named below serve to inform only. The author has no knowledge of the reliability of the stated contents, their clinical purity or value or safety at any given dosage.

One popular supplement formulation is that of Dr. Mercola—L-arginine (1,000 mg serving size in two capsules). It also claims the advantage of Time-Sorb®, a time-release matrix. Another example is Solgar® L-arginine Free Form, 500 mg/capsule. Arginine also comes in the by-prescription-only hydrochloride form infused under medical supervision, often to treat metabolic alkalosis. The latter is not a dietary supplement.

8.7 L-ARGININE SUPPLEMENTATION IN A FRUIT BAR

This clinical trial published by the journal *Cardiovascular Drugs & Therapy* aimed to determine the vascular and biochemical effects of a formulated product in the form of a food bar (HeartBar, Cooke Pharma, Inc., Belmont, CA) enriched with a combination of nutrients known to enhance the synthesis or activity of endothelium-derived NO (Sidebar 8.2).

Sidebar 8.2

"HeartBar is a medical food that should be used under the supervision of a physician. HeartBar contains L-arginine and other important ingredients to promote better vascular health."

Source: *Grocery Coupon Network. HeartBar Original Bar. http://www.grocerycouponnetwork. com/foodproducts/products.php?Id=25233. (Accessed November 14, 2013).*

Individuals with hypercholesterolemia manifest impaired FMD to which reduction in endothelium-derived NO is a contributing factor. Oral supplementation with large amounts of L-arginine (6 to 21 g/day) has been shown to improve endothelium-mediated vasodilation in hypercholesterolemia, but what amounts to clinical dosages may be impractical to take in capsule form. Therefore, the investigators developed a nutrient bar enriched with L-arginine as well as other ingredients that additively enhance NO/cGMP activity.

In a pilot study, hypercholesterolemic participants indicated that the bar was well tolerated. It had no adverse effects on serum chemistries or lipid profile. To determine if the nutrient bar normalizes endothelial function, a double-blind, placebo-controlled study was performed: FMD was assessed by high-resolution ultrasonography before and after 1 week of bar use (2 bars/day) in an additional group of hypercholesterolemic participants ranging in age between 47 and 67 years, showing impaired FMD before intervention. Vasodilator function in the "active bar" group improved to within a normal range and was significantly better than in the placebo bar group.

These findings support use of a nutrient bar[33] designed to enhance NO/cGMP activity to improve FMD in hypercholesterolemic individuals.[34]

8.8 L-CITRULLINE—SUPPLEMENTATION RAISES L-ARGININE LEVELS

L-Citrulline is an amino acid naturally present in the body. It is a key intermediate in the urea cycle. In the kidneys, vascular endothelium, and other tissues, it is converted into L-arginine, raising tissue levels of L-arginine promoting NO biosynthesis. Conversely, eNOS oxidizes L-arginine to L-citrulline and NO in the presence of L-arginine, and the essential cofactor BH4.[35] It has been shown that just like L-arginine, citrulline promotes cardiovascular and erectile function by raising NO formation:

Writing in *Cardiovascular Drug Review*, investigators report that while supplementing L-arginine has been shown to improve NO production and cardiovascular function in cardiovascular diseases associated with endothelial dysfunction, the benefit of supplementation decline over time: intestinal and hepatic metabolism of L-arginine to ornithine and urea by arginase ultimately makes oral delivery ineffective. In contrast, L-citrulline is not metabolized in the intestine or liver and does not induce tissue arginase, but rather inhibits its activity. L-citrulline entering the kidney, vascular endothelium, and other tissues can be readily converted to L-arginine, thus raising plasma and tissue levels of L-arginine and enhancing NO production. Supplemental L-citrulline may be a therapeutic adjunct in disease states associated with "L-arginine deficiencies."[36]

Recycling of L-citrulline to L-arginine is essential for NO production in endothelial cells. However, there is no direct evidence demonstrating the degree to which the recycling of L-citrulline to L-arginine is coupled to NO production. The journal *Nitric Oxide*, reported a study aimed at testing the hypothesis that the amount of NO formed would be significantly higher than the amount of L-citrulline formed due to the efficiency of L-citrulline recycling via the citrulline−NO cycle. Endothelial cells were incubated with [^{14}C]-L-arginine and stimulated by various agents to produce NO. The extent of NO and [^{14}C]-L-citrulline formation were simultaneously determined.

NO production exceeded apparent L-citrulline formation on the order of 8 to 1 under both basal and stimulated conditions. As further support, α-methyl-DL-aspartate, an inhibitor of argininosuccinate synthase (AS), a component of the citrulline−NO cycle, inhibited NO production in a dose-dependent manner. The results of this study provide evidence for the essential and efficient coupling of L-citrulline recycling via the citrulline−NO cycle to endothelial NO production.[37]

In the following paragraphs are examples of clinical reports that demonstrate effects of citrulline and citrulline/arginine combination in selected cardiovascular disorders.

Hypertension: The *Cardiology Journal* (formerly *Folia Cardiologica*) reported a study aimed to evaluate the effects of L-arginine and L-citrulline on blood pressure and right ventricular function in heart failure patients with unknown preserved ejection fraction (HFpEF). All patients underwent an echocardiogram and radioisotopic ventriculography rest/exercise, and were randomly assigned to an L-arginine group and a citrulline malate group for a two-month period. The principal echocardiographic finding was a statistically significant decrease in pulmonary artery pressure in the L-arginine group and in the citrulline group. Duration on treadmill and right ventricular ejection fraction post-exercise increased while diastolic and systolic artery pressure decreased significantly in both groups. There were no other statistically significant differences between the groups.[38]

Atherosclerosis: This study appearing in *Proceedings of the National Academy of Science USA*, aimed to determine the effect of ingested L-arginine, L-citrulline, and the antioxidants vitamins C and E on the progress of atherosclerosis in rabbits fed a high-cholesterol diet. The diet resulted in significant impairment of endothelium-dependent vasorelaxation in isolated thoracic aorta and blood flow in rabbit ear artery *in vivo*, in development of atheromas, and increased superoxide anion production in the thoracic aorta. The rabbits were then given L-arginine, L-citrulline, and/or antioxidants orally for 12 weeks.

It was found that L-arginine plus L-citrulline, either alone or in combination with antioxidants, significantly improved endothelium-dependent vasorelaxation and blood flow, there was marked regression in atheromatous lesions. These therapeutic effects were associated with concomitant increases in aortic endothelial NOS expression and plasma $NO(2)(-) + NO(3)(-)$ and cGMP levels. These results strongly suggested that consumption of these NO-boosting and antioxidant substances can reduce oxidative stress and reverse the progression of atherosclerosis.[39]

ED: Oral supplementation of L-arginine is subject to extensive pre-systemic metabolism, whereas L-citrulline escapes it and is converted to L-arginine. The journal *Urology* reports a study that aimed to test the efficacy and safety of oral L-citrulline supplementation as a donor for the L-arginine−NO pathway of penile erection to treat patients with mild ED.

Men with an erection hardness score of 3 received a placebo for 1 month, and 1.5 g/day L-citrulline for another month. The erection hardness score, frequency of intercourse per month, treatment satisfaction, and adverse events were recorded.

Patients' (mean age 56.5 ± 9.8 years) improvement in the erection hardness score from 3 (mild ED) to 4 (normal erectile function) occurred in 50% of those on L-citrulline and 8.3% of those on placebo. The mean number of occurrence of intercourse per month increased from 1.37 ± 0.93 at baseline to 2.3 ± 1.37 at the end of the treatment phase and 1.53 ± 1.00 at the end of the placebo phase. No adverse reaction to citrulline was noted.

The investigators averred that L-citrulline supplementation is less effective than phosphodiesterase type-5 (PDE5) enzyme inhibitors, at least in the short term, but it was shown to be safe and well accepted by patients.[40]

A number of distributors of vitamins and other supplements offer L-citrulline to consumers, including the NEO40 Daily available from Neogenis® Laboratories, Inc.[41] In addition to the products, the website also offers an opportunity to view references to the scientific basis to enhancing NO for health and wellness, a service rarely provided by companies that market health products.

8.9 L-ARGININE ORAL SUPPLEMENTS—NUTRACEUTICAL APPROACH TO SUPPORT ERECTILE FUNCTION

While most L-arginine-based products are readily available in pharmacies and health food stores across the country, the Real Health Laboratories, Inc., L-arginine-based VasoRectUltra™[42] formulation was designed specifically to support erectile function (See Note 1 and Note 2 at the end of the chapter).

Regarding quality assurance: The VasoRectUltra formulation is manufactured in USFDA-registered manufacturing plants. In addition to pre-process and in-process controls, Real Health Laboratories, Inc. contracts a USFDA-registered analytical chemistry laboratory and microbiology laboratory to test for safety, potency, and disintegration, using USP and AOAC methodology on each lot manufactured. Each lot of finished product undergoes this analytical chemistry and microbiological testing.

VasoRectUltra suggests a 2,000 mg/day serving in three capsules, fortified with calcium (as calcium carbonate), zinc (as zinc picolinate), Standardized Japanese Knotweed Extract (*Polygonum cuspidatum*; root; standardized to contain 50% resveratrol), Stinging Nettle Root Powder (*Urtica dioica*; root), Standardized Korean Ginseng Extract (*Panax ginseng*; root; standardized to contain 5% ginsenosides), Maca Powder (*Lepidium meyenii*; tuberous root), Grapeseed Extract (*Vitis vinifera I.*; seed; standardized to contain 95% proanthocyanidin), and Standardized *Gingko Biloba* extract (leaf; standardized to contain 24% ginkgoflavonglycosides, 6% terpene lactones). The role of some of these substances in supporting erectile function is described below mostly to demonstrate that ED has a basis in body chemistry. Others are detailed in later chapters of this monograph.

Resveratrol from standardized Japanese Knotweed extract (*Polygonum cuspidatum*) root: The formulation is boosted with *trans*-resveratrol (t-Resv), a high-yield NO-donor (see Chapter 9). Resveratrol occurs naturally in the skin of red grapes, whereas there are little more than traces of it in red wine. It is a powerful antioxidant that helps protect the cardiovascular system and the heart from oxidized LDL cholesterol (LDL-C), a well-known hazard to erectile function. Numerous clinical studies have shown the merits of resveratrol in supporting erectile function.

In a clinical study published in *Molecular Nutrition & Food Research*, conducted *in vitro* on human endothelial cells, it was reported that t-Resv (a purified form of resveratrol that remains active only when sheltered from the sunlight and from oxygen) increases NO production, improving endothelium function. The authors concluded that t-Resv induces a concentration-dependent, simultaneous increase in $[Ca^{2+}]$ and NO biosynthesis. Assuming that t-Resv exhibits the same properties in human blood vessels *in vivo*, its pharmacological properties may improve endothelial function.[43]

A study published in the *British Journal of Pharmacology*, concluded that t-Resv enhances NO production by increasing the activity of eNOS, even in existing unhealthy conditions.[44]

Boosting phosphodiesterase type-5 inhibitors: *The Journal of Sexual Medicine* reported that resveratrol in combination therapy with vardenafil (Levitra®) can improve erectile function where NO release is impaired. In other words, where Levitra alone may not improve erections, enhancing it with resveratrol may: Vardenafil and resveratrol synergistically enhance the ACh/NO/cGMP pathway in corpus cavernosal smooth muscle cells.[45]

Stinging nettle (*Urtica dioica*): While stinging nettle is not known to promote NO/cGMP formation, its addition to the VasoRectUltra formulation is intended to promote prostate health. It does this by inhibiting the growth activity of prostate cells that lead to benign prostate hyperplagia (BPH), a condition that interferes with erectile vitality. A number of studies have shown the beneficial effects of that herb.

Elderly male patients with lower urinary tract symptoms (LUTS) caused by BPH were given a combination of 160 mg sabal fruit extract WS 1473, and 120 mg urtica root extract WS 1031 per capsule (PRO 160/120). Following a single-blind placebo run-in phase of 2 weeks, the patients received 2 × 1 capsule/day of supplement under double-blind conditions over a period of 24 weeks. Double-blind treatment was followed by an open control period of 24 weeks during which all patients were administered PRO 160/120. Treatment efficacy measures included the assessment of lower urinary tract symptoms (LUTS) with the International Prostate Symptom Score (I-PSS)[46] self-rating questionnaire and a quality of life index, as well as uroflow and sonographic parameters.

Patients on PRO 160/120 had substantially higher total score reduction after 24 weeks of treatment than the placebo group in terms of obstructive as well as to irritation symptoms, and to patients with moderate or severe symptoms at baseline. After being switched to PRO 160/120, after the control period, patients on placebo showed a marked improvement in LUTS (as measured by the I-PSS). In conclusion, PRO 160/120 was clearly superior to the placebo for the amelioration of LUTS as measured by the I-PSS. PRO 160/120 is advantageous in obstructive and irritation urinary symptoms and in patients with moderate and severe symptoms. The herbal extract was well tolerated.[47]

A study appearing in *Andrologia* showed in an animal model that stinging nettle can be used as an effective drug for the management of BPH. The conclusion was based on measurement of prostate/body weight ratio, weekly urine output and serum testosterone levels, prostate-specific antigen levels on day 28, and histological examinations carried out on prostates from treatment and control groups.[48]

Zinc is included in the formulation for the same reason: Investigators reported in the *Indian Journal of Urology* that in their patients with BPH, there was 61% less mean tissue zinc as compared to normal tissues.[49]

Folic acid: The authors of a study by the Division of Cardiology, Johns Hopkins Medical Institutions, Baltimore, MD, report that folic acid regulates eNOS essential to the generation of NO, and it also protects the endothelium (see Chapter 10).[50] The authors state that given that eNOS uncoupling plays a major role in many cardiovascular disorders, the potential of folic acid supplement "as an inexpensive and safe oral therapy is intriguing and is stimulating ongoing investigations."

Grapeseed Extract (Vitis vinifera I.): The formulation also contains grapeseed extract, a proanthocyanidin antioxidant intended to help protect the endothelium from oxidative stress. Researchers from the Boston University School of Medicine published a study titled "Grape seed and skin extracts inhibit platelet function and release of reactive oxygen intermediates" in the *Journal of Cardiovascular Pharmacology*.

The investigators incubated blood platelets with seed or skin extract and found that it led to a decrease in platelet aggregation, to a marked decrease in superoxide release, as well as a significant increase in radical-scavenging activity and enhanced platelet NO. These effects were dose-dependent for both grape extracts. Co-incubation with seeds and skins led to enhanced NO release, and prevented superoxide production. Thus, the extracts from purple grape skins and seeds inhibit platelet function and platelet-dependent inflammatory responses at pharmacologically relevant concentrations, suggesting potentially beneficial platelet-dependent antithrombotic and anti-inflammatory properties of purple grape-derived flavonoids.[51]

Keeping in mind that insulin resistance leads to oxidative stress damage to the endothelium, the following study supports the value of grape seed extract as an antioxidant anti-inflammatory supplement: The journal *Diabetic Medicine* reported a study titled "Effects of grape seed extract in Type 2 diabetic subjects at high cardiovascular risk: a double blind randomized placebo controlled trial examining metabolic markers, vascular tone, inflammation, oxidative stress and insulin sensitivity."

It was found that grapeseed extract significantly improved markers of inflammation and glycemia and a sole marker of oxidative stress in obese type-2 diabetic subjects at high risk of cardiovascular events over a 4-week period, which suggests it may have a therapeutic role in decreasing cardiovascular risk.[52]

Hawthorn (*Crataegus* species): Hawthorn preparations have long been known for their benefit in heart ailments. Investigators tested their inhibitory effects on endothelin-1 (ET-1) synthesis by cultured endothelial cells. These actions were compared with that of grapeseed extract because the vasoactive components of both of these herbal remedies are mainly oligomeric flavan-3-ols called procyanidins.

Reporting in the journal *Clinical Science* (London), it was shown that extracts of Hawthorn and grapeseed were equally effective as inhibitors of ET-1 synthesis. ET-1is an inflammatory marker. Grapeseed extract also produced a potent endothelium-dependent vasodilator response on preparations of isolated aorta. Based on these results and previous findings, the investigators concluded that through their pharmacological properties procyanidins stimulate a response in endothelial cells which helps restore endothelial function.[53]

Ginkgo biloba: Adding Ginkgo to a formula targeting ED is consistent with supporting unimpeded blood flow through a healthy cardiovascular system. Ginkgo leaves contain flavonoids and terpenoids believed to have potent antioxidant properties that may reduce or even help prevent some of the damage of oxidative stress.

In the following report published in *Arzneimittelforschung*, meta-analysis was applied to controlled clinical trials where Ginkgo biloba extract EGb 761 was given to patients with peripheral arterial disease (PAD) in five placebo-controlled clinical trials with similar design and inclusion criteria. Treatment outcome was measured by the increased walking distance in a standardized treadmill exercise. The mean increase in walking distance achieved by patients given EGb 761 (a Ginkgo biloba preparation) was significantly higher than that achieved by the placebo group. The meta-analysis revealed a highly significant therapeutic effect of EGb 761 for the treatment of PAD.[54] PAD is linked to endothelium impairment and its treatment involves restoring endothelium health and function.

There are many clinical studies on Ginkgo biloba in connection with erectile function that are primarily aimed at treating ED coupled with depression, and they do not involve L-arginine. However, it is worth listing a few of these references here because both depression and treatments for depression are known to interfere with erectile vitality both directly and indirectly:

In the *Journal of Sexual and Marital Therapy*: Ginkgo biloba for antidepressant-induced sexual dysfunction. Among patients with sexual dysfunction secondary to a variety of antidepressant medications, including selective serotonin reuptake inhibitor (SSRIs), serotonin and norepinephrine reuptake inhibitors (SNRIs), monoamine oxidase inhibitors (MAOIs), and tricyclics, Ginkgo biloba was found to be 84% effective in treating antidepressant-induced sexual dysfunction predominately caused by selective serotonin reuptake inhibitors. Dosages of Ginkgo biloba extract ranged from 60 mg qd to 120 mg bid (average = 209 mg/day). The common undesirable side effects were gastrointestinal disturbances, headache, and general central nervous system activation.[55]

On the other hand, another study in *Human Psychopharmacology: Clinical and Experimental* reported no such efficacy. The results of this 2-month trial were that there was no statistically significant difference from placebo at weeks 2, 4, and 8 after medication; that in comparison with baseline, both the Ginkgo biloba group and the placebo group showed improvement in some aspect of sexual function, which is suggestive of the importance of the placebo effect in assessing sexual function.[56] This study did not replicate a prior positive finding supporting the use of Ginkgo biloba for antidepressant-induced, especially SSRI, sexual dysfunction.

It is possible that the critical element in comparing these studies is diagnosis and homogeneity of etiology. A noticeable improvement with placebo raises the question whether the ED was homogeneously of vascular origin. There is no other reason why Ginkgo would be effective unless Ginkgo itself is inert, and there is reason to doubt that. There is no way to know from the information given. Furthermore, some authorities hold that SSRIs may affect libido and delay orgasm, but do not affect erectile function,[57] further muddying the field when it comes to comparing clinical trial outcome from studies where participants' etiology is at best vague and most likely heterogeneous.

There are many supplements claiming to promote health, youthfulness, and improved sexual function. Only L-arginine and citrulline, backed by conventional medical scientific and clinical trials, can rightly claim that it supports both improved health and increased erectile strength because it is the basic essential nutrient "substrate" from which the body derives NO.

A number of studies report treatment with oral L-arginine in combination with other substances believed to promote erection, libido, or both. Many supplements include substances that have known sexual function-enhancing activity. These added components usually serve sexual performance vitality by promoting NO formation or through their antioxidant property, though in one instance, yohimbe, it addresses libido.

Korean Ginseng (Panax ginseng): Korean ginseng is likewise a useful addition to a nutraceutical formulation to support erectile function. It is a deciduous perennial herb with typical light-colored fleshy root. The taste of the Korean ginseng root is sweetish at first, but with a bitter aftertaste.

It grows on moist and shaded mountainsides in China, Korea, and Russia. The root is long and slender and sometimes resembles the shape of the human body. In China, the ginseng roots are called Jin-chen, meaning "like a man."

Korean ginseng is available as powder of the whole root, as cut and dried roots, and as liquid or powder extracts. The total ginsenosides (the active components) content of a six-year-old root varies between 0.7% and 3%. Ginsenosides have been shown to be anti-inflammatory and antioxidant.[58] They have also been shown to be anti-atherosclerotic, antihypertensive, and anti-diabetic, and to enhance sexual performance.

The *International Journal of Impotence Research* reported a study titled "Clinical efficacy of Korean red ginseng for erectile dysfunction" that compared ginseng treatment to placebo and trazodone (Desyrel®) in order to find a natural "drug" to treat ED without medical complications. In the group receiving ginseng, changes in early detumescence and erectile parameters, such as penile rigidity and girth, libido, and patient satisfaction, were significantly higher than in the other groups.

The overall therapeutic efficacies on ED were 60% for ginseng and 30% for placebo and trazodone-treated groups, statistically confirming the effect of ginseng. No complete remission of ED was noted, but partial responses were reported. There were no adverse effects.[59]

A study published in the *Journal of Urology* reported on the efficacy of Korean red ginseng for ED based on changes in the International Index of Erectile Function (IIEF), RigiScan® (UroHealth Systems, Laguna Niguel, California), hormonal levels, and penile duplex ultrasonography with audio-visual sexual stimulation. Patients with clinically diagnosed ED were enrolled in a study comparing the effects of Korean red ginseng (900 mg, 3 times daily) to placebo.

Mean IIEF scores were significantly higher in patients treated with Korean red ginseng than placebo. Scores on Questions 3 (penetration) and 4 (maintenance) were significantly higher in the ginseng than in the placebo group. In response to the global efficacy question, 60% of the patients answered that Korean red ginseng improved erection. Also, penile tip rigidity on RigiScan showed significant improvement in the ginseng group vs. placebo.[60]

Maca (*Lepidium meyenii*), another substance in the formulation, is a cultivated root in the *Brassica* family. It is consumed as a vegetable food, and it is considered to have aphrodisiac properties by inhabitants of the Andean region of South America. In this study published in *Andrologia*, men with mild ED were treated with Maca dry extract (2,400 mg/day) or placebo. Pre- and post-treatment changes in ED and subjective well-being were obtained with the IIEF-5 and the Satisfaction Profile (SAT-P).

After 12 weeks of treatment, both Maca- and placebo-treated patients experienced a significant increase in IIEF-5 score. However, Maca patients experienced a greater increase than those taking placebo. Both Maca- and placebo-treated subjects experienced a significant improvement in psycho-logical performance-related SAT-P score, but the Maca group scored signifi-cantly higher than that of the placebo group. Only Maca-treated patients experienced a significant improvement in physical and social performance-related SAT-P score compared with the baseline. The results suggest a small but significant effect of Maca supplementation on subjective perception of general and sexual well-being in adult patients with mild ED.[61]

Another study appearing also in *Andrologia*, aimed to assess the effect of Maca on subjective report of sexual desire. Men aged 21−56 years were given (gelatinized) Maca in either 1,500 mg or 3,000 mg dosages, or placebo. Measures were obtained on self-perception of sexual desire, scores for Hamilton test for depression, and Hamilton test for anxiety at 4, 8, and 12 weeks of treatment.

An improvement in sexual desire was observed with Maca after 8 weeks of treatment. Serum testosterone and estradiol levels were not dif-ferent in treatment and placebo groups, but Maca was found to have an independent effect on sexual desire, and this effect did not relate to either Hamilton scores for depression or anxiety or serum testosterone and estra-diol levels. The authors concluded that treatment with Maca improved sexual desire.[62]

8.10 L-ARGININE AND YOHIMBE/YOHIMBINE

Yohimbe is an evergreen tree found in Zaire, Cameroon, and Gabon. The bark holds a chemical called yohimbine, used in folk medicine to arouse sex-ual excitement. In Western medicine it is prescribed for sexual problems caused by medications for depression (SSRIs). Yohimbine blocks the pre- and post- synaptic alpha-2 adrenoceptors.[63]

Natural Medicines Comprehensive Database[64] rates effectiveness based on scientific evidence according to the following scale: Effective, Likely Effective, Possibly Effective, Possibly Ineffective, Likely Ineffective, Ineffective, and Insufficient Evidence to Rate. The effectiveness ratings for YOHIMBE are as follows. Possibly effective for:

- ED: There is evidence that the active ingredient, yohimbine, can be help-ful for ED. Some herbalists suggest that the yohimbe bark actually works better than the yohimbine ingredient alone (Yohimbine Hydrochloride, USP—a standardized form of yohimbine—is a prescription medicine that has been used to treat ED).[65] This has not been verified by any research studies.
- Sexual problems caused by SSRIs: There is evidence from many studies that the active ingredient, yohimbine, can improve sexual problems asso-ciated with this class of medications used for depression. However, this benefit has not been described specifically for the yohimbe bark.

A clinical trial reported in the journal *European Urology* aimed to compare the efficacy and safety for the treatment of ED of the combination of 6 g of L-arginine glutamate, and 6 mg of yohimbine hydrochloride, with that of 6 mg of yohimbine hydrochloride alone, and that of placebo alone. On-demand oral administration of the L-arginine glutamate (6 g) and 6 mg yohimbine combination was shown to be effective in improving erectile function in patients with mild to moderate ED.[66]

The aim of this study published in the *Iran Journal of Psychiatry* was to assess the efficacy and safety of a combination of yohimbine and L-arginine (designated as treatment SX) in the treatment of ED. Married patients aged 25−50 reporting a minimum of 3 months of ED were enrolled in this study. The severity of ED was based on EF domain scores on the IIEF. The scores of 15−25 were considered as mild-to-moderate ED. Patients received one capsule of SX or placebo on demand in a 1:1 ratio using a computer-generated randomization code.

The difference between the two groups was significant at the fourth week.[67] The investigators reported that, "Four adverse events were observed over the study. The difference between the SX and placebo was not significant in the frequency of adverse events." They do not say what the events were. However, they concluded that the formulation SX is safe and effective for the treatment of mild-to-moderate ED, at least in the short-term.

Archives of Sexual Behavior reports four independent but convergent meta-analyses of the efficacy of yohimbine in the treatment of erectile disorder. These analyses integrated data from controlled clinical trials of yohimbine (when used alone), uncontrolled trials examining yohimbine (alone), controlled trials of yohimbine when used in combination with other drugs, and uncontrolled trials of yohimbine plus other drugs. The authors document a consistent tendency for yohimbine and for other medications containing yohimbine to enhance erectile functioning relative to placebo.[68]

A clinical study titled "Double-blind, placebo-controlled safety and efficacy trial with yohimbine hydrochloride in the treatment of nonorganic erectile dysfunction" was reported in the *International Journal of Impotence Research*.

Patients with ED and without clearly detectable organic or psychological causes were given an oral dosage of 30 mg a day (two 5 mg tablets three times daily) for eight weeks. After four weeks, efficacy evaluation was based on both subjective and objective criteria. Subjective criteria included improvement in sexual desire, sexual satisfaction, frequency of sexual contacts, and quality of erection as penile rigidity during sexual contact/intercourse. Objective criteria of outcome were based on improvement in penile rigidity determined by use of polysomnography in the sleep laboratory.

Overall, yohimbine was found significantly more effective than placebo in response rate: 71% vs. 45%. Yohimbine was well tolerated, as only 7% of patients rated tolerability "fair" or "poor," and most adverse experiences were mild. There was no serious adverse event.[69]

The *Journal of Urology* published a meta-analysis titled "Yohimbine for erectile dysfunction: a systematic review and meta-analysis of randomized clinical trials." The authors concluded that, "The benefit of yohimbine medication for erectile dysfunction seems to outweigh its risks. Therefore, yohimbine is believed to be a reasonable therapeutic option for erectile dysfunction that should be considered as initial pharmacological intervention."[70]

8.11 CAUTIONARY NOTE

Yohimbe/Yohimbine is not detailed in any other chapter of this monograph, and for that reason, side effects and cautions are given here. From NCCAM: Yohimbe has been associated with high blood pressure, increased heart rate, headache, anxiety, dizziness, nausea, vomiting, tremors, and sleeplessness. It can be dangerous if taken in large doses or for long periods of time. It should not be taken in combination with monoamine oxidase (MAO) inhibitors, as effects may be additive.

Yohimbe should be used with caution when taken with medicines for high blood pressure, tricyclic antidepressants, or phenothiazines. People with kidney problems and people with psychiatric conditions should not use yohimbe. Women who are pregnant or breastfeeding should not take yohimbe.[71]

The *Journal of Medical Toxicology* reports an instance of refractory priapism associated with ingestion of yohimbe extract,[72] and there are other scattered reports of adverse reactions. However, may studies report it to be generally well tolerated.[73]

8.12 L-ARGININE AND PYCNOGENOL®

Pycnogenol (PYC) is an extract from the bark of the French maritime pine, *Pinus pinaster* (detailed in Chapter 9). It was shown that when patients suffering from ED ranging from moderate-to-mild (determined with IIEF-5), were treated with PYC (120 mg/day) for 3 months, it significantly reduced ED. Furthermore, a significant increase of plasma antioxidant activity was observed. Three months after PYC administration, the level of total cholesterol decreased significantly, as did also LDL-C. Placebo had no effect.[74]

In a study published in the *Journal of Sexual and Marital Therapy*, the investigators reported testing the possibility of overcoming ED by increasing the amounts of endogenous NO. They proposed oral administration of Pycnogenol because it promotes increased production of NO by NOS together with L-arginine. Men without confirmed organic ED received 3 ampoules of Sargenor/day (also Medapharma; an arginine aspartate), and a drinkable solution of the dipeptide arginyl aspartate (equivalent to 1.7 g L-arginine/day) over 3 months. During the second month, patients were additionally supplemented with 40 mg Pycnogenol two times per day; during the third month, the daily dosage was increased to three 40 mg Pycnogenol tablets.

After one month of treatment with L-arginine, there was no discernible change. The combination of L-arginine and Pycnogenol for the following month restored erectile function to 80%, and after three months of treatment, 92.5% of the men experienced a normal erection. No side effects were noted.[75]

The journal *Phytotherapy Research* reported a clinical study with Japanese patients with mild-to-moderate ED to determine the efficacy of a supplement regimen of Pycnogenol (60 mg/day), L-arginine (690 mg/day), and aspartic acid (552 mg/day). The results were assessed with a five-item erectile domain (IIEF-5). Blood biochemistry, urinalysis, and salivary testosterone were also measured. Eight weeks of supplements improved the total score of the IIEF-5, and a marked improvement was observed in hardness of erection and satisfaction with sexual intercourse. Blood pressure declined, and there was a slight increase in salivary testosterone in the supplement group as compared to controls. No adverse reactions were observed during the study period. It was concluded that Pycnogenol in combination with L-arginine as a dietary supplement is effective and safe in these patients with mild-to-moderate ED.[76]

8.13 CAVEAT: MAYO CLINIC CAUTIONS

The following is reprinted with permission of the copyright holder, Natural Standard[77,78]:

"Arginine may cause bloating; diarrhea; endocrine changes; gastrointestinal discomfort; hives; increased blood urea nitrogen, serum creatine, and serum creatinine; increased inflammatory response; leg restlessness, lower back pain; nausea, numbness (with arginine injection); rash; reduction in hematocrit; severe tissue necrosis with extravasation; systemic acidosis; or venous irritation. In heart disease patients, arginine may cause high white blood cell count, increased post-heart attack deaths, lack of energy and strength, and vertigo or increased blood pressure (in heart transplant patients).

"Arginine may increase the risk of bleeding. Caution is advised in patients with bleeding disorders or those taking drugs that may increase the risk of bleeding. Dosing adjustments may be necessary.

"Arginine may change blood sugar levels. Caution is advised in patients with diabetes or hypoglycemia, and in those taking drugs, herbs, or supplements that affect blood sugar. Blood glucose levels may need to be monitored by a qualified healthcare professional, including a pharmacist, and medication adjustments may be necessary.

"Use cautiously in patients with impaired kidney function or those at risk for hyperkalemia (abnormally high levels of blood potassium), including those with

diabetes or using drugs that elevate potassium levels, such as potassium-sparing diuretics and potassium supplements, as arginine may cause hyperkalemia. Fatal cardiac arrhythmia occurred in one patient.

"Use caution with phosphodiesterase inhibitors (e.g., sildenafil [Viagra®]), due to a theoretical risk of additive blood vessel widening and blood pressure lowering.

"Use with caution in postmenopausal patients, as night sweats and flushing have been reported.

"Use with caution in patients with herpes virus, as L-arginine may worsen this condition. L-arginine may increase the risk of herpes simplex cold sores.

"Use with caution in individuals at risk for headaches, as headache has been a reported side effect. In mountain climbers, L-arginine increased the risk of developing a headache.

"Use with caution in patients with immunological disorders.

"Use cautiously in patients with acrocyanosis, sickle cell anemia, and hyperchloremic acidosis, as arginine may cause worsening of symptoms.

"Use cautiously in patients with acrocyanosis, sickle cell anemia, and hyperchloremic acidosis, as arginine may cause worsening of symptoms.

"Use cautiously in patients with guanidinoacetate methyltransferase (GAMT) deficiency. This enzyme is involved in the conversion of amino acids such as arginine to creatine. . . .

"Avoid use in those with low blood pressure or those using blood pressure-lowering agents, due to the reported blood vessel-widening and blood pressure-lowering effects of L-arginine.

"Avoid with nitrates, as concurrent use may result in additive blood pressure-lowering and blood vessel-widening effects.

"Avoid use in patients given spironolactone, because arginine monohydrochloride has resulted in abnormally high potassium levels and fatal cardiac arrhythmia.

"Avoid use in patients with asthma, as arginine may cause an allergic and response, aggravate airway inflammation, and amplify inflammatory airway response. In human research, L-arginine increased exhaled nitric oxide, suggesting increased inflammatory response in asthmatic and cystic fibrosis subjects.

"Avoid use in patients at risk for or with a history of heart attack, as arginine may worsen outcomes and increase the risk of mortality.

"Avoid use in breast cancer patients.

"Avoid with known allergy or hypersensitivity to arginine. Symptoms may include rash, itching, or shortness of breath. Anaphylaxis has occurred after arginine injections. In clinical research, one patient experienced a mild allergic skin reaction to intravenous L-arginine. Hives have been reported."

L-arginine can cause an outbreak of dormant Herpes and/or exacerbate such an outbreak.

A number of other conditions including certain forms of cancer have been reported to accelerate with L-arginine supplementation.

NOTES

1. The author is consultant to Real Health Laboratories, Div. of PharmaCare US.
2. The L-arginine contained in the VasoRectUltra formulation is considered vegetarian, with a starting material of corn. It is vegetarian if removed from the capsule, which is gelatin (bovine source, not porcine).

REFERENCES

1. Uslu N, Gorgulu S, Alper AT, Eren M, Nurkalem Z, Yildirim A, et al. Erectile dysfunction as a generalized vascular dysfunction. *J Am Soc Echocardiogr* 2006;**19**:341−6.
2. Kaiser DR, Billups K, Mason C, Wetterling R, Lundberg JL, Bank AJ. Impaired brachial artery endothelium-dependent and -independent vasodilation in men with erectile dysfunction and no other clinical cardiovascular disease. *J Am Coll Cardiol* 2004;**43**:179−84.
3. Hale TM, Okabe H, Bushfield TL, Heaton JP, Adams MA. Recovery of erectile function after brief aggressive antihypertensive therapy. *J Urol* 2002;**168**:348−54.
4. Schwartz BG, Kloner RA. Cardiovascular implications of erectile dysfunction. *Circulation* 2011;**123**:e609−11.
5. Cama E, Colleluori DM, Emig FA, Shin H, Kim SW, Kim NN, et al. Human arginase II: crystal structure and physiological role in male and female sexual arousal. *Biochem* 2003;**42**:8445−51.
6. Cox JD, Kim NN, Traish AM, Christianson DW. Arginase-boronic acid complex highlights a physiological role in erectile function. *Nature Struct Biol* 1999;**6**:1043−7.
7. Morris Jr. SM. Recent advances in arginine metabolism: roles and regulation of the arginases. *Br J Pharmacol* 2009;**157**:922−30.
8. Li H, Meininger CJ, Hawker Jr. JR, Haynes TE, Kepka-Lenhart D, Mistry SK, et al. Regulatory role of arginase I and II in nitric oxide, polyamine, and proline syntheses in endothelial cells. *Am J Physiol Endocrinol Metab* 2001;**280**:E75−82.
9. Moody JA, Vernet D, Laidlaw S, Rajfer J, Gonzalez-Cadavid NF. Effects of long-term oral administration of L-arginine on the rat erectile response. *J Urol* 1997;**158**(3 Pt 1):942−7.
10. Shaul PW. Regulation of endothelial nitric oxide synthase: location, location, location. *Annu Rev Physiol* 2002;**64**:749−74.
11. Yang Y-M, Huang A, Kaley G, Sun D. eNOS uncoupling and endothelial dysfunction in aged vessels. *Am J Physiol Heart Circ Physiol* 2009;**297**:H1829−36.
12. Schmidt TS, Nicholas J. Mechanisms for the role of tetrahydrobiopterin in endothelial function and vascular disease. *Clin Sci* 2007;**113**:47−63.
13. Landmesser U, Dikalov S, Russ Price S, McCann L, Fukai T, Holland SM, et al. Oxidation of tetrahydrobiopterin leads to uncoupling of endothelial cell nitric oxide synthase in hypertension. *J Clin Invest* 2003;**111**:1201−9.
14. Heller R, Unbehaun A, Schellenberg B, Mayer B, Werner-Felmayer G, Werner ER. L-ascorbic acid potentiates endothelial nitric oxide synthesis via a chemical stabilization of tetrahydrobiopterin. *J. Biol. Chem* 2001;**276**:40−7.
15. Duffy SJ, Gokce N, Holbrook M, Huang A, Frei B, Keaney Jr JF, et al. Treatment of hypertension with ascorbic acid. *Lancet* 1999;**354**:2048−9.

16. Kaya C, Uslu Z, Karaman I. Is endothelial function impaired in erectile dysfunction patients? *Int J Impot Res* 2006;**18**:55−60.

17. Tang WH, Wang Z, Cho L, Brennan DM, Hazen SL. Diminished global arginine bioavailability and increased arginine catabolism as metabolic profile of increased cardiovascular risk. *J Am Coll Cardiol* 2009;**53**:2061−7.

18. Morris Jr. SM. Arginases and arginine deficiency syndromes. *Curr Opin Clin Nutr Metab Care* 2012;**15**:64−70.

19. Yin W-H, Chen J-W, Tsai C, Chiang M-C, Young MS, Lin S-J. L-arginine improves endothelial function and reduces LDL oxidation in patients with stable coronary artery disease. *Clin Nutr* 2005;**24**:988−97.

20. Gur S, Kadowitz PJ, Trost L, Hellstrom WJ. Optimizing nitric oxide production by time dependent L-arginine administration in isolated human corpus cavernosum. *J Urol* 2007;**178**(4 Pt 1):1543−8.

21. O'Leary ME, Fowler FJ, Lenderking WR, Barber B, Sagnier PP, Guess HA, et al. A brief male sexual function inventory for urology. *Urol* 1995;**46**:697−706.

22. Chen J, Wollman Y, Chernichovsky T, Iaina A, Sofer M, Matzkin H. Effect of oral administration of high-dose nitric oxide donor L-arginine in men with organic erectile dysfunction: results of a double-blind, randomized, placebo-controlled study. *BJU Int* 1999;**83**:269−73.

23. Zorgniotti AW, Lizza EF. Effect of large doses of the nitric oxide precursor, L-arginine, on erectile dysfunction. *Int J Impot Res* 1994;**6**:33−5.

24. Klotz T, Mathers MJ, Braun M, Bloch W, Engelmann U. Effectiveness of oral L-arginine in first-line treatment of erectile dysfunction in a controlled crossover study. *Urol Int* 1999;**63**:220−3.

25. Rosano GMC, Panina G, Cerquetani E, Leonardo F, Pelliccia F, Bonfigli B, et al. L-arginine improves endothelial function in newly diagnosed hypertensives. *J Am Coll Cardiol* 1998;**31**(2s1):262A.

26. Rector TS, Bank AJ, Mullen KA, Tschumperlin LK, Sih R, Pillai K, et al. Randomized, double-blind, placebo-controlled study of supplemental oral L-arginine in patients with heart failure. *Circulation* 1996;**93**:2135−41.

27. Adams MR, McCredie R, Jessup W, Robinson J, Sullivan D, Celermajer DS. Oral L-arginine improves endothelium-dependent dilatation and reduces monocyte adhesion to endothelial cells in young men with coronary artery disease. *Atherosclerosis* 1997;**129**:261−9.

28. Lucotti P, Setola E, Monti LD, Galluccio E, Costa S, Sandoli EP, et al. Beneficial effects of a long-term oral L-arginine treatment added to a hypocaloric diet and exercise training program in obese, insulin-resistant type 2 diabetic patients. *Amer J Physiol Endocrinol Metab* 2006;**291**:E906−12.

29. Dong JY, Qin LQ, Zhang Z, Zhao Y, Wang J, Arigoni F, et al. Effect of oral L-arginine supplementation on blood pressure: a meta-analysis of randomized, double-blind, placebo-controlled trials. *Am Heart J* 2011;**162**:959−65.

30. <http://www.nccam.nih.gov/health/supplements/wiseuse.htm>; 2013 [accessed 16.11.13].

31. <http://www.fda.gov/Food/ResourcesForYou/Consumers/ucm109760.htm>; 2013 [accessed 16.11.13].

32. <http://fnic.nal.usda.gov/dietary-supplements>; 2013 [accessed 16.11.13].

33. US Patent No. 606432 granted May 16, 2000 [Arginine- or lysine-containing fruit healthbar formulation] was assigned to Cooke Pharma, Belmont, CA.

34. Maxwell AJ, Anderson B, Zapien MP, Cooke JP. Endothelial dysfunction in hypercholesterolemia is reversed by a nutritional product designed to enhance nitric oxide activity. *Cardiovasc Drugs Ther* 2000;**14**:309–16.

35. Förstermann U, Münzel T. Endothelial nitric oxide synthase in vascular disease. From marvel to menace. *Circulation* 2006;**113**:1708–14.

36. Romero MJ, Platt DH, Caldwell RB, Caldwell RW. Therapeutic use of citrulline in cardiovascular disease. *Cardiovasc Drug Rev* 2006;**24**:275–90.

37. Flam BR, Eichler DC, Solomonson LP. Endothelial nitric oxide production is tightly coupled to the citrulline–NO cycle. *Nitric Oxide* 2007;**17**:115–21.

38. Orozco-Gutiérrez JJ, Castillo-Martínez L, Orea-Tejeda A, Vázquez-Díaz O, Valdespino-Trejo A, Narváez-David R, et al. Effect of L-arginine or L-citrulline oral supplementation on blood pressure and right ventricular function in heart failure patients with preserved ejection fraction. *Cardiol J* 2010;**17**:612–8.

39. Hayashi T, Juliet PA, Matsui-Hirai H, Miyazaki A, Fukatsu A, Funami J, Iguchi A. L-Citrulline and L-arginine supplementation retards the progression of high-cholesteroL-diet-induced atherosclerosis in rabbits. *Proc Natl Acad Sci USA* 2005;**102**:13681–6.

40. Cormio L, De Siati M, Lorusso F, Selvaggio O, Mirabella L, Sanguedolce F, et al. Oral L-citrulline supplementation improves erection hardness in men with mild erectile dysfunction. *Urol* 2011;**77**:119–22.

41. Neogenis® Laboratories, Inc., Austin, Texas, 78746. <http://www.neogenis.com>; 2013 [accessed 17.11.13].

42. <http://www.realhealthlabs.com/store/the-vasorect-ultra-formula-30-day-supply.html>; 2013 [accessed 17.11.13].

43. Elíes J, Cuíñas A, García-Morales V, Orallo F, Campos-Toimil M. Trans-resveratrol simultaneously increases cytoplasmic Ca^{2+} levels and nitric oxide release in human endothelial cells. *Mol Nutr Food Res (Special Issue: Resveratrol – Current Status and Outlook)* 2011;**55**:1237–48.

44. Förstermann U, Li H. Therapeutic effect of enhancing endothelial nitric oxide synthase (eNOS) expression and preventing eNOS uncoupling. *Br J Pharmacol* 2011;**164**:213–23.

45. Fukuhara S, Tsujimura A, Okuda H, Yamamoto K, Takao T, Miyagawa Y, et al. Vardenafil and resveratrol synergistically enhance the nitric oxide/cyclic guanosine monophosphate pathway in corpus cavernosal smooth muscle cells and its therapeutic potential for erectile dysfunction in the streptozotocin-induced diabetic rat: preliminary findings. *J Sex Med* 2011;**8**:1061–71.

46. <http://www.urospec.com/uro/Forms/ipss.pdf>; 2013 [accessed 17.11.13].

47. Lopatkin N, Sivkov A, Walther C, Schläfke S, Medvedev A, Avdeichuk J, et al. Long-term efficacy and safety of a combination of sabal and urtica extract for lower urinary tract symptoms—a placebo-controlled, double-blind, multicenter trial. *World J Urol* 2005;**23**:139–46.

48. Nahata A, Dixit VK. Ameliorative effects of stinging nettle (Urtica dioica) on testosterone-induced prostatic hyperplasia in rats. *Andrologia* 2012;**44**(Suppl 1):396–409.

49. Christudoss P, Selvakumar R, Fleming JJ, Gopalakrishnan G. Zinc status of patients with benign prostatic hyperplasia and prostate carcinoma. *Indian J Urol* 2011;**27**:14–8.

50. Moens AL, Vrints CJ, Claeys MJ, Timmermans P-P, Champion HC, Kass DA. Mechanisms and potential therapeutic targets for folic acid in cardiovascular disease. *Am J Physiol–Heart C* 2008;**294**:H1971–7.

51. Vitseva O, Varghese S, Chakrabarti S, Folts JD, Freedman JE. Grape seed and skin extracts inhibit platelet function and release of reactive oxygen intermediates. *J Cardiovasc Pharmaco* 2005;**46**:445−51.

52. Kar P, Laight D, Rooprai HK, Shaw KM, Cummings M. Effects of grape seed extract in Type 2 diabetic subjects at high cardiovascular risk: a double blind randomized placebo controlled trial examining metabolic markers, vascular tone, inflammation, oxidative stress and insulin sensitivity. *Diabet Med* 2009;**26**:526−31.

53. Corder R, Warburton RC, Khan NQ, Brown RE, Wood EG, Lees DM. The procyanidin-induced pseudo laminar shear stress response: a new concept for the reversal of endothelial dysfunction. *Clin Sci (Lond)* 2004;**107**:513−7.

54. Schneider B. Ginkgo biloba extract in peripheral arterial diseases. Meta-analysis of controlled clinical studies. *Arzneimittelforschung* 1992;**42**:428−36.

55. Cohen AJ, Bartlik B. Ginkgo biloba for antidepressant-induced sexual dysfunction. *J Sex Marital Ther* 1998;**24**:139−43.

56. Kang B-J, Lee S-J, Kim M-D, Cho M-J. A placebo-controlled, double-blind trial of Ginkgo bilboa for antidepressant-induced sexual dysfunction. *Hum Psychopharmacol Clin Exp* 2002;**17**:279−84.

57. Madeo B, Bettica P, Milleri S, Balestrieri A, Granata AR, Carani C, et al. The effects of citalopram and fluoxetine on sexual behavior in healthy men: evidence of delayed ejaculation and unaffected sexual desire. A randomized, placebo-controlled, double-blind, double-dummy, parallel group study. *J Sex Med* 2008;**5**:2431−41.

58. Kiefer D, Pantuso T (2003). Panax ginseng. *American Family Physician*, Oct. 15: <http://www.aafp.org/afp/2003/1015/p1539.html>; 2013 [accessed 14.11.13].

59. Choi HK, Seong DH, Rha KH. Clinical efficacy of Korean red ginseng for erectile dysfunction. *Int J Impot Res* 1995;**7**:181−6.

60. Hong B, Ji YH, Hong JH, Nam KY, Ahn TY. A double-blind crossover study evaluating the efficacy of Korean red ginseng in patients with erectile dysfunction: a preliminary report. *J Urol* 2002;**168**:2070−3.

61. Zenico T, Cicero AF, Valmorri L, Mercuriali M, Bercovich E. Subjective effects of Lepidium meyenii (Maca) extract on welL-being and sexual performances in patients with mild erectile dysfunction: a randomised, double-blind clinical trial. *Andrologia* 2009;**41**:95−9.

62. Gonzales GF, Córdova A, Vega K, Chung A, Villena A, Góñez C, et al. Effect of Lepidium meyenii (MACA) on sexual desire and its absent relationship with serum testosterone levels in adult healthy men. *Andrologia* 2002;**34**:367−72.

63. Shepperson NB, Duval N, Massingham R, Langer SZ. Pre- and postsynaptic alpha-adrenoceptor selectivity studies with yohimbine and its two diastereoisomers rauwolscine and corynanthine in the anesthetized dog. *J Pharmacol Exp Ther* 1981;**219**:540−6.

64. *Natural Medicines Comprehensive Database*. Published by Pharmacists Letter, 13th ed., 2012).

65. <http://pharmrev.aspetjournals.org/cgi/pmidlookup?view=long&pmid=11546836>; 2013 [accessed 17.11.13].

66. Lebret T, Hervé JM, Gorny P, Worcel M, Botto H. Efficacy and safety of a novel combination of L-arginine glutamate and yohimbine hydrochloride: a new oral therapy for erectile dysfunction. *Eur Urol* 2002;**41**:608−13.

67. Akhondzadeh S, Amiri A, Bagheri AH. Efficacy and safety of oral combination of Yohimbine and L-arginine (SX) for the treatment of erectile dysfunction: a multicenter, randomized, double blind, placebo-controlled clinical trial. *Iran J Psychiat* 2010;**5**:1−3.

68. Carey MP, Johnson BT. Effectiveness of yohimbine in the treatment of erectile disorder: four meta-analytic integrations. *Arch Sex Behav* 1996;**25**:341−60.

69. Vogt HJ, Brandl P, Kockott G, Schmitz JR, Wiegand MH, Schadrack J, et al. Double-blind, placebo-controlled safety and efficacy trial with yohimbine hydrochloride in the treatment of nonorganic erectile dysfunction. *Int J Impot Res* 1997;**9**:155−1561.

70. Ernst E, Pittler MH. Yohimbine for erectile dysfunction: a systematic review and meta-analysis of randomized clinical trials. *J Urol* 1998;**159**:433−6.

71. <http://nccam.nih.gov/health/yohimbe>; 2013 [accessed 17.11.13].

72. Myers A, Barrueto Jr. F. Refractory priapism associated with ingestion of yohimbe extract. *J Med Toxicol* 2009;**5**:223−5.

73. Riley AJ. Yohimbine in the treatment of erectile disorder. *Br J Clin Pract* 1994;**48**:133−6.

74. Ďuračková Z, Trebatický B, Novotný V, Žitňanová I, Breza J. Lipid metabolism and erectile function improvement by pycnogenol extract from the bark of Pinus pinaster in patients suffering from erectile dysfunction-a pilot study. *Nutr Res* 2003;**23**:1189−98.

75. Stanislavov R, Nikolova V. Treatment of erectile dysfunction with pycnogenol and L-arginine. *J Sex Marital Ther* 2003;**29**:207−13.

76. Aoki H, Nagao J, Ueda T, Strong JM, Schonlau. F, Yu-Jing S, et al. Clinical assessment of a supplement of Pycnogenol® and L-arginine in Japanese patients with mild to moderate erectile dysfunction. *Phytother Res* 2012;**26**:204−7.

77. <http://www.mayoclinic.com/health/L-arginine/NS_patient-arginine/DSECTION=safety>; 2013 [accessed 17.11.13].

78. <http://www.naturalstandard.com>; 2013 [accessed 17.11.13].

Chapter 9

The Polyphenolic Antioxidant Resveratrol, the Carotinoid Lycopene, and the Proanthocyanidin Pycnogenol

9.1 INTRODUCTION

The previous two chapters detailed the role of the amino acid L-arginine in cardiovascular health and erectile function (ED). It also has other functions, both metabolic and in connection with wound healing, kidney function, and many more. However, foremost is that it is the principal substrate for the formation of nitric oxide (NO) in the first blood vessel-control signal system, the nitric oxide/cyclic guanosine monophosphate (NO/cGMP) pathway. In addition, it is also an antioxidant, preventing peroxidation, and a major element of the immune system. In fact, NO is the "bullet" that macrophages fire at invading microorganisms.

The plant world is not immune to predatory attacks by microorganisms, as noted in a previous chapter, and many different varieties of flora also have their own version of an immune system which consists, not uncommonly, of powerful antioxidants. This chapter addresses a few of these substances in connection with their contribution to understanding the purported contribution of exogenous, plant-based, antioxidants in reducing the impact of oxidative stress on our cardiovascular health and erectile function.

A number of plant products recently gained the attention of researchers for their unique benefit in cardiovascular health and sexual vitality. That plant products can also serve as palliatives is long known: Roman Legionnaires knew to chew on willow stem (*Salix*) to relieve pain and inflammation, and they knew that gum mastic (*Pistacia lentiscus*), relieved stomach disorders. Today, we know that the former holds salicylic acid, and the latter *polymeric myrcene*, a bactericide that targets *Helicobacter pylori*, the principal bacterial cause of stomach ulcers. There are, of course many more, but this monograph is not about plants. Nevertheless, the idea is

Robert Fried: Erectile Dysfunction as a Cardiovascular Impairment.
DOI: http://dx.doi.org/10.1016/B978-0-12-420046-3.00009-3

259

gaining ground that plant products may be our principal source of readily available exogenous antioxidants.

As we have come to know more and more about oxidative stress and its detrimental effects on endothelial function and cardio-sexual health, we recognize our dependence on these free-radical scavengers as essential to health … perhaps even our survival. There is much that we don't know about antioxidants, and what's worse, much that we believe that we know is actually wrong. For this reason, the substances described in the following sections and chapters are limited to those that conventional science has scrutinized and validated, and declared safe for consumption.

9.2 RESVERATROL

Resveratrol is a powerful antioxidant produced by plants. It belongs to a class of polyphenolic compounds called stilbenes, and it is essentially the plant immune system. Polyphenols also protect human body cells from free radicals that otherwise contribute to tissue damage.

Resveratrol may be found in many common vegetables, especially cruciferous vegetables like broccoli, and in fruits such as blueberries, blackberries, etc., and also in certain wines. The resveratrol constituent of wine, varying in concentration with the cultivar, generally contains relatively little of it— red wine holds on the order of 0.2 and 5.8 mg/L. In a review titled "Wine, resveratrol and health," the authors contend that resveratrol naturally present in wine proffers many health benefits, including cardioprotection. However, its bioavailability in certain foods and beverages may be quite low.[1,2,3] More commonly, resveratrol is available as a nutritional supplement primarily from Japanese *knotweed* (*Polygonum cuspidatum*).

9.2.1 Total Antioxidant Status

Much has been made of the damaging effect of free radicals and reactive oxygen species (ROSs) in connection with endothelial as well as overall body health, and the pitfalls of evaluation of antioxidant capacity were underscored in a previous chapter in connection with misleading claims of the health benefits and other commercial abuses of oxygen radical absorbance capacity (ORAC). Nevertheless, clinical use of measures of antioxidant status is well established. It is from such studies about total antioxidant status (TAS) that we learn the value of antioxidants in our daily life, even though existing technology does not lead to practical self-assessment.

Even if there were a readily available methodology, absent adequate information about the consequences of increasing total antioxidant status, or the implications of TAS lower than expected in any given individual, it cannot be determined what to make of any TAS value. Furthermore, as noted in a previous chapter, even the most enthusiastic scientific research sources writing about the value of antioxidants in cardio-sexual health never fail to caution about possible harmful effects of excessive dietary antioxidant intake subverting endogenous production.

A number of techniques now exist to measure TAS, and it is from use of these methods that we derive guidelines for nutrition and dietary supplementation. The *International Journal of Food Science & Nutrition* reports a study aimed to correlate the free radical scavenging and antioxidant activity of two known substances, Trolox and α-tocopherol, using three *in vitro* methods, linoleic acid emulsion, brain homogenate and 2,2-diphenyl-1-picryl-hydrazyl (DPPH).

9.2.2 Trolox and α-Tocopherol

A steady state α-tocopherol shows a greater inhibition of spontaneous oxidation of brain homogenate (a slurry of tissue often used to assay unbound compounds or drugs) than does Trolox (Sidebar 9.1), while the latter shows a better antioxidant activity performance in inhibition of linoleic acid peroxidation and free radical scavenging activity.

Sidebar 9.1

Note: Trolox is the Hoffman-LaRoche trade name for a water-soluble derivative of vitamin E. It is an antioxidant, like vitamin E, and is used in biological or biochemical applications to reduce oxidative stress or damage. As noted in a previous chapter, Trolox equivalent antioxidant capacity (TEAC) is a measurement of antioxidant strength based on Trolox, measured in units called Trolox Equivalents (TE; e.g. μmolTE/100g). The TEAC assay is used to measure antioxidant capacity of foods, beverages, and supplements

Antioxidant activity significantly changes according to the substrate, the parameter adopted to compare the substances in the same method, and the form used to express antioxidant concentration.[4] A detailed review of "The chemistry behind antioxidant capacity assays" may be found in, Huang, Ou, & Prior (2005) *Journal of Agricultural Food Chemistry*.[5] One of many applications of this technology to determining antioxidant properties of foods is reported in the *Journal of Nutrition*.

The authors contend that epidemiologic studies have pointed to an inverse relationship between the quantity of fruits and vegetables consumed and statistics about disease and deaths. Furthermore, it is thought that the antioxidant content of fruits and vegetables may contribute to protection from disease. Because plant foods contain many different kinds of antioxidants, it would be useful to know the cumulative capacity of their components to scavenge free radicals—total antioxidant capacity (TAC).

A variety of foods commonly consumed in Italy, including 34 vegetables, 30 fruits, 34 beverages, and 6 vegetable oils, were analyzed using three different assays: i.e. Trolox equivalent antioxidant capacity (TEAC), total radical-trapping antioxidant parameter (TRAP), and ferric reducing-antioxidant power (FRAP). These assays, based on different chemical mechanisms, were selected to take into account the wide variety and range of action of antioxidant compounds present in actual foods.

Among vegetables, spinach had the highest antioxidant capacity in the TEAC and FRAP assays. This may be due in part to the fact that it does not hold as much iron as is commonly thought: A misplaced decimal initially raised its iron content ten-fold. Besides, it is a high-oxalate food and oxalate inhibits iron absorption—Popeye not withstanding. Next came peppers, whereas asparagus had the greatest antioxidant capacity in the TRAP assay.

- Among fruits, the highest antioxidant activities were found in berries (i.e. blackberry, red currant, and raspberry), regardless of the assay used.
- Among beverages, coffee had the greatest TAC, regardless of the method of preparation or analysis, followed by citrus juices, which exhibited the highest value among soft beverages.
- Finally, soybean oil had the highest antioxidant capacity of the oils followed by extra virgin olive oil, whereas peanut oil was less effective.

The authors propose that the use of an appropriate questionnaire to estimate antioxidant intake will allow the determination of the relation between dietary antioxidants and oxidative stress-induced diseases.[6]

In a clinical trial published in *Diabetes Care*, TAS of serum was measured using the TEAC assay in type-1 diabetic and non-diabetic participants. The presence of coronary artery calcification (CAC) was assessed in the diabetic participants with electron beam computed tomography. TAS was found to be reduced in type-1 diabetic participants compared with non-diabetic participants. There were associations between TAS and HbA1c and duration of diabetes. Significant CAC was considered present if the *Agatston score* (indicates the amount of calcium in coronary arteries)[7] was greater than 10.

The diabetic participants with significant CAC were older; they had longer duration of diabetes, and were more likely to have high blood pressure; they had higher total cholesterol concentration, serum creatinine concentration, and urinary albumin-to-creatinine ratio. They had lower serum TAS compared with those without significant calcification. The power of TAS to predict CAC was independent of many of the traditional coronary heart disease (CHD) risk factors. The authors concluded that TAS is reduced in type-1 diabetes and is associated with the presence of CAC.[8]

The following scientific reports support the various pathways by which resveratrol—especially *trans*-resveratrol—has been shown to benefit cardio-sexual function.

9.2.3 Resveratrol Enhances Nitric Oxide Biosynthesis and Protects the Endothelium

trans-Resveratrol increases NO production: In a study in *Molecular Nutrition & Food Research*, it was reported that *trans*-resveratrol, a purified form of resveratrol that remains active only when sheltered from the sunlight and from oxygen, increases NO production, thus improving endothelium function. The aim of this study was to investigate whether the dietary polyphenol *trans*-resveratrol raises $[Ca^{2+}](c)$ in endothelial cells, leading to a simultaneous augmentation of NO biosynthesis.

The researchers simultaneously and separately measured $[Ca^{2+}](c)$ and NO in human endothelial cells using the Ca^{2+} indicator fura-2 and the NO-sensitive fluorescent probe 4,5-diaminofluorescein. In approximately 30% of the cells, *trans*-resveratrol induced an increase in $[Ca^{2+}](c)$ with both a transient and a sustained component, and a simultaneous increase in NO biosynthesis. This effect was reduced by non-selective Ca^{2+}-channel blockers, inhibition of intracellular Ca^{2+} release, inhibition of endothelial nitric oxide synthase (eNOS) and, to a lesser extent, inhibition of extracellular signal-regulated kinase 1/2 (ERK 1/2) or 5′-adenosine monophosphate-activated protein kinase (AMPK). *trans*-Resveratrol did not modify *in vitro* eNOS activity, suggesting that the observed NO formation takes place by mobilization of Ca^{2+} and not through direct effects on eNOS.

The authors report that they show for the first time, that *trans*-resveratrol induces a concentration-dependent simultaneous increase in $[Ca^{2+}](c)$ and NO biosynthesis that could be linked to its endothelium-dependent vaso-relaxant (EDR) effect. They assume that if it would act in the same way in human blood vessels *in vivo*, the pharmacological properties observed might contribute to the beneficial cardiovascular effects of this polyphenol by improving endothelial function.[9]

The polyphenolic natural product resveratrol is considered to have potential as a drug for prevention and treatment of cardiovascular diseases (CVDs). A review published in *Biofactors* summarizes the molecular effects of resveratrol on endothelial cells that play a key role in the development of those diseases: Resveratrol enhances endothelial NO production, improves endothelial redox balance, and inhibits endothelial activation in response to pro-inflammatory and metabolic factors. The action of resveratrol appears to be mediated by an enzyme, sirtuin 1 (SIRT₁).[10]

9.2.4 Resveratrol Induces Mitochondrial Biogenesis and Protects against Metabolic Decline

The ability of resveratrol to improve mitochondrial function requires SIRT₁. That enzyme is now thought to be a major therapeutic target for the treatment of metabolic and inflammatory disorders.[11]

9.2.5 Resveratrol Is Antioxidant

Researchers writing in the *Journal of Physiology & Pharmacology* report finding that resveratrol also regulates the gene expression of pro-oxidative and anti-oxidative enzymes in human endothelial cells: NADPH oxidases (Nox) are the predominant producers of superoxide in the vasculature, whereas superoxide dismutase (SOD) and glutathione peroxidase 1 (GPx1) are the major enzymes responsible for the inactivation of superoxide and hydrogen peroxide (H_2O_2), respectively, as noted in a previous chapter.

Incubation of human umbilical vein endothelial cells (HUVEC) and HUVEC-derived EA.hy 926 cells with resveratrol resulted in a concentration- and time-dependent downregulation of Nox4, the most abundant NADPH oxidase catalytic subunit (quantitative real-time RT-PCR). The same resveratrol regimen upregulated the mRNA expression of SOD1 and GPx1. Protein levels of SOD1 and GPx1 were enhanced by resveratrol in a concentration-dependent manner (by Western blot analyses). Pretreatment of EA.hy926 cells with resveratrol completely abolished DMNQ-induced oxidative stress.

The authors conclude that the suppression of the expression of pro-oxidative genes (such as NADPH oxidase) and induction of anti-oxidative enzymes (such as SOD1 and GPx1) might be an important component of the vascular protective effect of resveratrol.[12]

9.2.6 Resveratrol Promotes Endothelial Nitric Oxide Synthase Activity

The following study in the *British Journal of Nutrition* emphasizes the beneficial effect of the pattern of wine consumption on NO formation. Investigators examined the effect of repeated and long-term treatment with resveratrol on NO production in endothelial cells as a model of routine wine consumption. Repeated treatment with resveratrol for five days resulted in an increase in endothelial eNOS protein content and NO production in the HUVEC in a concentration-dependent manner. A significant increase in functional eNOS protein content was observed even at 50 nm.

These results demonstrate that resveratrol within the physiological range, even in small quantities, increases eNOS, thereby promoting NO production, suggesting that eNOS induction might result from the accumulation of nanomolar concentrations of resveratrol. The authors contend that the study results may explain the observation that cardiovascular benefits of red wine are experienced with routine consumption, but not with acute consumption.[13]

9.2.7 Resveratrol Promotes Nitric Oxide/Cyclic Guanosine Monophosphate Formation

According to the authors of a review in the journal *Nitric Oxide*, resveratrol (3,5,4′-trihydroxy-*trans*-stilbene), a polyphenol phytoalexin found in many varieties of plant species, stimulates endothelial production of NO, and reduces oxidative stress; it inhibits vascular inflammation and prevents platelet aggregation. It has been shown to reduce blood pressure and left ventricle hypertrophy in experimentally induced hypertension in animals, and to slow the progression of atherosclerosis.

Resveratrol protects NO/cGMP formation: According to an article in the *British Journal of Pharmacology*, *trans*-resveratrol is one of several compounds that was found to prevent eNOS uncoupling and, at the same time, enhance eNOS expression.[14]

9.2.8 Resveratrol Enhances Tetrahydrobiopterin Biosynthesis, thus Reversing Endothelial Nitric Oxide Synthase Uncoupling

One study appearing in the *Journal of Pharmacology and Experimental Therapeutics* attributed the protective effect of *trans*-resveratrol to the elevation of tetrahydrobiopterin (BH4) levels. Resveratrol decreases BH4 oxidation by reducing superoxide production and enhancing ROS inactivation (upregulation of SOD1, SOD2, SOD3, catalase, and GPx1) (in ApoE-KO mice). This novel mechanism (reversal of eNOS uncoupling) might contribute to the protective effects of resveratrol.[15]

9.2.9 The Sirtuin "Family"

Recent research reveals that one of the pathways by which resveratrol exerts a biological influence is through the so-called "silent information regulator genes" (sirtuins), proteins present in all species from bacteria to mammals. Seven of these have been found to be present in mammals. It can at best be said that though identified, their importance underscored, their responses to environmental factors and their role in health and disease are currently poorly understood.

A report in *Alternative Medicine Review* discusses the dietary, lifestyle, and environmental factors that are known to influence sirtuin activity, and summarizes research on the importance of vitamin B_3 in supporting sirtuin enzyme activity, as well as the role specifically of the amide form of this vitamin (nicotinamide) to inhibit sirtuin enzyme activity. It appears that polyphenols, especially resveratrol, influence sirtuins. Resveratrol is said to be a sirtuin activator:

In Part 2 of that review, the author links resveratrol-spurred sirtuins to a number of health concerns. Human $SIRT_1$ promotes endothelium dependent vasodilation. It affects insulin sensitivity, it has anti-obesity activity, and it appears to be sensitive to calorie intake changes. Due to its activation of sirtuins, resveratrol is most likely to produce a noticeable physiological effect under stressful circumstances or those involving unhealthy lifestyle habits.[16,17]

9.2.10 Dose-Dependent Non-Monotonic Response of Resveratrol

There appears to be a nonlinear, dose-dependent health benefit of resveratrol. At a lower dose, it is cardioprotective by increasing formation of cell-survival proteins, and reducing myocardial infarct size. At higher doses, resveratrol exerts signal-inducing apoptosis in cancer cells. At ever-higher doses, resveratrol depresses cardiac function, elevates levels of apoptotic protein expressions, results in an unstable redox environment, increases myocardial infarct size, and number of self-destructing cells.

At high dose, resveratrol not only hinders tumor growth, but also inhibits the synthesis of RNA, DNA, and protein, causes structural chromosome aberrations, chromatin breaks, chromatin exchanges, blocks cell proliferation, decreases wound healing, decreases endothelial cell growth by fibroblast growth factor-2 (FGF-2) and vascular endothelial growth factor, and angiogenesis in healthy tissue cells leading to cell death. Thus, at lower dose, resveratrol can be very useful in maintaining human health, whereas at higher dose, resveratrol has pro-apoptotic actions on healthy cells as well as on tumor cells.[18]

Additional references:

Red wine significantly enhances endothelial cell function by modifying NO bioavailability. A moderate intake of red wine can increase NO availability.[19] Intake of red wine increases the number and functional capacity of circulating endothelial progenitor cells by enhancing NO bioavailability.

Because resveratrol acts as an agonist at the estrogen receptor level, this study tested its effect on eNOS expression in human endothelial cells. eNOS expression and eNOS-derived NO production were increased after long-term incubation with resveratrol. The stimulation of eNOS expression and activity may contribute to the cardiovascular protective effects attributed to resveratrol.[20] Resveratrol enhances expression and activity of eNOS.

9.2.11 Atherosclerosis

Resveratrol supplementation as complementary treatments in primary prevention of CVD: One-year consumption of a resveratrol-rich grape supplement reduced inflammation in patients on statins for primary prevention of CVD and at high CVD risk (i.e. with diabetes or hypercholesterolemia). This study showed for the first time that a dietary intervention with grape resveratrol could complement the gold standard therapy in the primary prevention of CVD.[21]

9.2.12 Resveratrol Decreases Oxidized Low-Density Lipoprotein and Inflammatory Markers

The aim of a follow-up study titled "Consumption of a grape extract supplement containing resveratrol decreases oxidized low-density lipoprotein (oxLDL) and apolipoprotein B (ApoB) in patients undergoing primary prevention of cardiovascular disease: a triple-blind, 6-month follow-up, placebo-controlled, randomized trial," published in *Molecular Nutrition & Food Research*, was to investigate the effect of a grape supplement containing resveratrol on oxLDL, ApoB, and serum lipids on statin-treated patients in primary CVD prevention.

Patients in three parallel arms consumed one capsule (350 mg) daily for 6 months containing resveratrol-enriched grape extract (GE-RES, Stilvid®), grape extract (GE, similar polyphenolic content but no resveratrol), or placebo (maltodextrin). After 6 months, no changes were observed in the placebo group, and only LDL cholesterol (LDL-C) decreased in the GE group.

In contrast, LDL-C, ApoB , oxLDL, and oxLDL/ApoB decreased in the Stilvid group, whereas the ratio non-high-density lipoprotein cholesterol (HDL-C) (total atherogenic cholesterol load)/ApoB) increased. No adverse effects were observed in any of the patients: The treatment lowered atherogenic markers and is thought to exert additional cardioprotection beyond the "gold-standard" medication in such patients. Resveratrol was found to be necessary to achieve these effects.[22]

The Research Group on Quality, Safety, and Bioactivity of Plant Foods, CEBAS-CSIC, reported a triple-blind, randomized, parallel, dose—response, placebo-controlled, 1-year follow-up trial in the *American Journal of Cardiology* aimed at ascertaining the effects of a dietary resveratrol-rich grape supplement on the inflammatory and fibrinolytic status of patients at high risk of CVD and treated according to current guidelines for primary prevention of CVD. Patients were assigned to three groups, consumed placebo (maltodextrin), a resveratrol-rich grape supplement (resveratrol 8 mg), or a conventional grape supplement lacking resveratrol, for the first 6 months, and a double dose for the next 6 months.

In contrast to placebo and conventional grape supplement, the resveratrol-rich grape supplement significantly decreased high-sensitivity C-reactive protein (CRP), as well as other inflammatory markers, and increased anti-inflammatory interleukin-10 (IL-10). Adiponectin and soluble intercellular adhesion molecule-1 (sICAM-1) tended to increase and decrease, respectively. No adverse effects were observed in any patient.

In conclusion, 1-year consumption of a resveratrol-rich grape supplement improved the inflammatory and fibrinolytic status in patients who were on statins for primary prevention of, and at high risk of, CVD with diabetes or hypercholesterolemia, plus one or more cardiovascular risk factors. The results show for the first time that a dietary intervention with grape resveratrol could complement the "gold standard" therapy in the primary prevention of CVD.[22]

Additional references:

Resveratrol modulates vascular cell function, inhibits LDL oxidation, and suppresses platelet aggregation. Further studies of its benefits as a cardiovascular protective agent would help establish its bioavailability and *in vivo* cardioprotective effects in humans.[23]

Resveratrol may prevent lipid oxidation, platelet aggregation, promote arterial vasodilation and modulate the levels of lipids and lipoproteins. A potent anti-oxidant, it reduces oxidative stress and regenerates α-tocopherol, strengthening the anti-oxidant defense mechanism. It has been found to be safe, as no significant toxic effects have been identified, even when consumed at higher concentrations. It is an effective anti-atherogenic agent, which could be used in the prevention and treatment of CVD.[24]

9.2.13 Hypertension

Resveratrol improves flow-mediated dilation (FMD): *Nutrition, Metabolism & Cardiovascular Diseases* reported the outcome of a study of the effects of consuming resveratrol in red wine on FMD. Overweight/obese (BMI $25-35$ kg m(-2) men, or post-menopausal women with untreated borderline hypertension (systolic BP: $130-160$ mm Hg or diastolic BP: $85-100$ mm Hg), consumed three doses of resveratrol (resVida™ 30, 90, and 270 mg) and a placebo at weekly intervals. Plasma resveratrol and FMD were measured one hour after treatment.

There was a significant dose effect of resveratrol on plasma resveratrol concentration and on FMD, increased in comparison to placebo. FMD was found to be linearly related to log plasma resveratrol concentration.[10] Resveratrol treatment resulted in a dose-related increase in plasma resveratrol and in FMD.[25]

Previous cell culture-based studies have shown potential health benefit effects of dietary polyphenols, but such studies typically relied on higher concentrations than those commonly observed in blood following consumption of polyphenol-rich foods or beverages. This study examined the effects in cultured HUVECs. Resveratrol and quercetin increased eNOS in the absence of H_2O_2 and decreased H_2O_2-induced endothelin 1 (ET-1) mRNA expression. Resveratrol and quercetin decreased endothelin secretion into the media, blocking the stimulatory effect of H_2O_2.[26]

9.2.14 Resveratrol Lowers Blood Pressure

The *Journal of Pharmacy & Pharmacology* reported the antihypertensive effect of an alcohol-free hydro-alcoholic grape skin extract (GSE) obtained from skins of a vinifera grape (*Vitis labrusca*) in a rat model of hypertension. Oral administration of GSE significantly reduced systolic, mean, and diastolic arterial pressure in Wistar rats with desoxycorticosterone acetate-salt and N(G)-nitro-L-arginine methylester (L-NAME)-induced experimental hypertension. The antihypertensive effects were attributed to a combination of the vasodilator and the antioxidant actions of GSE.[27]

9.2.15 Resveratrol Improves Heart Function

Polyphenolic compounds from red grapes acutely improve endothelial function in patients with CHD as shown by a positive effect on brachial artery FMD. These results, reported in the *European Journal of Cardiovascular Prevention & Rehabilitation*, might well explain the favorable effects of red wine on the cardiovascular system.[28]

The aim of a study titled "Cardioprotection by resveratrol: A human clinical trial in patients with stable coronary artery disease" was designed to determine whether resveratrol had a clinically measurable cardioprotective effect in patients after myocardial infarction. In this clinical trial, post-infarction patients received 10 mg of resveratrol daily for 3 months. Systolic and diastolic left ventricular function, FMD, and several other laboratory and hemorheological parameters were measured before and after the treatment.

Left ventricular diastolic function improved significantly with resveratrol, so did endothelial function (FMD), and so did red blood cell deformability and platelet aggregation vs. no such findings in a placebo group. The authors concluded that resveratrol improved left ventricle diastolic function, improved endothelial function, lowered LDL-C level, and protected against unfavorable hemorheological changes measured in patients with coronary artery disease (CAD).[29]

A study in *Pharmacological Research* reported the associations between total urinary resveratrol metabolites (TRMs) as biomarkers of wine and resveratrol consumption, and cardiovascular risk factors in a large cross-sectional study including high cardiovascular-risk individuals in the PREDIMED Study in Spain. TRMs were analyzed by liquid chromatography—tandem mass spectrometry (LC-MS/MS) with a previous solid-phase extraction.

TRMs improved the mean HDL-C, triglyceride (TG), and plasma concentrations, and heart rate. TRMs adjusted for a biomarker of resveratrol intake decreased fasting blood glucose, TG concentrations, and heart rate. Both resveratrol and wine intake, evaluated as TRMs, were associated with beneficial changes in blood lipid profiles, fasting blood glucose (only resveratrol), and heart rate, suggesting that resveratrol intake via wine consumption might help to decrease cardiovascular risk factors.[30]

Additional reference:

Red wine extract, as well as resveratrol and proanthocyanidins found in grape seeds, are equally effective in reducing myocardial ischemic reperfusion injury, which suggests that these red wine polyphenolic antioxidants play a crucial role in cardioprotection.[31]

9.2.16 Diabetes—Resveratrol Improves Glycemic Control

This clinical study titled "Resveratrol supplementation improves glycemic control in type 2 diabetes mellitus" was carried out to test the hypothesis that oral supplementation of resveratrol would improve glycemic control and the associated risk factors in patients with type-2 diabetes mellitus (T2DM). Patients with T2DM were enrolled from Government Headquarters Hospital, Ootacamund, India, and assigned to intervention and control groups. The control group received only oral hypoglycemic agents, whereas the intervention group received resveratrol (250 mg/day) along with their oral hypoglycemic agents for a period of 3 months.

Hemoglobin A1c (HbA1c), lipid profile, urea nitrogen, creatinine, and protein were measured at the baseline and at the end of 3 months: supplementation of resveratrol for 3 months significantly improves the mean HbA1c, systolic blood pressure, total cholesterol, and total protein in T2DM patients. No significant changes in body weight and HDL-C and LDL-C were observed. No significant changes were noted in control participants.

The authors propose that oral supplementation of resveratrol may be effective in improving glycemic control and may possibly provide a potential adjuvant for the treatment and management of diabetes.[32]

The Journals of Gerontology Series A: Biological Sciences & Medical Sciences reported a pilot study to determine whether resveratrol could improve glucose metabolism and vascular function in older adults with impaired glucose tolerance (IGT). Participants aged 72 ± 3 years with IGT were enrolled in a 4-week open-label study of resveratrol (daily dose 1, 1.5, or 2 g). Following a standard mixed meal (110 g carbohydrate, 20 g protein, 20 g fat), the researchers measured 3-hour glucose and insulin area under the curve (AUC), insulin sensitivity (Matsuda index), and secretion (corrected insulin response at 30 minutes). Endothelial function was assessed by reactive hyperemia peripheral arterial tonometry (RH-PAT; reactive hyperemia index) before and 90 minutes post-meal. Results did not differ by dose, so data were combined for analysis.

It was found that after 4 weeks of resveratrol, fasting plasma glucose was unchanged, but peak post-meal and 3-hour glucose AUC declined. Matsuda index improved, and corrected insulin response at 30 minutes was unchanged. There was a trend toward improved post-meal reactive hyperemia index. Weight, blood pressure, and lipids were unchanged. It was concluded that at doses between 1 and 2 g/day, resveratrol improves insulin sensitivity and post-meal plasma glucose in subjects with IGT.[33]

The aim of a study published in the *British Journal of Nutrition* was to determine whether resveratrol could improve insulin sensitivity in T2DM patients. Patients were enrolled in a 4-week-long study and were given initial general examination including blood chemistry. The treatment group received oral 2×5 mg resveratrol, and a control group received placebo. Before and after the second and fourth weeks of the trial, insulin resistance/sensitivity, creatinine-normalized *ortho*-tyrosine level in urine samples (as a measure of oxidative stress), incretin levels, and phosphorylated protein kinase B (pAkt): protein kinase B (Akt) ratio in platelets were assessed. After the fourth week, resveratrol significantly decreased insulin resistance (homeostasis model of assessment for insulin resistance), and urinary *ortho*-tyrosine excretion, while it increased the pAkt:Akt ratio in platelets. It seemed to have no effect on β-cell function (i.e. homeostasis model of assessment of β-cell function).

The study demonstrated for the first time that resveratrol improves insulin sensitivity in humans, and that may be due to a resveratrol-induced decrease in oxidative stress that leads to a more efficient insulin signaling via the Akt pathway.[34]

The *American Journal of Physiology, Heart & Circulatory Physiology* published a study titled "Resveratrol improves left ventricular diastolic relaxation in type 2 diabetes by inhibiting oxidative/nitrative stress: *in vivo* demonstration with magnetic resonance imaging." This study conducted on an animal model of T2DM was designed to examine how resveratrol protects against diabetes-induced cardiac dysfunction. Normal control (m-Lepr(db)) mice and T2DM (Lepr(db)) mice were treated with resveratrol orally for 4 weeks.

In vivo magnetic resonance imaging (MRI) showed that resveratrol improved cardiac function by increasing the left ventricular diastolic peak filling rate in Lepr(db) mice. This protective role is partially explained by improvement in NO production and inhibition of oxidative/nitrative stress in cardiac tissue. Among the effects of resveratrol was increased NO production by enhancement of eNOS expression. Resveratrol seemingly protects against cardiac dysfunction by inhibiting oxidative/nitrative stress and improving NO availability.[35]

Additional reference:

Insulin and resveratrol can prevent cardiac dysfunction in diabetes (animal model).[36]

9.2.17 Resveratrol Promotes Erectile Function and Potentiates Phosphodiesterase Type-5 Inhibitors

Resveratrol supports erectile function: A study in *Archives of Pharmacological Research* reported that resveratrol triggered penile erection, as well as enhancing blood testosterone levels (in rats).[37]

Parenthetically, it may be small consolation to American men to learn that diabetic rats have the same ED as they themselves may have due also to damaged endothelium, and therefore impaired NO release. On the bright side, researchers found that resveratrol in combination therapy with vardenafil (Levitra®) improved erectile function where NO-formation is impaired: Levitra, for instance, may not improve erections in some men, but enhancing it with resveratrol may do so.[38] The phosphodiesterase type-5 (PDE5) inhibitors, including Viagra®, Cialis®, and Levitra do not invariably work for all men with ED. It is thought that they fail when the endothelium cannot deliver even a minimum required volume of NO per unit of time to cause and maintain erection. Supplementing a prescription PDE5 inhibitor with a compound that enhances endothelium function might be the answer.

The contribution of resveratrol to supporting erectile function was shown in connection with fortifying L-arginine supplement in the last chapter. It promotes healthy endothelial function in a number of cardiovascular disorders where the endothelium is otherwise in jeopardy. Parenthetically, the white rabbit is often the model of choice for the study of diet and atherosclerosis, as the response of the rabbit arterial vascular system, especially the aorta, bears a good similarity to the diet and atherogenesis response in people. Rats and mice come in second, head-to-head.

In a study published in the *International Journal of Impotence Research*, rabbits were fed a 2% w/w cholesterol and 2% w/w high cholesterol plus resveratrol (4 mg kg(−1) per day) diet for 6 weeks. Total cholesterol levels in the plasma were measured, as well as in the thoracic aorta (TA), mesenteric artery (MA), renal artery (RA), and corpus cavernosum (CC).

Vascular and endothelial functions in renal artery, thoracic aorta, mesenteric artery, and corpus cavernosum were assessed by isolated tissue bath with cumulative doses of acetylcholine (ACh) and sodium nitroprusside. There were no significant changes on plasma total cholesterol levels between cholesterol and (cholesterol + resveratrol)-treated groups. Vaso-relaxation responses to ACh in the resveratrol-treated group showed significant changes when compared with the hypercholesterolemic group.

The authors concluded that resveratrol might be an effective treatment in prevention of atherosclerotic changes in arteries and corpus cavernosum, and that the initial effects of hypercholesterolemia on ED and endothelial dysfunction may be prevented by resveratrol.[39]

9.2.18 Resveratrol Increases Testosterone

A two-part study published in *Archives of Pharmacological Research* reported that (a) the relaxation effects of resveratrol were measured on isolated New Zealand white rabbit corpus cavernosum, pre-contracted by phenylephrine (PE) (5 × 10(−5) M), and (b) measures of reproductive organ weight, blood testosterone levels, testicular histopathology, sperm counts, as well as the sperm motility and deformity in male ICR mice given an oral dose of resveratrol (50 mg/kg) for 28 days were obtained.

Resveratrol elicited a concentration-dependent relaxing effect on corpus cavernosum, leading to a median effective concentration of 0.29 mg/mL. Repeated treatment with resveratrol (50 mg/kg) did not cause an increase in body weight, reproductive organ weight, or testicular microscopic findings. However, resveratrol did elicit an increase in blood testosterone concentration, testicular sperm counts, and sperm motility by 51.6%, 15.8%, and 23.3%, respectively, without influence on sperm deformity.

The authors concluded that resveratrol had a positive effect on male reproductive function by triggering penile erection, as well as enhancing blood testosterone levels, testicular sperm counts, and sperm motility.[37]

A study published in the *Journal of Sexual Medicine* aimed to determine why some patients with ED do not respond to treatment with PDE5 inhibitors (vardenafil/Levitra, in this case). The authors contended that typically low NO availability may reduce the efficacy of PDE5 inhibitors and that perhaps resveratrol, reported to activate eNOS through activation of $SIRT_1$, may produce improvement.

Rats were chosen as their corpus cavernosal smooth muscle cells (CCSMCs) also form eNOS and synthesize cGMP. Diabetic rats in different groups were treated with resveratrol and/or vardenafil or both in combination for the last 4 weeks of an 8-week period of diabetes induction. Conventional controls were in place. Intracellular cGMP levels rose with resveratrol treatment in CCSMCs.

Diabetic rats showed impairment of erectile function. The combination treatment of resveratrol and vardenafil had a synergistic effect: treatment with either resveratrol or vardenafil improved intracavernous pressure/mean arterial pressure (ICP/MAP) ratio, and combination therapy with resveratrol and vardenafil had a synergistic effect in improvement of ICP/MAP. Treatment with either resveratrol or vardenafil elevated cGMP levels in CCSMCs and improved erectile function in the diabetic rats. Resveratrol or combination therapy of resveratrol and vardenafil was found to improve erectile function where NO release is known to be impaired.[38]

The point should not be lost here that the idea of inducing diabetes in an animal model in order to study ED is predicated on the knowledge, *a priori*, that diabetes induces ED. The point should also not be lost here that while PDE5 inhibitors can prolong duration of formation of ACh/NO/cGMP in marginally functional endothelium, it is nevertheless an antioxidant that tips the balance. The last point that should also not be lost here is that while we're talking "erectile dysfunction," this is all about impaired endothelial and cardiovascular health, albeit due in this case to diabetes.

9.2.19 Adverse Effects

Resveratrol in diet is not known to be toxic or to have adverse effects in humans, but there have been only a few controlled clinical trials to date.

A recent trial evaluating the safety of oral resveratrol in ten subjects found a single dose up to 2 g/day resulted in no serious adverse effects.[40,41] As noted earlier, it is not without health hazard in very large doses.

According to a note in the journal *Drug Metabolism Reviews*: Resveratrol is well tolerated in healthy people without any co-medication. However, supplemental doses of resveratrol in the range of 1 g/day or more by far exceed the natural intake through food. Whether resveratrol-drug interactions can be harmful in patients taking additional medications remains unknown. Recent *in vivo* studies and clinical trials indicate a possible drug—drug interaction potential using high-dosage formulations.

In this review, the known *in vitro* and *in vivo* effects of resveratrol on various cytochrome P450 (CYP) isoenzymes are summarized. Resveratrol may lead to interactions with various CYPs especially when taken in high doses. Aside from systemic CYP inhibition, intestinal interactions must also be considered. They can potentially lead to reduced first-pass metabolism, resulting in higher systemic exposure to certain co-administered CYP substrates (including lipids and steroidal hormones; see the US Food and Drug Administration (USFDA) website for more information).[42] Therefore, patients who ingest high doses of this food supplement combined with additional medications may be at risk of experiencing clinically relevant drug—drug interactions.[43]

9.3 LYCOPENE

9.3.1 Carotinoids

Carotenoids are pigments found mostly in plants and some algae, bacteria, and fungi. They are not made by animal species, so we get them from plants. Probably the one best known for its bright orange color is carotene, found in carrots and apricots. It has been said that consuming a diet rich in carotenoids from natural foods, such as fruits and vegetables, is a boost to health.

Lycopene is a naturally occurring carotinoid found in many fruits and vegetables, but in greatest concentration in tomatoes. In the US, 85% of dietary lycopene comes from tomato products such as tomato juice or paste. One cup (240 mL) of tomato juice provides about 23 mg of lycopene. Cooking raw tomatoes actually converts the lycopene into a form that is more bioavailable for the body. By comparison, the concentration of lycopene in three forms of tomato (raw, paste, and ketchup) exceeds the concentration in many other natural sources, thus making the case that dousing the hamburger and fries with ketchup is likely a good idea.

With the notable exception of crude palm oil and Gac fruit (*Momordica cochinchinensis*) found in South East Asia, but mostly in Vietnam, most carotenoid-rich fruits and vegetables are low in lipids, thus also lowering their bioavailability. The addition of oil significantly enhances the absorption of all carotenoids (α-carotene, β-carotene, lycopene, and lutein). We are mostly incapable of synthesizing carotenoids, and must obtain them through our diet.

9.3.2 Lycopene Promotes Endothelium Health

A study published in the journal *Nutrition Research* reported the effects of 70 g/day of tomato paste containing 33.3 mg of lycopene as measured by FMD of the brachial artery by ultrasonography. The regimen was intended to estimate endothelial function at day 1 (acute response) and day 15 (midterm response). Plasma lipid peroxides were measured with a photometric enzyme-linked immunosorbent assay (ELISA) as an index of total oxidative status. Tomato supplementation led to an overall increased FMD-mediated arterial blood vessel dilation by day 15. Daily tomato paste consumption exerts a beneficial effect on endothelial function.[44] Until now, there has been no well-established definition of "lycopene deficiency," and there was no direct evidence that repletion of low lycopene levels has any benefit. That may be changing.

9.3.3 Cook the Food or Use a Synthetic Form

The aim of a study reported in the *Journal of the Association of Physicians of India* was to assess the efficacy of various forms of lycopene in combating oxidative stress. It was concluded that lycopene supplementation in natural and synthetic form is protective against oxidative stress as shown by decrease in concentration of oxidative stress biomarkers. However, the synthetic form is more effective as it needs no heat transformation to be directly absorbed by cells.[45]

Adding cooked tomatoes in a meal is good for the endothelium, and cooking them with olive oil is even better: Plasma lycopene concentrations in healthy subjects who consumed one meal per day of tomatoes (470 g), cooked with or without extra virgin olive oil (25 ml) for five days, were compared to those on a low lycopene diet. There was an 82% increase in plasma *trans*-lycopene and a 40% increase in *cis*-lycopene in those who consumed tomatoes cooked in olive oil. The addition of olive oil to diced tomatoes during cooking greatly increases the absorption of lycopene. The results highlight the importance of how a food is prepared and consumed in determining the bioavailability of dietary carotenoids such as lycopene.[46]

Curiously, a study published in the *American Journal of Clinical Nutrition* found that consuming a resveratrol regimen equivalent to 32–50 mg lycopene/day, or lycopene supplements (10 mg/day) for four weeks, had no effect on carotenoid and lipid profiles, inflammatory markers, blood pressure, and arterial stiffness: The aim of this study was to determine whether consuming tomato-based foods by healthy middle-aged volunteers affects known biomarkers of CVD risk.

After a 4-week run-in period with a low-tomato diet, volunteer men and women 40−65 years old were randomly assigned to 1 of 3 dietary intervention groups and asked to consume a control diet low in tomato-based foods, a high tomato-based diet, or a control diet supplemented with lycopene capsules (10 mg/day) for 12 wk. Blood samples were collected at baseline, at 6 weeks, and after the intervention and were analyzed for carotenoid and lipid profiles and inflammatory markers. Blood pressure, weight, and arterial stiffness were also measured. Dietary intake was also noted during the intervention

Their data showed that a relatively high daily consumption of tomato-based products is ineffective at reducing conventional CVD risk markers in moderately overweight, healthy, middle-aged individuals.[47] This trial was registered as ISRCTN34203810.[48]

9.3.4 Lycopene Attenuates Oxidative Stress

Every meal consumed results in the normal metabolic formation of free radicals. Including cooked tomatoes or tomato products in a meal reduces postprandial oxidative stress: The journal *Molecular Nutrition & Food Research* reported a study where healthy weight men and women consumed high-fat meals known to induce postprandial oxidative stress on two separate occasions containing either a processed tomato product or non-tomato alternative. Blood samples were collected for changes in glucose, insulin, lipids, oxLDL, inflammatory cytokines, and measures of FMD. Both meals induced increases in plasma glucose, insulin, and lipid concentrations, and a trend for higher TGs was observed after the tomato meal. Tomato significantly reduced high-fat meal-induced LDL oxidation and rise in interleukin-6.[49]

9.3.5 Reactive Hyperemia Peripheral Arterial Tonometry— The Finger-Tip Hyperemia Test for Endothelial Function

There are two forms of hyperemia, *active* or *functional* hyperemia, and *reactive* or *passive* hyperemia. Active hyperemia is dictated by metabolic demand, causing increased blood flow to an active body organ. Reactive hyperemia occurs when blood circulation to a region of the body is blocked.

The *Journal of the American College of Cardiology* published an article describing digital pulse volume changes during reactive hyperemia in patients without obstructive CAD and either normal or abnormal coronary microvascular endothelial function. The investigators called the test the RH-PAT index. It is a measure of reactive hyperemia calculated as the ratio of the digital pulse volume during reactive hyperemia divided by that pulse volume at baseline. An RH-PAT index less than 1.35 was found to have a sensitivity of 80% and a specificity of 85% to identify patients with coronary endothelial dysfunction.

In their study, they report that the average RH-PAT index was lower in patients with coronary endothelial dysfunction compared with those with normal coronary endothelial function.[50]

The objective of a study appearing in the *Journal of Diabetes and its Complications*, was to assess acute postprandial and post long-term weight loss effects of a low-carbohydrate diet vs. a low-fat diet on subclinical markers of CVD in adults with T2DM. Results are illustrated in Figure 9.1.

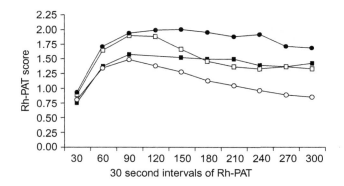

FIGURE 9.1 Fasting and postprandial changes in RH-PAT. Open circle represents low-carbohydrate fasting RH-PAT scores at 2 weeks, open squares represent low-fat fasting RH-PAT scores, black circles are low-carbohydrate post-meal RH-PAT scores, and black squares are low-fat post-meal RH-PAT scores. All data are from the 2-week meal test. Source: *Davis, N. J., Crandall, J. P., Gajavellic, S., Berman, J. W., Tomutae, N., Wylie-Rosett, J., et al. (2011). Differential effects of low-carbohydrate and low-fat diets on inflammation and endothelial function in diabetes.* J Diabetes Complicat, *25, 371−376. Reproduced with kind permission from the* Journal of Diabetes Complications.

Baseline and 6 month measures of CRP, interleukin-6 (IL-6), soluble sICAM, and soluble E-selectin, were drawn from archived samples of participants randomized in a clinical trial comparing a low-carbohydrate and a low-fat diet. In a subset of participants, postprandial measures of these markers were obtained 3 hours after a low-carbohydrate or low-fat liquid meal. Endothelial function was also measured by RH-PAT during the meal test.

After 6 months, CRP declined in the low-fat group; sICAM declined in the low-carbohydrate group; and soluble E-selectin decreased. All differences were statistically significant. A significant negative correlation between change in HDL and change in soluble E-selectin was obtained and also with the change in ICAM.

Low-carbohydrate and low-fat diets both have beneficial effects on CVD markers. There may be different mechanisms through which weight loss with these diets potentially reduces CVD risk.[51]

9.3.6 Supplementing Lycopene Can Reduce Oxidative Stress

Serum lycopene rises as a dose-dependent function after 8 weeks of supplementation. A 15 mg/day group showed a greater increase in plasma SOD activity than a placebo group in a finger-tip hyperemia test for endothelial function (RH-PAT index). In the 15 mg/day group after 8 weeks, high-sensitivity CRP (hs-CRP), systolic blood pressure, sICAM-1 and sVCAM-1 (tests for vascular adhesion factors) significantly decreased, and β-carotene and LDL-particle size significantly increased.

The authors also noted that the beneficial effect of lycopene supplementation on endothelial function (i.e. RH-PAT and sVCAM-1) were "remarkable" in participants with relatively impaired endothelial function at baseline. Changes in RH-PAT index correlated with SOD activity especially in the 15 mg lycopene/day group and hs-CRP. In addition, changes in lycopene correlated with hs-CRP and SOD activity. The authors concluded that increased serum lycopene after supplementation can reduce oxidative stress.[52]

Lycopene taken in doses equal to or greater than 25 mg daily can reduce LDL-C by about 10% (comparable to the effect of low doses of statins) in patients with slightly elevated cholesterol levels: A review published in *Maturitas* summarizes current evidence on the effect of lycopene on serum lipid concentrations and blood pressure. The authors conducted meta-analyses on studies in the PubMed and Cochrane databases for intervention studies investigating the effect of lycopene on blood lipids or blood pressure for a minimum duration of 2 weeks. The period covered was 1955 to September 2010.

Lycopene was shown to have a significant cholesterol-lowering effect for total serum cholesterol and LDL-C in the subgroup of trials using lycopene dosages greater than 25 mg daily. Subgroup meta-analysis of trials using lower lycopene dosages showed no significant effect. The effect of lycopene on systolic blood pressure of all trials suggested a significant blood pressure-reducing effect.[53]

This study, appearing in *Atherosclerosis*, aimed to investigate the relationship between the antioxidant effects of resveratrol and its ability to promote cholesterol efflux in incubated macrophages. Resveratrol was found to significantly promote cholesterol efflux in a dose-dependent manner. The authors concluded that resveratrol appears to be a natural antioxidant that enhances cholesterol efflux, and propose that it has potential as a natural antioxidant that could be used to prevent and treat CVD.[54]

9.3.7 Lycopene Helps Avert Atherosclerosis

Insufficient serum lycopene levels may be a marker for atherosclerosis, whereas high levels are associated with reduced prevalence of CVD: The *Journal of Biological Regulators & Homeostatic Agents* reported a comparison of plasma concentrations of lycopene in subjects who had neither symptoms nor ultrasonic evidence of carotid artery atherosclerosis to participants with ultrasonic evidence of carotid atherosclerosis. The latter exhibited significantly higher serum concentrations of total cholesterol, LDL-C, and TGs, and significantly lower plasma concentrations of lycopene. Higher serum levels of lycopene may protect against carotid atherosclerosis.[55]

Lycopene reduces foam cell formation, thus impeding the time-course of atherosclerosis: A study published in the *Journal of Nutritional Biochemistry* reports that lycopene may exert its anti-atherogenic role through changes in cholesterol metabolism that attenuate foam cell formation. Invasion of arterial cell wall by cholesterol-laden foam cells is thought to be the first step in endothelium impairment and atherosclerosis.[56] oxLDL can cause inflammation in arterial blood vessels, causing macrophages to gravitate to the site of inflammation where they engulf LDL-C, giving them a "foamy" appearance.

The *American Journal of Physiology, Endocrinology & Metabolism* reports a study where human monocyte-derived macrophages (HMDM) were incubated with lycopene in the presence or absence of native LDL (nLDL) or LDL modified by oxidation (oxLDL), aggregation (aggLDL), or acetylation (acLDL). The cholesterol content, lipid synthesis, scavenger receptor activity, and the secretion of inflammatory (IL-1β and tumor necrosis factor (TNF)-α) and anti-inflammatory (IL-10) cytokines were determined.

It was found that lycopene may reduce macrophage foam cell formation induced by modified LDL by decreasing lipid synthesis and downregulating the activity and expression of scavenger receptor A (SR-A). However, these effects are accompanied by impaired secretion of the anti-inflammatory cytokine IL-10, suggesting that lycopene can also exert a concomitant pro-inflammatory effect.[57]

9.3.8 Lycopene Protects the Heart

Epidemiological studies show that high lycopene levels lower CVD risk: The journal *Experimental Biology and Medicine* (Maywood) reported a two-part study conducted in Finland: Study 1 examined the role of lycopene as a risk-lowering factor with regard to acute coronary events and stroke. The participants were middle-aged men free of CHD and stroke at the study baseline enrolled in the prospective Kuopio Ischemic Heart Disease Risk Factor (KIHD) Study. Men in the lowest quartile of serum levels of lycopene had a significant (3.3-fold) risk of the acute coronary event or stroke as compared with others.

Study 2 assessed the association between plasma lycopene and intima-media thickness of the common carotid artery wall (CCA-IMT) in an analysis of the Antioxidant Supplementation in the Atherosclerosis Prevention (ASAP) study in asymptomatic men and women. Adjusting for common cardiovascular risk factors, low plasma levels of lycopene were associated with an 18% increase of IMT in men, compared to men with higher than median plasma.[58]

9.3.9 Lycopene May Not Lower Lipid Levels but It Reduces Lipid Peroxidation

It was reported in a previous chapter in connection with theories of atherosclerosis that LDL-C may not, *per se*, present much of a hazard to endothelial, cardiovascular, and erectile function, but that it stops being relatively benign when it oxidizes (oxLDL). Evidence that beneficial effects of lycopene emerge without reduction in LDL-C but with reduction in oxLDL shift the emphasis from avoiding cholesterol to reducing oxidative stress: Researchers from the Department of Biochemistry, People's College of Medical Sciences and Research Centre, Bhanpur, Bhopal, India, reported the results of a study aimed to evaluate the beneficial effect of lycopene in tomatoes. In a CHD group, and an age-matched control group, they examined the lipid peroxidation rate by assessing levels of malondialdehyde (MDA), and serum enzymes involved in antioxidant activities such as SOD, GPx, glutathione reductase, reduced glutathione and lipid profile, which includes total cholesterol, TGs, HDL, LDL, and very low density lipoprotein (VLDL), after 60 days of tomato supplementation.

There were significantly lower levels of serum antioxidant enzymes and very high lipid peroxidation rates in the CHD group, as compared to the controls, and significantly higher levels of lipids in the CHD group, when compared to the controls. After 60 days of tomato supplementation, the CHD group showed a significant improvement in the levels of serum enzymes involved in antioxidant activities and decreased lipid peroxidation rate, but there were no significant changes in lipid profile.[59]

9.3.10 Diabetes

Spanish investigators reported a meta-analysis of clinical studies of the role of lycopene in T2DM in the medical journal *Nutricion Hospitalaria*. A search in Medline and the Cochrane Library, using the MeSH terms "carotenoids" OR "lycopene" AND "type 2 diabetes mellitus," assessed for quality using the JADAD and STROBE scales, yielded ten articles.

Analysis showed that after adjusting for other risk factors, the "carotenoids" OR "lycopene" ("OR" is a meta-analysis search term, as in OR or AND) for developing T2DM was similar among the different levels of lycopene intake. Plasma levels of lycopene increase in the intervention groups. Lycopene decreases the MDA and lipid peroxidation. Lycopene may have a beneficial effect on oxidative stress in diabetic patients.[60]

9.3.11 Sexual Health and Benign Prostate Hyperplasia

Not only does resveratrol restore erectile function in streptozotocin (STZ) diabetic rats, but so does lycopene: The present study published in

Pharmazie aimed to determine whether lycopene could lower oxidative stress and attenuate ED in diabetic rats. Lycopene (10, 30, 60 mg/kg/day) was administered via intragastric intubation for 8 weeks to STZ-induced (50 mg/kg, i.v.) diabetic rats. Lycopene treatment resulted in significant dose-related restoration of erectile function by lowering blood glucose, reducing oxidative stress, and upregulating eNOS expression.[61]

In a study appearing in the *Journal of Nutrition*, investigators examined the effects of lycopene on normal human prostate epithelial cells (PrEC) by treating them with synthetic all-E-lycopene (up to 5 μmol/L) and assessing proliferation via [3H]thymidine incorporation. The effects of lycopene on cell cycle progression were investigated by flow cytometry. Protein expressions of cyclins D1 and E were analyzed to determine whether lycopene affects cell cycle progression. It was found that lycopene inhibits growth of non-neoplastic PrEC *in vitro*, and it was hypothesized that lycopene might likewise inhibit the growth of PrEC *in vivo*.[62]

The group that reported the results in the previously cited study of the effects of lycopene on PrEC in culture conducted the following pilot study to investigate the effects of lycopene supplementation in elderly men with histologically proven benign prostate hyperplasia (BPH) free of prostate cancer. Each one of two groups took either lycopene at a dose of 15 mg/day, or placebo for 6 months. The effects of the intervention on carotenoid status, clinical diagnostic markers of prostate proliferation, and symptoms of the disease, were assessed. The goal of the trial was inhibition, or even reduction, of increased serum prostate-specific antigen (PSA) levels.

After six months of lycopene supplementation, PSA levels in men decreased significantly, whereas there was no change in the placebo group. Plasma lycopene concentration increased in the group taking lycopene, however, other plasma carotenoids did not change. On the other hand, prostate enlargement progressed in the placebo group as assessed by trans-rectal ultrasonography and digital rectal examination. The prostate did not enlarge in the lycopene group. Symptoms of the disease, as assessed by the International Prostate Symptom Score questionnaire, improved in both groups with a significantly greater effect in men taking lycopene supplements.[63]

9.4 PYCNOGENOL

In 1534, French explorer Jacques Cartier led a ship into what we know now as the "mouth of the St. Lawrence River" in Canada. It became stranded, and all of Cartier's crew became ill with scurvy, a very common plague to seafarers then. They were dying at an alarming rate when fortuitously, Quebec Indians made them a brew from red pine bark. It saved the lives of all but those at death's door.

Meticulously recorded in Cartier's journal, this miraculous medicine was duly entered in the archives of the French Royal Library upon his return home, and was thus basically lost to history until the late 1940s. Then, a medical scientist, Dr. Jacques Masquelier, identified the palliative constituent, an oligomeric proanthocyanidin (OPC) now commercially known as Pycnogenol, the US registered trademark name for a patented product derived from the pine bark of the tree *Pinus pinaster*. The patented extract contains 65–75% proanthocyanidins (procyanidins): 100 mg contains 65 to 75 mg of proanthocyanidins (procyanidins).[64]

A powerful antioxidant, Pycnogenol holds 65 to 75% proanthocyanidins. Thus, a 100 mg serving would contain 65 to 75 mg of proanthocyanidins. The active ingredients in Pycnogenol can also be extracted from other sources, including peanut skin, grape seed oil, and witch hazel bark. Pycnogenol consists of a concentrate of water-soluble polyphenols, including bioflavonoids catechin, and taxifolin, as well as phenolcarbonic acids. Antioxidants such as bioflavonoids enhance eNOS expression and subsequent NO release from endothelial cells.

Pycnogenol is reported to have vaso-relaxant cardiovascular benefits, angiotensin-converting enzyme (ACE)-inhibiting activity, and it can enhance the microcirculation by increasing capillary permeability. These properties are attributed to its antioxidant activity.[65]

Likewise, a review in the *International Journal of Clinical Pharmacology & Therapeutics* informs us that Pycnogenol is primarily composed of procyanidins and phenolic acids. Procyanidins are biopolymers of catechin and epicatechin subunits, which are recognized as important constituents in human nutrition. Pycnogenol contains a wide variety of procyanidins. Clinical studies indicate that Pycnogenol protects against oxidative stress in several cell systems by doubling the intracellular synthesis of anti-oxidative enzymes and by acting as a potent scavenger of free radicals. Other anti-oxidant effects involve a role in the regeneration and protection of vitamin C and E. Anti-inflammatory activity has been demonstrated *in vitro* and *in vivo* in animals.

Dilation of the small blood vessels has been observed in patients with CVD. The ability to inhibit ACE is associated with a mild antihypertensive effect. Pycnogenol has low acute and chronic toxicity with mild unwanted effects occurring in a small percentage of patients following oral administration.[64]

9.4.1 Pycnogenol Promotes Endothelial Function and Lowers Blood Pressure

In this study published in the *Journal of Cardiovascular Pharmacology & Therapeutics*, the investigators tested the possibility that Pycnogenol could stimulate NO/cGMP. In the *in vitro* experiments, Pycnogenol (1–10 μg/mL) relaxed epinephrine (E)-, norepinephrine (NE)-, and PE-contracted intact rat aortic ring preparations in a concentration-dependent manner.

When the endothelial lining of the aortic ring was obliterated, Pycnogenol had no effect, indicating a NO/cGMP pathway relaxation effect. This endothelium-dependent response was clearly caused by enhanced NO levels, because the NOS inhibitor N-methyl-L-arginine (NMA) reversed or prevented the relaxation. This response, in turn, was reversed by addition of L-arginine.

Pycnogenol-induced relaxation persisted after exposure of intact rings to high levels of SOD, suggesting that the mechanism of EDR did not involve scavenging of superoxide anion. In addition to causing relaxation, preincubation of aortic rings with Pycnogenol (1−10 μg/mL) inhibited subsequent endothelium independent E- and NE-induced contractions in a concentration-dependent manner. Fractionation of Pycnogenol by Sephadex LH-20 chromatography resulted in three fractions, one of which, fraction 3, oligomeric procyanidins, exhibited potent NO/cGMP activity.

These results indicate that Pycnogenol, in addition to its antioxidant activity, stimulates eNOS activity to increase NO levels, which could counteract the vasoconstrictor effects of NE and PE.[66]

A study published in the journal *Hypertension Research* reported the effects of Pycnogenol on endothelium-dependent (NO/cGMP) vasodilation in humans: forearm blood flow (FBF) responses to ACh and to sodium nitroprusside (SNP—cyanide, an NO-donor), an endothelium-independent vasodilator, was evaluated in healthy young men before and after two weeks of daily oral administration of Pycnogenol (180 mg/day) or placebo. The findings suggest that Pycnogenol augments endothelium-dependent vasodilation by increasing NO production.[67]

Hypertension is a known hazard to sexual function, and most who suffer from it are treated with prescription antihypertensive medications intended to relax chronically constricted blood vessels. Some of these meds unfortunately have ED as a listed potential adverse side effect. A clinical trial published in *Nutrition Research* aimed to determine the possible protective effect of oral Pycnogenol administrated for eight weeks to non-smoking, mildly hypertensive patients. Men and women with systolic blood pressure between 140 and 159 mm Hg and/or diastolic blood pressure of 90 and 99 mm Hg were given 200 mg/day of Pycnogenol, or placebo. Blood pressure was taken during supplementation, and the serum level of thromboxane was measured.

A significant decrease in the systolic blood pressure was observed during Pycnogenol supplementation, and serum thromboxane concentration was significantly decreased during treatment. It was concluded that supplementing Pycnogenol can effectively decrease systolic blood pressure in mildly hypertensive patients.[68]

Pycnogenol lowers medication requirements and lowers blood pressure in hypertensives with T2DM: The journal *Nutritional Research* reported a study on the clinical effectiveness of 125 mg/day of Pycnogenol supplement in reducing antihypertensive medication use and CVD risk factors in patients with T2DM and mild-to-moderate hypertension. Pycnogenol treatment achieved blood pressure control in 58.3% of patients at the end of the 12 weeks, with 50% reduction in individual pretrial dose of ACE-inhibitors.

Plasma endothelin-1 decreased by 3.9 pg/mL. Mean HbA1c dropped significantly, and fasting plasma glucose declined by 23.7 mg/dL. LDL-C declined by 12.7 mg/dL, and urinary albumin level decreased by week 8. After 12 weeks of supplementation, Pycnogenol resulted in improved diabetes control, lowered CVD risk factors, and reduced antihypertensive medicine use as compared to controls.[69] Similar anti-diabetic and antihypertensive effects of Pycnogenol supplementation were reported in other publications.[70]

9.4.2 Pycnogenol and Heart Function

Patients with CAD received 200 mg/day Pycnogenol for eight weeks followed by placebo or vice versa, on top of standard cardiovascular therapy. There was a 2-week washout period between the two treatment periods. At baseline and after each treatment period, endothelial function, non-invasively assessed by brachial artery FMD using high-resolution ultrasound, biomarkers of oxidative stress and inflammation, platelet adhesion, and 24-hour blood pressure monitoring, were evaluated.

Pycnogenol treatment was associated with improvement in brachial artery FMD. 15-F(2t)-Isoprostane, an index of oxidative stress, significantly decreased after Pycnogenol treatment, though inflammation markers, platelet adhesion, and blood pressure did not change after treatment or placebo. The authors report that, "This study provides the first evidence that the antioxidant Pycnogenol improves endothelial function in patients with CAD by reducing oxidative stress."[71]

9.4.3 Pycnogenol and Coenzyme Q10 in Treatment of Heart Failure

The journal of internal medicine, *Panminerva Medica*, reported on the benefits of a Pycnogenol & coenzyme Q10 combination (PycnoQ10) as an adjunct to medical treatment in stable heart failure patients. The aim of this 12-week observational study was to provide functional parameters such as exercise capacity, ejection fraction, and distal edema. Patients with a stable level of heart failure within the past three months and stable NYHA class II or III (6 months) were included in the study. Heart failure management was in accordance with American Heart Association (AHA) guidelines for "best treatment."

In the treatment group, systolic and diastolic pressure, as well as heart rate and respiratory rate, were significantly lowered with PycnoQ10. Heart ejection fraction increased by 22.4%. Walking distance on a treadmill increased 3.3-fold. Distal edema decreased significantly in PycnoQ10-treated patients.[72]

9.4.4 Pycnogenol is Antioxidant

The aim of this study was to test the effect of Pycnogenol supplementation on measures of oxidative stress and the lipid profile in humans. Healthy participants received Pycnogenol (150 mg/day) for 6 weeks. Fasting blood was collected at baseline, and after 3 and 6 weeks of supplementation, and again after a 4-week washout period.

After 6 weeks of supplementation with Pycnogenol, a significant increase in plasma polyphenol levels was detectable. It was reversed after the 4-week washout phase. The antioxidant effect of Pycnogenol was demonstrated by a significant increase in ORAC in plasma throughout the supplementation period. The ORAC value returned to baseline after the 4-week washout period. Pycnogenol also significantly reduced LDL-C levels and increased HDL-C levels in plasma in two-thirds of the participants.

Whereas LDL changes reversed during washout, HDL increase did not. It was concluded that Pycnogenol significantly increases antioxidant capacity of plasma, as determined by ORAC, and exerts favorable effects on the lipid profile.[73]

A study from the journal *Nutrition* aimed to determine whether commonly used antioxidants alter MDA modification of proteins, a known mechanism of free radical-related tissue injury. This study examined the effects of adding 1 mg/mL of Pycnogenol, 5 mM of α-tocopherol, 5 mM of ascorbate, and 0.2 mg/mL of an ethanol equivalent of red and white wine to MDA-protein content of endothelial cells in culture. The addition of Pycnogenol, but not of the other antioxidants, was associated with significant reduction in MDA-protein content compared with controls, indicating an additional mechanism by which Pycnogenol protects against oxidative stress.[74]

Phytotherapy Research published a study aimed to evaluate the effects of 6-month supplementation with Pycnogenol on health risk factors in subjects with metabolic syndrome. Pycnogenol was used with the aim of improving risk factors associated also with central obesity, elevated TGs, low HDL-C, high blood pressure, and fasting blood glucose. Participants with all five risk factors of metabolic syndrome, ranged in age between 45 and 55 years, received Pycnogenol for 6 months.

In the 6-month study period, in comparison to control participants, Pycnogenol supplementation with 150 mg/day decreased waist circumference, total cholesterol levels, blood pressure, and increased the HDL-C levels. Pycnogenol lowered fasting glucose from baseline 123 $\pm$ 8.6 mg/dL to 106.4 $\pm$ 5.3 after 3 months, and to 105.3 $\pm$ 2.5 at the end of the study. Men's waist circumference decreased after 3 months, and again after

6 months. Plasma free radicals decrease in the Pycnogenol group was greater than in the control group.[75]

9.4.5 Pycnogenol and Type-2 Diabetes

A report in the journal *Diabetes Research & Clinical Practice* informs us that fractions of Pycnogenol-containing compounds of diverse molecular masses from polyphenolic monomers, dimers, and higher oligomers exhibited a pronounced inhibitory activity on α-glucosidase: Pycnogenol exhibited the most potent inhibition (IC(50) about 5 μg/mL) on α-glucosidase compared to green tea extract (IC(50) about 20 μg/mL) and acarbose (IC(50) about 1 mg/mL).

The inhibitory action of Pycnogenol was stronger in extract fractions containing higher procyanidin oligomers. Oligomeric procyanidins contribute to its glucose-lowering effects observed in clinical trials with diabetic patients.[76]

9.4.6 Pycnogenol and Erectile Function

Japanese patients with mild-to-moderate ED were instructed to take a supplement (Pycnogenol/60 mg/day, L-arginine/690 mg/day, and aspartic acid/552 mg/day) or placebo for 8 weeks. Results were assessed with the IIEF-5 plus blood biochemistry, urinalysis, and salivary testosterone.

Eight weeks of supplement intake improved the total IIEF-5 score, with marked improvement in "hardness of erection" and "satisfaction with sexual intercourse." Decreased blood pressure was noted along with improved liver function markers (aspartate transaminase and γ-glutamyl transpeptidase), and also a slight increase in salivary testosterone. No adverse reactions were observed during the study period.[77]

A clinical trial of the efficacy of Prelox® and L-arginine aspartate (Prelox is a proprietary blend of L-arginine and Pycnogenol; Horphag Research UK Ltd, London, UK) on ED in men 30 to 50 years old was reported in the *British Journal of Urology International*. The formulation, Prelox, was administered over a period of 6 months. The IIEF was used to quantify changes in sexual function.

The erectile domain of the IIEF (questions 1−5 plus 15) improved significantly from baseline after 3 months, and even more after 6 months of treatment. Some increase was noted in the placebo group. The treatment effect was significantly different from placebo. Total plasma testosterone levels increased significantly after 6 months of treatment compared to the placebo group.[78]

Similar findings were reported in the following sources: Oral administration of L-arginine in combination with Pycnogenol causes a significant improvement in sexual function in men with ED without any side effects.[79]

Pycnogenol for 1 month restored erectile function to normal. Intercourse frequency doubled. eNOS levels rose significantly in spermatozoa and serum testosterone. Cholesterol levels and blood pressure declined. No undesirable effects were reported.[80]

9.4.7 Caveat

Pycnogenol is reportedly generally safe when taken in doses of 50 mg to 450 mg daily for up to 6 months. However, it has been reported to cause dizziness, gut problems, headache, and mouth ulcers. It might cause the immune system to become more active, possibly increasing symptoms of autoimmune diseases such as multiple sclerosis (MS), lupus (systemic lupus erythematosus, SLE), and rheumatoid arthritis (RA).

REFERENCES

1. Guerrero RF, García-Parrilla MC, Puertas B, Cantos-Villar E. Wine, resveratrol and health: a review. *Nat Prod Commun* 2009;**4**:635−58.
2. Baur JA, Sinclair DA. Therapeutic potential of resveratrol: the in vivo evidence. *Nat Rev Drug Discov* June 2006;**5**:493−506.
3. Gu X, Creasy L, Kester A, Zeece M. Capillary electrophoretic determination of resveratrol in wines. *J. Agric. Food Chem* 1999;**47**:3223−7.
4. Castro IA, Rogero MM, Junqueira RM, Carrapeiro MM. Free radical scavenger and antioxidant capacity correlation of α-tocopherol and Trolox measured by three in vitro methodologies. *Int J Food Sci Nutr* 2006;**57**:75−82.
5. Huang D, Ou B, Prior RL. The chemistry behind antioxidant capacity assays. *J Agric Food Chem* 2005;**53**:1841−56.
6. Pellegrini N, Serafini M, Colombi B, Del Rio D, Salvatore S, Bianchi M, et al. Total antioxidant capacity of plant foods, beverages and oils consumed in Italy assessed by three different in vitro assays. *J. Nutr.* 2003;**133**:2812−9.
7. <http://www.nhlbi.nih.gov/health//dci/Diseases/cscan/cscan_whatdoes.html>; 2013 [accessed 17.11.13].
8. Valabhji J, McColl AJ, Richmond W, Schachter M, Rubens MB, Elkeles RS. Total antioxidant status and coronary artery calcification in type-1 diabetes. *Diabetes Care* 2001;**24**:1608−13.
9. Elíes J, Cuíñas A, García-Morales V, Orallo F, Campos-Toimil M. Trans-resveratrol simultaneously increases cytoplasmic Ca(2 +) levels and nitric oxide release in human endothelial cells. *Mol Nutr Food Res* 2011;**55**:1237−48.
10. Schmitt CA, Heiss EH, Dirsch VM. Effect of resveratrol on endothelial cell function: molecular mechanisms. *Biofactors* 2010;**36**:342−9.
11. Price NL, Gomes AP, Ling AJ-Y, Duarte FV, Martin-Montalvo A, North BJ, et al. SIRT[1] is required for AMPK activation and the beneficial effects of resveratrol on mitochondrial function. *Cell Metab* 2012;**15**:675−90.
12. Spanier G, Xu H, Xia N, Tobias S, Deng S, Wojnowski L, et al. Resveratrol reduces endothelial oxidative stress by modulating the gene expression of superoxide dismutase 1 (SOD1), glutathione peroxidase 1 (GPx1) and NADPH oxidase subunit (Nox4). *J Physiol Pharmacol* 2009;**60**(Suppl 4):111−6.

13. Takahashi S, Nakashima Y. Repeated and long-term treatment with physiological concentrations of resveratrol promotes NO production in vascular endothelial cells. *Br J Nutr* 2012;**107**:774−80.

14. Förstermann U, Li H. Therapeutic effect of enhancing endothelial nitric oxide synthase (eNOS) expression and preventing eNOS uncoupling. *Brit J Pharmacol* 2011;**164**:213−23.

15. Xia N, Daiber A, Habermeier A, Closs EI, Thum T, Spanier G, et al. Resveratrol reverses endothelial nitric-oxide synthase uncoupling in apolipoprotein E knockout mice. *J Pharmacol Exp Ther* 2010;**335**:149−54.

16. Kelly G. A review of the sirtuin system, its clinical implications, and the potential role of dietary activators like resveratrol: part 1. *Altern Med Rev* 2010;**15**:245−63.

17. Kelly GS. A review of the sirtuin system, its clinical implications, and the potential role of dietary activators like resveratrol: part 2. *Altern Med Rev* 2010;**15**:313−28.

18. Mukherjee S, Dudley JI, Das DK. Dose-dependency of resveratrol in providing health benefits. *Dose Response* 2010;**8**:478−500.

19. Huang PH, Chen YH, Tsai HY, Chen JS, Wu TC, Lin FY, et al. Intake of red wine increases the number and functional capacity of circulating endothelial progenitor cells by enhancing nitric oxide bioavailability. *Arterioscler Thromb Vasc Biol* 2010;**30**:869−77.

20. Wallerath T, Deckert G, Ternes T, Anderson H, Li H, Witte K, et al. Resveratrol, a polyphenolic phytoalexin present in red wine, enhances expression and activity of endothelial nitric oxide synthase. *Circulation* 2002;**106**:1652−8.

21. Tomé-Carneiro J, Gonzálvez M, Larrosa M, García-Almagro FJ, Avilés-Plaza F, Parra S, et al. Consumption of a grape extract supplement containing resveratrol decreases oxidized LDL and ApoB in patients undergoing primary prevention of cardiovascular disease: a triple-blind, 6-month follow-up, placebo-controlled, randomized trial. *Mol Nutr Food Res* 2012;**56**:810−21.

22. Tomé-Carneiro J, Gonzálvez M, Larrosa M, Yáñez-Gascón MJ, García-Almagro FJ, Ruiz-Ros JA, et al. One-year consumption of a grape nutraceutical containing resveratrol improves the inflammatory and fibrinolytic status of patients in primary prevention of cardiovascular disease. *Am J Cardiol* 2012;**110**:356−63.

23. Bradamante S, Barenghi L, Villa A. Cardiovascular protective effects of resveratrol. *Cardiovasc Drug Rev* 2004;**22**:169−88.

24. Ramprasath VR, Jones PJ. Anti-atherogenic effects of resveratrol. *Eur J Clin Nutr* 2010;**64**:660−8.

25. Wong RH, Howe PR, Buckley JD, Coates AM, Kunz I, Berry NM. Acute resveratrol supplementation improves flow-mediated dilatation in overweight/obese individuals with mildly elevated blood pressure. *Nutr Metab Cardiovasc Dis* 2011;**21**:851−6.

26. Nicholson SK, Tucker GA, Brameld JM. Physiological concentrations of dietary polyphenols regulate vascular endothelial cell expression of genes important in cardiovascular health. *Br J Nutr* 2010;**103**:1398−403.

27. Soares De Moura R, Costa Viana FS, Souza MA, Kovary K, Guedes DC, Oliveira EP, et al. Antihypertensive, vasodilator and antioxidant effects of a vinifera grape skin extract. *J Pharm Pharmacol* 2002;**54**:1515−20.

28. Lekakis J, Rallidis LS, Andreadou I, Vamvakou G, Kazantzoglou G, Magiatis P, et al. Polyphenolic compounds from red grapes acutely improve endothelial function in patients with coronary heart disease. *Eur J Cardiovasc Prev Rehabil* 2005;**12**:596−600.

29. Magyar K, Halmosi R, Palfi A, Feher G, Czopf L, Fulop A, et al. Cardioprotection by resveratrol: A human clinical trial in patients with stable coronary artery disease. *Clin Hemorheol Microcirc* 2012;**50**:179−87.

30. Zamora-Ros R, Urpi-Sarda M, Lamuela-Raventós RM, Martínez-González MÁ, Salas-Salvadó J, Arós F, et al. High urinary levels of resveratrol metabolites are associated with a reduction in the prevalence of cardiovascular risk factors in high-risk patients. *Pharmacol Res* 2012;**65**:615–20.

31. Das DK, Sato M, Ray PS, Maulik G, Engelman RM, Bertelli AA, et al. Review: cardioprotection of red wine: role of polyphenolic antioxidants. *Drugs Exp Clin Res* 1999;**25**:115–20.

32. Bhatt JK, Thomas S, Nanjan MJ. Resveratrol supplementation improves glycemic control in type 2 diabetes mellitus. *Nutr Res* 2012;**32**:537–41.

33. Crandall JP, Oram V, Trandafirescu G, Reid M, Kishore P, Hawkins M, et al. Pilot study of resveratrol in older adults with impaired glucose tolerance. *J Gerontol A Biol Sci Med Sci* 2012;**67**:1307–12.

34. Brasnyó P, Molnár GA, Mohás M, Markó L, Laczy B, Cseh J, et al. Resveratrol improves insulin sensitivity, reduces oxidative stress and activates the Akt pathway in type 2 diabetic patients. *Br J Nutr* 2011;**106**:383–9.

35. Zhang H, Morgan B, Potter BJ, Ma L, Dellsperger KC, Ungvari Z, et al. Resveratrol improves left ventricular diastolic relaxation in type 2 diabetes by inhibiting oxidative/nitrative stress: in vivo demonstration with magnetic resonance imaging. *Am J Physiol Heart Circ Physiol* 2010;**299**:H985–94.

36. Huang JP, Huang SS, Deng JY, Chang CC, Day YJ, Hung LM. Insulin and resveratrol act synergistically, preventing cardiac dysfunction in diabetes, but the advantage of resveratrol in diabetics with acute heart attack is antagonized by insulin. *Free Radic Biol Med* 2010;**49**:1710–21.

37. Shin S, Jeon JH, Park D, Jang MJ, Choi JH, Choi BH, et al. trans-Resveratrol relaxes the corpus cavernosum ex vivo and enhances testosterone levels and sperm quality in vivo. *Arch Pharm Res* 2008;**31**:83–7.

38. Fukuhara S, Tsujimura A, Okuda H, Yamamoto K, Takao T, Miyagawa Y, et al. Vardenafil and resveratrol synergistically enhance the nitric oxide/cyclic guanosine monophosphate pathway in corpus cavernosal smooth muscle cells and its therapeutic potential for erectile dysfunction in the streptozotocin-induced diabetic rat: preliminary findings. *J Sex Med* 2011;**8**:1061–71.

39. Soner BC, Murat N, Demir O, Guven H, Esen A, Gidener S. Evaluation of vascular smooth muscle and corpus cavernosum on hypercholesterolemia. Is resveratrol promising on erectile dysfunction? *Int J Impot Res* 2010;**22**:227–33.

40. La Porte C, Voduc N, Zhang G, Seguin I, Tardiff D, Singhal N, et al. Steady-State pharmacokinetics and tolerability of trans-resveratrol 2000 mg twice daily with food, quercetin and alcohol (ethanol) in healthy human subjects. *Clin Pharmacokinet* 2010;**49**:449–54.

41. Walle T, Hsieh F, DeLegge MH, Oatis JE, Walle UK. High absorption but very low bioavailability of oral resveratrol in humans. *Drug Metab. Dispos* 2004;**32**:1377–82.

42. <http://www.fda.gov/drugs/developmentapprovalprocess/developmentresources/druginteractionslabeling/ucm093664.htm>; 2013 [accessed 17.11.13].

43. Detampel P, Beck M, Krähenbühl S, Huwyler J. Drug interaction potential of resveratrol. *Drug Metab Rev* 2012;**44**:253–65.

44. Xaplanteris P, Vlachopoulos C, Pietri P, Terentes-Printzios D, Kardara D, Alexopoulos N, et al. Tomato paste supplementation improves endothelial dynamics and reduces plasma total oxidative status in healthy subjects. *Nutr Res* 2012;**32**:390–4.

45. Sarkar PD, Gupta T, Sahu A. Comparative analysis of Lycopene in oxidative stress. *J Assoc of Physicians India* 2012;**60**:17–9.

46. Fielding JM, Rowley KG, Cooper P, O'Dea K. Increases in plasma lycopene concentration after consumption of tomatoes cooked with olive oil. *Asia Pac J Clin Nutr* 2005;**14**: 131−6.

47. Thies F, Masson LF, Rudd A, Vaughan N, Tsang C, Brittenden J, et al. Effect of a tomato-rich diet on markers of cardiovascular disease risk in moderately overweight, disease-free, middle-aged adults: a randomized controlled trial. *Am J Clin Nutr* 2012;**95**: 1013−22.

48. <http://www.isrctn.org>; 2013 [accessed 17.11.13].

49. Burton-Freeman B, Talbot J, Park E, Krishnankutty S, Edirisinghe I. Protective activity of processed tomato products on postprandial oxidation and inflammation: a clinical trial in healthy weight men and women. *Mol Nutr Food Res* 2011;**56**:622−31.

50. Bonetti PO, Pumper GM, Higano ST, Holmes Jr. DR, Kuvin JT, Lerman A. Noninvasive identification of patients with early coronary atherosclerosis by assessment of digital reactive hyperemia. *J Am Coll Cardiol* 2004;**44**:2137−41.

51. Davis NJ, Crandall JP, Gajavellic S, Berman JW, Tomutae N, Wylie-Rosett J, et al. Differential effects of low-carbohydrate and low-fat diets on inflammation and endothelial function in diabetes. *J Diabetes Complicat* 2011;**25**:371−6.

52. Kim JY, Paik JK, Kim OY, Park HW, Lee JH, Jang Y, et al. Effects of lycopene supplementation on oxidative stress and markers of endothelial function in healthy men. *Atherosclerosis* 2011;**215**:189−95.

53. Ried K, Fakler P. Protective effect of lycopene on serum cholesterol and blood pressure: meta-analyses of intervention trials. *Maturitas* 2011;**68**:299−310.

54. Berrougui H, Grenier G, Loued S, Drouin G, Khalil A. A new insight into resveratrol as an atheroprotective compound: inhibition of lipid peroxidation and enhancement of cholesterol efflux. *Atherosclerosis* 2009;**207**:420−7.

55. Riccioni G, Scotti L, Di Ilio E, Bucciarelli V, Ballone E, De Girolamo M, et al. Lycopene and preclinical carotid atherosclerosis. *J Biol Regul Homeost Agents* 2011;**25**:435−41.

56. Palozza P, Parrone N, Simone RE, Catalano A. Lycopene in atherosclerosis prevention: an integrated scheme of the potential mechanisms of action from cell culture studies. *Arch Biochem Biophys* 2010;**504**:26−33.

57. Napolitano M, De Pascale C, Wheeler-Jones C, Botham KM, Bravo E. Effects of lycopene on the induction of foam cell formation by modified LDL. *Am J Physiol Endocrinol Metab* 2007;**293**:E1820−7.

58. Rissanen T, Voutilainen S, Nyyssönen K, Salonen JT. Lycopene, atherosclerosis, and coronary heart disease. *Exp Biol Med (Maywood)* 2002;**227**:900−7.

59. Bose KS, Agrawal BK. Effect of lycopene from cooked tomatoes on serum antioxidant enzymes, lipid peroxidation rate and lipid profile in coronary heart disease. *Singapore Med J* 2007;**48**:415−20.

60. Valero MA, Vidal A, Burgos R, Calvo FL, Martínez C, Luengo LM, et al. Meta-analysis on the role of lycopene in type 2 diabetes mellitus. *Nutr Hosp* 2011;**26**:1236−41.

61. Gao JX, Li Y, Zhang HY, He XL, Bai AS. Lycopene ameliorates erectile dysfunction in streptozotocin-induced diabetic rats. *Pharmazie* 2012;**67**:256−9.

62. Obermüller-Jevic UC, Olano-Martin E, Corbacho AM, Eiserich JP, van der Vliet A, Valacchi G, et al. Lycopene inhibits the growth of normal human prostate epithelial cells in vitro. *J Nutr* 2003;**133**:3356−60.

63. Schwarz S, Obermüller-Jevic UC, Hellmis E, Koch W, Jacobi G, Biesalski H-K. Lycopene inhibits disease progression in patients with benign prostate hyperplasia. *J Nutr* 2008;**138**:49−53.

64. Rohdewald P. A review of the French maritime pine bark extract (Pycnogenol), a herbal medication with a diverse clinical pharmacology. *Int J clin pharmacol ther* 2002;**40**: 158–68.

65. Packer L, Rimbach G, Virgili F. Antioxidant activity and biologic properties of a procyanidin-rich extract from pine (Pinus maritima) bark, pycnogenol. *Free Radic Biol Med* 1999; **27**:704–24.

66. Fitzpatrick DF, Bing B, Rohdewald P. Endothelium-dependent vascular effects of pycnogenol. *J Cardiovasc Pharmacol Ther* 1998;**32**:509–15.

67. Nishioka K, Hidaka T, Nakamura S, Umemura T, Jitsuiki D, Soga J, et al. Pycnogenol, French maritime pine bark extract, augments endothelium-dependent vasodilation in humans. *Hypertens Res* 2007;**9**:775–80.

68. Hosseini S, Lee J, Sepulveda RT, Fagan T, Rohdewald P, Watson RR. A randomized, double blind, placebo controlled, prospective, 16 week crossover study to determine the role of pycnogenol in modifying blood pressure in mildly hypertensive patients. *Nutr Res* 2001;**21**: 67–76.

69. Zibadi S, Rohdewald PJ, Park D, Watson RR. Reduction of cardiovascular risk factors in subjects with type 2 diabetes by pycnogenol supplementation. *Nutr Res* 2008;**28**:315–20.

70. Liu X, Wei J, Tan F, Zhou S, Würthwein G, Rohdewald P. Antidiabetic effect of pycnogenol French maritime pine bark extract in patients with diabetes type II. *Life Sci* 2004;**75**:2505–13.

71. Enseleit F, Sudano I, Périat D, Winnik S, Wolfrum M, Flammer AJ, et al. Effects of Pycnogenol on endothelial function in patients with stable coronary artery disease: a double-blind, randomized, placebo-controlled, cross-over study. *Eur Heart J* 2012;**33**: 1589–97.

72. Belcaro G, Cesarone MR, Dugall M, Hosoi M, Ippolito E, Bavera P, et al. Investigation of Pycnogenol® in combination with coenzymeQ10 in heart failure patients (NYHA II/III). *Panminerva Med* 2010;**52**(2 Suppl 1):21–5.

73. Devaraj S, Vega-López S, Kaul N, Schönlau F, Rohdewald P, Jialal I. Supplementation with a pine bark extract rich in polyphenols increases plasma antioxidant capacity and alters the plasma lipoprotein profile. *Lipids* 2002;**37**:931–4.

74. Kim J, Chehade J, Pinnas JL, Mooradian AD. Effect of select antioxidants on malondialdehyde modification of proteins. *Nutr* 2000;**16**:1079–81.

75. Belcaro G, Cornell U, Luzzi R, Cesarone MR, Dugall M, Feragalli B, et al. Pycnogenol® supplementation improves health risk factors in subjects with metabolic syndrome. *Phytother Res* 2013.10.1002/ptr.4883

76. Chäfer A, Högger P. Oligomeric procyanidins of French maritime pine bark extract (Pycnogenol) effectively inhibit alpha-glucosidase. *Diabetes Res Clin Pract* 2007;**77**:41–6.

77. Aoki H, Nagao J, Ueda T, Strong JM, Schonlau F, Yu-Jing S, et al. Clinical assessment of a supplement of Pycnogenol® and L-arginine in Japanese patients with mild to moderate erectile dysfunction. *Phytother Res* 2012;**26**:204–7.

78. Ledda A, Belcaro G, Cesarone MR, Dugall M, Schönlau F. Investigation of a complex plant extract for mild to moderate erectile dysfunction in a randomized, double-blind, placebo-controlled, parallel-arm study. *BJU Int* 2010;**106**:1030–3.

79. Stanislavov R, Nikolova V. Treatment of erectile dysfunction with pycnogenol and L-arginine. *J Sex Marital Ther* 2003;**29**:207–13.

80. Stanislavov R, Nikolova V, Rohdewald P. Improvement of erectile function with prelox: a randomized, double-blind, placebo-controlled, crossover trial. *Int J Impot Res* 2008;**20**: 173–80.

Chapter 10

Selected Micronutrients and the Metabolic Basis for Their Support of Endothelial Health and Erectile Function

10.1 INTRODUCTION

The last chapter presented evidence that certain plant-based antioxidants benefit cardiovascular and heart health and erectile function. The evidence supporting these benefits goes to the core of the theme of this monograph that:

(A) the constellation of cardiovascular and heart disorders such as hypertension, atherosclerosis, coronary heart disease (CHD), including metabolic syndrome as well as type-2 diabetes, includes also vasculogenic erectile dysfunction (ED) with equal standing, and that

(B) what they all have etiologically in common is endothelium impairment, and that

(C) endothelium impairment is principally due, for a number of means, to oxidative stress interfering with nitric oxide (NO) formation and the nitric oxide/cyclic guanosine monophosphate (NO/cGMP) vasodilation pathway.

The logic of examining the antioxidant activity of resveratrol, lycopene, and Pycnogenol®, for example, is that by strong inference it supports the etiological role of oxidative stress in the development and progress of cardiovascular and erectile calamities.

Despite overwhelming scientific and clinical trial outcome evidence, there is still considerable controversy surrounding supplementing antioxidants, except in vague media terms, suggesting that they might somehow be "good for your health." There is as much controversy in medical and health care circles.

The controversy surrounding the therapeutic value of supplementing micronutrients hinges on whether they do in fact represent "complementar(ies)" or "alternatives" to conventional prevention of endothelial impairment and ED

Robert Fried: Erectile Dysfunction as a Cardiovascular Impairment.
DOI: http://dx.doi.org/10.1016/B978-0-12-420046-3.00010-X

and treatment means. This monograph proposes that when complementary or alternative treatments work, it is for exactly the same reason that successful conventional treatments work, and that proof of their efficacy is based on the same replicable controlled procedures as those followed in conventional clinical trials.

"Alternative" is often an arbitrary and even more often, pejorative, term applied to any stuff that is not a regular part of conventional medical therapeutics. It is mostly lost to history that the term *vitamin* is derived from the research of Kazimierz Funk (1884–1967), who in 1912 described a nutrient component he termed a "vital amine"—vital to life. The current trend of assigning negligible worth to vitamin and other micronutrient supplementation may in part be due to the fact that we have for the most part eliminated such common diseases as those that faced Funk: scurvy, beriberi, and pellagra, to mention just a few. These, along with iron and zinc deficiencies, are still common in developing countries—albeit mostly out of the American public eye.

There are many micronutrients and foods, including tea and cocoa, that have been shown beneficial to endothelial health and erectile function, but this chapter is concerned only with select substances where the controversy about use centers on a forum of science vs. science findings, rather than science vs. hearsay or folk-medicine … add a heaping teaspoon of wishful thinking, stir well … ..

10.2 CHROMIUM REDUCES INSULIN RESISTANCE IN TYPE-2 DIABETES

Chromium is crucial to the function of insulin, the "master hormone" of metabolism. Insulin controls blood sugar levels and many other aspects of carbohydrate breakdown and storage, and it also modulates fat, protein, and energy metabolism. There are several different forms of chromium, and to function properly, insulin requires adequate availability of trivalent chromium ($Cr3+$). The US Food and Drug Administration (FDA) designated chromium an essential nutritional trace element.

Although insulin mainly works in muscle, fat, and the liver, it also exerts profound effects on many other body tissues. It is the primary hormone that controls how the body cells absorb, use, and store nutrients and energy. Besides regulating the cellular absorption and utilization of glucose, amino acids, and fatty acids, insulin also activates and inactivates enzymes and directly affects certain processes, including protein synthesis.

10.2.1 Dietary Intake and Chromium Deficiency

US government sources report that most Americans get less chromium in their daily diets than the amount recommended by the RDA Committee: 50 to 200 μg of chromium/day. The vast majority of Americans get less than 50 μg/day, and therefore, according to the RDA 10th Edition, 1989, "In the majority of all chromium supplementation studies in the United States, at least half the subjects with impaired glucose tolerance improved upon chromium supplementation, suggesting that the lower ranges of chromium intake from typical US diet are not optimal with regard to chromium nutriture."[1]

Note: An article in *Proceedings of the Nutrition Society* made an interesting point that lies at the basis of chromium supplementation in connection with blood sugar metabolism: The symptoms of chromium deficiency are similar to those for metabolic syndrome, and supplemental chromium has been shown to improve that condition. The article also cites cinnamon, and that is why it is detailed in the next section.[2]

However, chromium supplementation may be specific to conditions involving insulin resistance (IR) even present obesity because, according to an article in *Metabolic Syndrome and Related Disorders*, "Chromium picolinate does not improve key features of metabolic syndrome in obese non-diabetic adults."[3] This means that chromium will not help control weight if weight gain is not linked to IR.

10.2.2 Chromium, Insulin Resistance, and Diabetes

Most of the research on chromium is linked to diabetes or to non-diabetic conditions of high blood glucose levels after ingesting simple sugar. A study on the link between chromium supplementation and type-2 diabetes was presented in June 1996 at the Annual Scientific Sessions of the American Diabetes Association held in San Francisco, CA. Researchers from the Human Nutrition Research Center of the United States Department of Agriculture collaborated with Chinese researchers from the Beijing Medical University: The results deemed "impressive" are summarized below [see Reference 4].

Insulin resistance as in type-2 diabetes means that the insulin circulating in blood is relatively ineffective. Individuals with type-2 diabetes are the most likely to have chromium deficiency.

The following is the ending to the chapter on chromium nutrition in the latest edition of the standard reference textbook on medical nutrition:

"Based on current knowledge of chromium function and nutrition, the possibility cannot be ignored that inadequate chromium status may be responsible in part for some cases of impaired glucose tolerance, hyperglycemia, hypoglycemia, glycosuria [sugar in urine], and refractoriness to insulin." [From,[5] abstracted from[6]]

One of the most definitive explanations of IR can be found in the journal *Diabetes*. It tells us that resistance to insulin-stimulated glucose uptake is common in most patients with impaired glucose tolerance (IGT) or non-insulin-dependent diabetes mellitus (NIDDM), as well as in approximately 25% of non-obese individuals with normal glucose tolerance. It concludes with the contention that resistance to insulin-stimulated glucose uptake and hyperinsuli-nemia are involved in the etiology and clinical course of three major related diseases—NIDDM, hypertension, and coronary artery disease (CAD).[7]

According to researchers writing in *Hormones & Metabolic Research*, tissue chromium levels are lower than normal in diabetes, and there is a link between low circulating levels of chromium and the incidence of type-2 diabetes. Although it is still somewhat controversial, there is support for supplementation with chromium picolinate, a stable and highly bioavailable form of chromium supplementation that has been shown to facilitate insulin release and improve insulin sensitivity. The authors recommend 200 μg per day.[8]

10.2.3 Chromium Reduces Insulin Resistance

The aim of the study appearing in *Folia Medica* was to evaluate the effect of chromium supplementation on IR in patients with type-2 diabetes. Overweight patients were assigned to two groups: One group had good metabolic control, whereas another had bad control. Both groups received oral supplements of 30 μg of chromium picolinate for six days. Serum concentration of chromium was recorded using atom-absorption methods. Immune-reactive insulin and the IR index at baseline, and at the end of the two-month period, were also recorded. Serum concentrations of chromium was significantly lower in diabetic patients than in the healthy individuals used as controls.

There was a significant decrease in immune-reactive insulin and the IR index after two months of supplementation of 30 μg chromium picolinate daily. Chromium included early in the complex therapy of diabetes is beneficial in the reduction of the degree of IR.[9]

The Dutch journal of nutrition, *Nederlands Tijdschrift voor Geneeskdunde*, reported a clinical trial involving supplementation with 1,000 μg of chromium in Chinese patients with type-2 diabetes. Supplementation with 1,000 μg of chromium led to a 2% fall in the glycosylated hemoglobin level (HbA1c).

The investigators reported that although toxic effects of chromium are seldom seen, nevertheless the safety of chromium picolinate has been questioned. They caution that individual patients with type-2 diabetes mellitus may have an increased risk of hypoglycemic episodes when taking chromium supplements as self-medication.[4]

10.2.4 Chromium Supplementation Reduces Oxidative Stress in Type-2 Diabetes

A study in the *Journal of Agricultural Food & Chemistry* aimed to determine the effects of chromium supplementations on oxidative stress in patients with normal blood glucose concentration (euglycemia) with HbA1C values less than 6.0%, and one group of type-2 diabetes patients with mild hyperglycemia (6.8 to 8.5%), and another group severely hyperglycemic (HbA1C greater than 8.5%). These were given 1,000 µg/day of chromium supplements as yeast for 6 months, or placebo.

At the start, plasma chromium level in the mild and the severe hypoglycemia groups were 25−30% lower than those of the normal levels group. The values of thiobarbituric acid reactive substances (TBARS, byproducts of damage due to oxidative stress) and total antioxidative status (TAS) of the mild and the severe groups were significantly higher than those of the normal group.

After supplementation, the levels of plasma TBARS in the mild and severe chromium supplementation groups declined significantly, whereas the opposite was found in the normal group, and showed no significant changes in the placebo group. The levels of plasma TAS in the normal and the mild chromium groups were significantly decreased, the opposite was found in the severe group, and it showed no significant changes in the placebo group. No significant difference was found in the antioxidant enzyme activity of superoxide dismutase, glutathione peroxidase, and catalase during supplementations. These data suggest that chromium supplementation was effective in reducing the increased oxidative stress in type-2 diabetes whose HbA1C levels were greater than 8.5%.[10]

10.2.5 Vitamins C and E Enhance Chromium Supplementation in Reducing Oxidative Stress and Improving Glucose Metabolism in Type-2 Diabetes

The *Journal of Clinical Biochemistry & Nutrition* reported a study of the effects of combining chromium and vitamins C and E supplementation on oxidative stress in adults with type-2 diabetes with HbA1c greater than 8.5%. Participants were divided into three groups: Gp 1, placebo; gp 2, 1 K µg chromium plus vitamin C; gp 3, 1 K µg chromium plus vitamin 800 IU vitamin E on a daily basis for six months. Baseline plasma chromium levels were not significantly different between the supplementation and placebo groups. TBARS and TAS were also not significantly different.

Following the study period, plasma TBARS levels, fasting glucose, HbA1c and IR declined significantly in the chromium, the chromium plus vitamin C, and the chromium plus vitamin E groups. Not so in the placebo group. Plasma TAS and glutathione peroxidase were significantly higher in the three treatment groups compared to the placebo group. The author concluded that chromium alone, and chromium plus vitamin C and E supplementation, was effective for reducing oxidative stress and improving glucose metabolism in type-2 diabetes patients.[11]

According to a clinical report in the *Journal of Trace Elements in Medicine & Biology*, patients who received 9 g brewer's yeast (42 µg chromium) supplementation for 3 months showed significantly improved glycemic control and lipid variables in new-onset type-2 diabetes.[12] Likewise, the results of a clinical trial appearing in the journal *Diabetes* indicates that supplemental chromium improves HbA1c, glucose, insulin, and cholesterol in type-2 diabetes. However, levels higher than the upper limit of the Estimated Safe and Adequate Daily Dietary Intake are required.[13]

10.2.6 Effects of Chromium Picolinate Supplementation on Body Composition

Chromium picolinate supplementation can lead to significant improvement in body composition: A clinical study appearing in *Current Therapeutic Research* aimed to determine the effects of chromium picolinate on body composition. Patients received either a placebo or 200 μg or 400 μg of chromium per day in the form of at least two servings of a protein/carbohydrate nutritional drink that held the different dosages of chromium picolinate. Participants were outpatients and they were not provided with weight loss, dietary, or exercise guidance. Body composition was measured before and after a 72-day test period by using water displacement with residual lung volumes determined by helium dilution.

On completion, post-test, a body composition improvement index (BCI) was calculated for each subject by adding the loss of body fat and gain in non-fat mass, and subtracting fat gained and lean mass lost. Analysis of the pre-study data revealed no significant differences in body composition between the three groups, but after the test period, both the 200 μg and 400 μg groups had significantly higher positive changes in BCIs compared with placebo. No significant differences in BCI were found between the 200 and 400 μg groups. It was concluded that supplementation with a minimum of 200 μg/day of chromium picolinate can lead to significant improvement in body composition.[14]

10.2.7 Efficacy of Chromium Challenged

A randomized, placebo-controlled, double-blind trial in subjects with impaired glucose tolerance appeared in *Diabetes Care* aimed at determining the effect of chromium picolinate supplementation on glucose tolerance, IR, and lipids in patients with IGT.

The participants, identified by history of abnormal glucose tolerance, including past gestational diabetes, were selected by their response to a 75 g oral glucose tolerance test (OGTT). In the fasting state, blood insulin, chromium, total cholesterol, and triacylglycerides were assessed. Chromium picolinate supplementation was administered in 400 μg twice-daily (800 μg/day) dosages (Bullivants Natural Health Products, Baulkham Hills, NSW, Australia, supplied the active tablets and the placebo). IR was calculated with the homeostasis model assessment (HOMA), using the formula HOMA of IR (HOMA-IR) = (fasting insulin × fasting glucose)/22.5.

There were no significant differences between the treatment and placebo groups at baseline and after 3 months of treatment; the only changes to achieve statistical significance were a small rise in serum chromium in the active treatment group. There was no difference in rate to progression or regression of IGT. The authors report finding no beneficial effect of chromium supplementation in participants with IGT despite increases in serum chromium levels.[15]

10.2.8 Komorowski and Juturu: Response to Gunton *et al.* (2005)

Quoted with permission from *Diabetes Care*[15]:

"We read the recent article by Gunton et al.[15] *with great interest and feel that it warrants comment. In this study, the authors stated that they 'found no beneficial effect of chromium supplementation in the treatment of people with IGT [impaired glucose tolerance]. The results are in conflict with other clinical studies that showed chromium picolinate can enhance or normalize impaired glucose metabolism, as described in a recent review.[17] The lack of effect described by the authors may be explained by the apparent low dose of elemental chromium used in the study.*

"The authors stated that the chromium picolinate 'dose (at 800 μg/day) was at the higher end of the ranges used in previous studies.'[15] However, chromium picolinate administered at 800 μg per day yields a daily dose of 100 μg per day of elemental chromium (i.e. chromium picolinate contains 12.4% elemental chromium). An elemental chromium dose of 100 μg a day is half of the suggested minimum amount (200 μg) of elemental chromium previously shown to exhibit efficacy in glucose and lipid metabolism.[17] A daily dose of 200–1,000 μg of elemental chromium, as chromium picolinate, is the efficacious dosage range used in previous studies.

"Bullivants Natural Health Products, the supplier of the study products used by the authors, stated that 400 μg of the chromium picolinate product they produce yields 50 μg of elemental chromium. The study was conducted in Australia, and the 50-μg elemental chromium dose is also the maximum daily dose allowed by the Australian Therapeutic Goods Administration.[18]

"It was also interesting to note that although the serum chromium levels significantly rose in the active group, the serum chromium levels were not significantly higher in the active group than in the placebo group after 3 months of supplementation (active group 5.2 ± 8.9 nmol/L, placebo group 4.4 ± 4.0 nmol/L). For these reasons, we believe study subjects in the active group may have been administered daily doses of 50 μg elemental chromium, twice daily.

"We recommend future studies be conducted in people with impaired glucose tolerance (following criteria defined by the American Diabetes Association) using daily doses of chromium picolinate providing ≥ 200–1,000 μg of elemental chromium for at least 90 days. We also recommend evaluating efficacy using the 75-g oral glucose tolerance test with calculation of the area under the curve using the trapezoidal method."

10.2.9 Insulin Resistance, Chromium, and Erectile Dysfunction

Men with ED have a high incidence of metabolic syndrome and IR. IR with or without the presence of metabolic syndrome significantly increases the risk of cardiovascular disease (CVD) and ED. In fact, investigators report a clinical study in the *Journal of Sexual Medicine* where men were evaluated for multiple cardiovascular risk factors and graded on severity of ED by the Sexual Health Inventory for Men (SHIM) questionnaire. The prevalence of metabolic syndrome was determined by NCEP/ATP III criteria. IR was measured by quantitative insulin-sensitivity check index (QUICKI).

It was found that the total cholesterol/HDL ratio was moderately and negatively correlated with QUICKI, and likewise the triglyceride (TG)/HDL ratio. Metabolic syndrome was present in 43% of their ED population as opposed to 24% in a matched patient population. Approximately 79% of their total population had IR and 73% of the non-diabetic portion had IR, compared to 26% in a general population study. Metabolic syndrome, IR, and fasting blood sugar (FBS) greater than 110 mg/dL correlated positively with increasing severity of ED as evidenced by SHIM score.

The authors concluded that men with ED have a high incidence of metabolic syndrome and IR. They recommend early detection of metabolic disease in patients with ED to prevent further endothelial dysfunction in younger men with increased cardiovascular risk but who present for treatment of ED alone.[19]

It is especially advisable to curb IR because it plays a major role in ED and it has an earlier onset in men with diabetes as shown in Figure 10.1.[20]

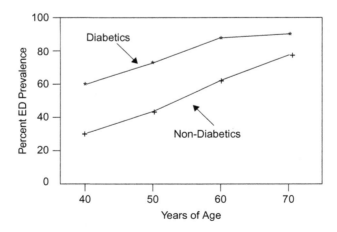

FIGURE 10.1 Frequency and age of onset of erectile dysfunction in type-2 diabetes and non-diabetics. Source: *Figure was redrawn from that provided by Dr. Culley C. Carson, University of North Carolina at Chapel Hill. Francis, S. H., & Corbin, J. D. (2011). PDE5 inhibitors: targeting erectile dysfunction in diabetics.* Curr. Opin. Pharmacol, *11, 683–688. Reproduced with kind permission from* Current Opinion in Pharmacology.

The authors of a meta-analysis-based review appearing in the journal *Fertility & Sterility* contend that diagnosing and treating IR should be part of the initial management plan for ED because IR is linked to defective vascular NO production and impaired insulin-induced vasodilation, both of which are likely to cause ED.[21]

10.2.10 Caveat

A clinical study appearing in the journal *BMC Endocrine Disorders* was conducted on non-obese normoglycemic subjects who were given 500 μg chromium picolinate twice a day. The authors found that after 16 weeks of chromium picolinate supplementation, there was no change in insulin sensitivity in non-obese normoglycemic individuals. Furthermore, those who had high serum chromium levels at the start paradoxically had a decline in insulin sensitivity. Caution is recommended with the use of this supplement.[22]

Writing in *Current Diabetes Reports*, researchers contend that a consistent, significant, and beneficial effect of chromium supplementation may not invariably be observed in individuals with type-2 diabetes. Patient selection may be an important factor in determining clinical response, as it was shown that decreased glucose and improved insulin sensitivity may be more likely in insulin-resistant individuals with type-2 diabetes who have more elevated fasting glucose and HbA1c levels.[23]

Note from the standard medical reference on nutrition, Shils, Olson & Shike, eds. (1994). *Modern nutrition in health and disease*:[5]

"Trivalent chromium has such a low order of toxicity that deleterious effects from excessive intake of this form of chromium do not occur readily. Trivalent chromium becomes toxic only at extremely high amounts. Chromium then acts as a gastric irritant rather than as a toxic element interfering with essential metabolism or biochemistry."

Nevertheless, *caveat emptor*.

10.3 CINNAMON—CLINICAL AND RESEARCH DATA ON BENEFITS IN TYPE-2 DIABETES

Cinnamon is a common spice originating in Southeast Asia. It has been known for some time to be effective in curbing blood sugar. One of its several constituents is an oligomeric proanthocyanidin (OPC), a powerful antioxidant previously encountered in Pycnogenol. The major active constituent is cinnamaldehyde. It has a relatively unknown, albeit reputable, research track record in connection with IR and treatment of type-2 diabetes, but there is little reason to believe that it will be a major player in its treatment. Nevertheless it deserves mention due to its demonstrated antidiabetic properties.

There is cinnamon, and then again, there is cinnamon: At the outset, a distinction needs to be made between *Cinnamomum zeylanicum* (Ceylon cinnamon), aka *Cinnamomum verum* (true cinnamon), and *Cinnamomum cassia* (Saigon cinnamon). They don't come from the same plant, though they are from the same Lauraceae family, and both types have shown blood sugar benefits. However, only *Cinnamomum zeylanicum* is "true cinnamon," unlike "Cassia," that is not "true cinnamon." Both varieties of cinnamon contain small amounts of coumarin, but it is found in greater concentration in Cassia cinnamon. Coumarin, not to be confused with coumadin, an anticoagulant, has a relatively low-level of toxicity. Nevertheless, organizations, including the Federal Institute for Risk Assessment in Berlin, Germany, have recommended avoiding consumption of "large amounts" of the cassia.

In powder form, it is not easy to tell these two types of cinnamon apart. However, when cinnamon sticks are rolled from the thick bark of the cassia plants, they look like a one-piece, thick bark layer that does not show multiple layers. In contrast, in Ceylon cinnamon sticks (true cinnamon) one can see multiple layers of a thinner bark.

10.3.1 What is the Active Constituent in Cinnamon?

According to investigators writing in the *Journal of the American College of Nutrition*, a (methyl)hydroxychalcone in cinnamon functions as an insulin mimetic in 3T3-LI adipocytes.[24] Chalcones are phenols, elements in formation of flavonoids. More recently, cinnamtannin B1, an antioxidant, has been linked to the insulin-like biological activity of cinnamon.[25] It has been shown that extracts from cinnamon enhance the activity of insulin.

The objective of this study appearing in the *Journal of Agriculture and Food Chemistry* was to isolate and characterize insulin-enhancing complexes from cinnamon that may be involved in the alleviation or possible prevention and control of glucose intolerance and diabetes. Water-soluble polyphenol polymers from cinnamon that increase insulin-dependent *in vitro* glucose metabolism roughly 20-fold and display antioxidant activity were isolated and characterized by nuclear magnetic resonance (NMR) and mass spectroscopy (MS). The polymers were composed of monomeric units with a molecular mass of 288. Two trimers with a molecular mass of 864 and a tetramer with a mass of 1,152 were isolated. Their protonated molecular masses indicated that they are A-type doubly linked procyanidin oligomers of the catechins and/or epicatechins. These polyphenolic polymers found in cinnamon may function as antioxidants, potentiate insulin action, and may be beneficial in the control of glucose intolerance and diabetes.[25]

Despite numerous clinical trials in humans and in animal models, and despite many clearly positive outcomes, there is still some controversy over whether cinnamon really has the ability to control blood sugar. Extract of

Cinnamomum zeylanicum, cinnamaldehyde, shows pronounced hypoglycemic and hypolipidemic effects (in streptozotocin (STZ)-induced diabetic rats): A report in *Phytomedicine* reiterates that *Cinnamomum zeylanicum* is a widely accepted traditional medicine to treat diabetes in India. For that reason, the study was designed to identify antidiabetic compounds, if any, by bioassay-guided fractionation.

The active compound that decreased plasma glucose levels was purified and found to be cinnamaldehyde, determined by chemical and physiochemical evidence. The LD(50) value of cinnamaldehyde was determined as 1850 ± 37 mg/kg bw. Cinnamaldehyde was then administered at different doses (5, 10, and 20 mg/kg bw) for 45 days to STZ (60 mg/kg bw)-induced male diabetic wistar rats. Dose-dependent plasma glucose concentration decreased significantly compared to the control; (oral) cinnamaldehyde (20 mg/kg bw) significantly decreased HbA1C; and there was significantly decreased serum total cholesterol, TG levels and significantly raised plasma insulin, hepatic glycogen, and high-density lipoprotein-cholesterol levels.

Cinnamaldehyde restored plasma aspartate aminotransferase, alanine aminotransferase, lactate dehydrogenase, alkaline phosphatase, and acid phosphatase levels to near normal, indicating that it can correct hypoglycemic and hypolipidemic effects in STZ-induced diabetic rats.[26]

It was demonstrated in another clinical study also involving chromium supplementation that cinnamon improved glucose, insulin, cholesterol, and HbA1c in patients with type-2 diabetes, and that the polyphenols improved insulin sensitivity *in vitro* in animals and in humans.

Cinnamon reduced mean fasting serum glucose, total cholesterol, and low-density lipoprotein cholesterol (LDL-C) in type-2 diabetes patients after 40 consecutive days of consumption of 1 to 6 g cinnamon. Participants with the metabolic syndrome who consumed an aqueous extract of cinnamon were shown to have improved fasting blood glucose, systolic blood pressure, percentage body fat, and increased lean body mass compared with the placebo group. However, not all studies have reported beneficial effects of either supplemental chromium or cinnamon, and the responses seem to be related to the duration of the study, the form of chromium or cinnamon used, and the extent of obesity and glucose intolerance of the participants in the study.[2]

The relative efficacy of *Cinnamomum cassia* and *Cinnamomum zeylanicum* was tested in an animal model (rat) to evaluate blood glucose and plasma insulin levels under various conditions. The *cassia* extract was found to be superior to the *zeylanicum* extract, and the *cassia* extract was slightly more powerful than an equivalent amount of *cassia bark*. A decrease in blood glucose level was observed in a glucose tolerance test, whereas it was not obvious in rats that were not challenged by a glucose load. Elevation in plasma insulin was observed: the *cassia* extract had a direct antidiabetic effect.[27]

Cinnamon increases fatty acid and glucose metabolism in adipose tissue. The title of a recent review in the journal *Diabetes, Obesity & Metabolism*, "Controversies surrounding the clinical potential of cinnamon for the management of diabetes," pretty much sums it up. The review tells us that cinnamon, a commonly consumed spice originating from Southeast Asia, is currently being investigated as a potential preventative nutraceutical supplement and treatment for IR, metabolic syndrome, and type-2 diabetes. It has been learned from *in vitro* studies that it may improve IR by preventing and reversing impairments in insulin signaling in skeletal muscle, and that it increases fatty acid and glucose metabolism in adipose tissue.

Studies have also shown that cinnamon has potent anti-inflammatory properties. However, numerous human clinical trials with cinnamon have been conducted with divergent findings, including some that showed no beneficial effect whatsoever. Other studies, on the other hand, have shown improved cholesterol levels, systolic blood pressure, insulin sensitivity, and postprandial glucose levels. However, the only function showing consistent improvement with cinnamon is fasting glucose levels. Therefore, the authors hold that it is premature to propose cinnamon supplementation based on the evidence.[28]

A review appearing in the journal *Diabetic Medicine* provided the following information:

In vitro C. zeylanicum can reduce postprandial intestinal glucose absorption by inhibiting pancreatic α-amylase and α-glucosidase, stimulating cellular glucose uptake by membrane translocation of glucose transporter-4, stimulating glucose metabolism and glycogen synthesis, inhibiting gluconeogenesis, and stimulating insulin release and potentiating insulin receptor activity.

In animal models, *Cinnamomum zeylanicum* reduces fasting blood glucose, LDL, and HbA1c, raises high-density lipoprotein cholesterol (HDL-C), and raises circulating insulin levels. It can also significantly improve metabolic problems associated with IR.[29]

Here are additional references:

- Meta-analysis revealed that cinnamon or cinnamon extract improves fasting blood glucose in type-2 diabetes or pre-diabetes. In the *Journal of Medical Foods*.[30]
- Cinnamon reduces serum glucose, TG, LDL-C, and total cholesterol in type-2 diabetes patients. In *Diabetes Care*.[31]
- Cinnamon reduces fasting plasma glucose concentrations in diabetic patients with poor glycemic control. In the *European Journal of Clinical Investigation*.[32]
- Cinnulin PF® (a proprietary water soluble extract of *Cinnamomum burmannii*)[33] supplementation reduces fasting blood glucose, systolic blood

pressure, and improves body composition in metabolic syndrome. In the journal *International Society of Sports Nutrition*.[34]

- Cinnamon may lower postprandial glucose response in normal weight and obese adults. In the *Journal of the Academy of Nutrition and Dietetics*.[35]
- Two grams of cinnamon for 12 weeks significantly reduced the HbA1c, systolic and diastolic blood pressure in poorly controlled type-2 diabetes patients. In *Diabetic Medicine*.[36]
- In contrast, one meta-analysis found no improvement in Hb1Ac, and the authors concluded that, "cinnamon should not be recommended for the improvement of glycemic control." In the journal *Nederlands Tijdschrift voor Geneeskunde*.[37]

10.4 THE HOMOCYSTEINE, ATHEROSCLEROSIS, AND VITAMIN CONUNDRUM

Chapter 3 included the homocysteine theory of atherosclerosis among the factors thought to be etiologically related to endothelium damage, CVD and heart disease, and ED. There was at first considerable resistance to the idea that something other than oxidized LDL-C could also cause atherosclerosis. The homocysteine theory gradually caught on, supported by many clinical studies gathering compelling evidence. Elevated levels of homocysteine have been consistently linked to endothelium impairment by a number of pathways, yet the impact is still controversial.

10.4.1 Homocysteine

Homocysteine is an amino acid formed during methionine metabolism. Methionine can convert to homocysteine, or bond with proteins to form so-called "protein-bound" homocysteine. About 80% of homocysteine is protein-bound in plasma. Metabolism of homocysteine by one pathway requires vitamin B_{12} and folic acid, and by another, it requires vitamin B_6. Levels of blood homocysteine depend on consumption of protein, as well as how much is metabolized, which in turn is said to depend on availability of B vitamins and folate.

10.4.2 Homocysteine Levels also Rise with Age

Concentrations of homocysteine above 14 μmol/L affect about 20% of any population studied, and a still higher proportion of elderly people.[38] According to a report in *The Lancet*, among more than 2,000 patients referred to a stroke prevention clinic or vascular disease prevention clinic, the proportion with concentrations of plasma total homocysteine (tHcy) of 14 μmol/L or more increased significantly with age (Figure 10.2).

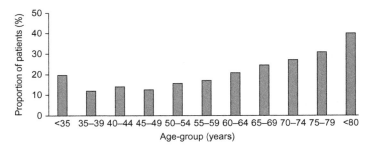

FIGURE 10.2 Age distribution of increased plasma tHcy ($\geq 14\,\mu$mol/L) among patients referred to vascular prevention clinics. Source: *Spence, D. (2009). Mechanisms of thrombogenesis in atrial fibrillation.* Lancet, *373, 1006. Reproduced with kind permission from Lancet.*

It is thought not to be coincidental that the proportion of vascular disease patients with metabolic B_{12} deficiency is 30% in those older than 71 years.[38,39]

10.4.3 Range and Severity of Hazard of Plasma Homocysteine Levels

The basic ranges and severities are as follows:

- Normal — 5 to 15 μmol/L
- Moderate — 16 to 30 μmol/L
- Intermediate — 31 to 100 μmol/L
- Severe — greater than 100 μmol/L

Research on the role of homocysteine in CVD and heart disease, and its purported treatment with folate and B vitamins, recently hit a major bump in the road. In summary, it all now stands at this:

- Elevated level of homocysteine is causally linked to atherosclerosis and endothelium impairment and CVD and heart disease. It should be treated with folate and B vitamins.
- Elevated level of homocysteine is not causally linked to atherosclerosis, but it is a "marker" of atherosclerosis.
- Elevated level of homocysteine is in no way causally linked to atherosclerosis and endothelium impairment and CVD and heart disease. No special supplementation of folate and B vitamins is called for. In fact, folate supplementation can destabilize atheromas and thus worsen atherosclerosis. However, one is advised to consume the RDA of folate and B vitamins anyway—but from foods, not supplements.
- Elevated level of homocysteine has nothing to do with atherosclerosis. In fact, it is causally linked to kidney dysfunction. Folate and B vitamins are not relevant to treatment.

10.4.4 Homocysteine Impairs Endothelium Function

Whichever of these turns out to be right is of no small concern to treatment of sexual dysfunction because elevated levels of homocysteine are also implicated in endothelium impairment. A study titled "Cavernosal dysfunction in a rabbit model of hyperhomocysteinaemia," published in the *British Journal of Urology International*, aimed to determine the effects of sustained elevated homocysteine levels on cavernosal smooth muscle function in rabbits, using a methionine-enriched diet.

Rabbit standard diet was supplemented with methionine (20 g/kg chow) for 4 weeks, unlike that of control animals. Cavernosal strips mounted in an organ bath relaxed when stimulated with carbachol, sodium nitroprusside (SNP), or noncholinergic, nonadrenergic (NANC)-mediated relaxation to electrical-field stimulation (EFS). Cavernosal tissue cGMP levels were assessed using an enzyme-linked immunosorbent assay (ELISA), and superoxide ($O2(.^-$), free radicals) production assessed using an assay of the superoxide dismutase (SOD)-inhibitable reduction of ferricytochrome c.

Among other results, there was a significant reduction in cavernosal cGMP levels in the experimental group, with a more than five-fold increase in cavernosal tissue free radical production. Supplementing the diet of rabbits with methionine for 4 weeks caused early and sustained elevated homocysteine levels and resulted in marked inhibitory effect on endothelium-dependent relaxation and NO formation in isolated corpus cavernosum. The effect was said to be mediated by reactive oxygen species (ROS).[40]

The journal *Metabolism* reported a study of the role of homocysteine in human ED in non-diabetic patients (excluding also CAD, and vitamin B_{12} or folate deficiency). The International Index of Erectile Function (IIEF) questionnaire was used to gauge erectile quality. Fasting plasma glucose, total cholesterol, TG, HDL-C, LDL-C, vitamin B_{12}, folic acid, and homocysteine levels of patients were measured. Penile color Duplex ultrasound was used to detect vascular abnormalities in non-diabetic patients with ED. Patients with ED were on average about 9 years older than the control subjects, and they had higher fasting plasma glucose, total cholesterol, LDL-C, and homocysteine levels.

There was a significant negative correlation between mean homocysteine level and mean IIEF domain score. The penile color Doppler ultrasound findings showed that there was a significant negative correlation between mean homocysteine level and the 1st, 5th, and 10th minute peak-systolic velocity (PSV). Age and homocysteine levels were the main determinants in ED. It was concluded that elevated homocysteine levels seem to be an important determinant in ED, and that it is likely that it is related to endothelial dysfunction.[41]

10.4.5 Homocysteine and Tetrahydrobiopterin: A Link to Endothelium Dysfunction

Tetrahydrobiopterin (BH4) is a naturally occurring essential cofactor for the conversion of a nitrogen component of L-arginine to NO by endothelial nitric oxide synthase (eNOS). Investigators report in *Arteriosclerosis, Thrombosis & Vascular Biology* that unavailability of BH4 results in attenuation of eNOS production of NO, instead generating superoxide. In vascular disease states, there is oxidative degradation of BH4 by ROS. Augmentation of BH4 concentrations in vascular disease by pharmacological supplementation, by enhancement of its rate of biosynthesis, or by measures to reduce its oxidation, have been shown to enhance NO bioavailability.[42]

It is therefore to the point that it was reported in the *American Journal of Physiology, Endocrinology & Metabolism* that, "Homocysteine impairs coronary artery endothelial function by inhibiting BH4 in patients with hyperhomocysteinemia." When BH4 levels are inadequate, eNOS is no longer coupled to L-arginine oxidation, causing ROS rather than NO production, and inducing vascular endothelial dysfunction.

Plasma levels of NO and BH4 were significantly lower in patients with elevated homocysteine (hyperhomocysteinemia, HHcy) than in controls. Coronary flow velocity reserve (CFVR) was significantly lower in the elevated HHcy patients than in the control group. Plasma level of homocysteine was negatively correlated with CFVR. The uncoupling of eNOS induced by HHcy in patients with chronic HHcy may explain CAD induced by dysfunction of the coronary artery endothelium.[43]

10.4.6 Consistently Elevated Plasma Homocysteine Level Is a Health Hazard

There is reason to believe that elevated levels of homocysteine can lead to atherosclerosis:

- Homocysteine generates superoxide and hydrogen peroxide, which have been linked to damage to arterial endothelium.
- Homocysteine changes coagulation factor levels so as to encourage clot formation.
- Homocysteine prevents small arteries from dilating, so they are more vulnerable to obstruction.
- Homocysteine causes smooth muscle cells in the arterial wall to multiply.
- Homocysteine, when infused into animal arteries, caused the linings to slough off and produce lesions, much like atheromas.
- Homocysteine has a reactive product, homocysteine thiolactone, which interacts with LDLs, causing them to precipitate and damage endothelial tissue.

- Homocysteine thiolactone causes platelet aggregation.

High homocysteine levels in a population were shown to be associated with higher levels of CVD mortality.[44]

10.4.7 Each Increase of 5 μmol/L in Homocysteine Level Raises the Risk of Coronary Heart Disease Events by Approximately 20 Percent Independent of Traditional Risk Factors

As reported in the *Mayo Clinic Proceedings*, a meta-analysis of homocysteine and CHD on MEDLINE, and homocysteine and Framingham risk factors, and the incidence of CHD in the general adult population without known CHD, found elevations of 20 to 50% in CHD risk for each increase of 5 μmol/L. That meta-analysis also yielded a risk ratio for coronary events of 1.18 for each increase of 5 μmol/L in homocysteine level. The association between homocysteine and CHD was similar when analyzed by sex, length of follow-up, outcome, study quality, and study design.[45]

In the British United Provident Association (BUPA) prospective study of men aged 35 to 64 years, homocysteine in stored samples was compared in those with no history of ischemic heart disease (IHD), to those who subsequently died of IHD, over a mean follow-up of 8.7 years, to age-matched controls.[46]

This study showed that men in the highest plasma homocysteine quartile level were nearly three times more likely to die of IHD, even after adjusting for factors like apolipoprotein levels and blood pressure than men with normal levels. The risk of CHD among men in the highest quartile of serum homocysteine (sHcy) levels was 3.7 times the risk among men in the lowest quartile. There was a continuous dose–response relationship, with risk increasing by 41% for each 5 μmol/L increase in the sHcy level. After adjustment for apolipoprotein B levels and blood pressure, this estimate was 33%.

In a meta-analysis of the retrospective studies of homocysteine level and myocardial infarction, the age-adjusted association was stronger, 84% increase in risk for a 5 μmol/L increase in the homocysteine level, possibly because the participants were younger; the relationship between sHcy level and IHD seems to be stronger in younger persons than in older persons.

The BUPA study also contained data from other prospective and retrospective studies that examined Homocysteine levels and fatal or nonfatal cardiac events. There was considerable consistency, with an odds ratio of IHD for each 5 μmol/L increase in sHcy of 1.8 in a total of about 2,300 participants.

Another prospective examination of the relationship between plasma homocysteine and mortality was carried out in patients with angiographically confirmed CAD.[47] The majority of the patients subsequently underwent coronary-artery bypass grafting or percutaneous transluminal coronary angio-plasty. The remaining patients were treated medically.

After a median follow-up of 4.6 years, there was a strong, graded relation between plasma homocysteine levels and overall mortality. After four years, 3.8% of patients with homocysteine levels below 9 μmol per liter had died, as compared with 24.7% of those with homocysteine levels of 15 μmol per liter or higher. Homocysteine levels were only weakly related to the extent of CAD, but were strongly related to the history with respect to myocardial infarction, the left ventricular ejection fraction, and the serum creatinine level.

10.4.8 Homocysteine and Vitamins

sHcy concentrations have been correlated with plasma vitamin concentration and vitamin intake in adults 67 to 96 years old, survivors of the Framingham study. Homocysteine was higher in men and women over 80 years of age, but after adjusting for age, gender and levels of other vitamins, there was a strong inverse correlation between plasma homocysteine and folate. Individuals with the lowest plasma folate were twelve times more likely to have an elevated homocysteine concentration, defined as being above 14 μmol/L, after adjusting for age, gender, and other B vitamins.

The mean sHcy concentration and the proportion of individuals with ele-vated plasma homocysteine greater than 14 μmol/L correlated positively with B vitamin intake. The quintile with the lowest B vitamin intake had a preva-lence of homocysteinemia of 53%. Similar results were found for a vitamin index expressing serum concentrations of B vitamins.[48]

10.4.9 Treatment with Vitamin B_6

It was reported in *Circulation* that risk for CAD rises with increasing plasma homocysteine, regardless of age and sex. Also, low pyridoxal-5'-phosphate (P-5-P, a more readily absorbable metabolite of vitamin B_6) confers an inde-pendent risk: Patients with CAD were tested for concentrations of plasma homocysteine, folate, vitamin B_{12}, and P-5-P. Homocysteine levels corre-lated negatively with all vitamins: Low P-5-P was seen in 10% of patients, but in only 2% of controls.[49]

Within the range currently considered to be normal, the risk for CAD rises with increasing plasma homocysteine, regardless of age and gender, and with no threshold effect. In addition to a link with homocysteine, low P-5-P confers an independent risk for CAD.

10.4.10 Is there Vitamin B_{12} Deficiency?

The results of a study reported in the *Journal of the American College of Nutrition* suggest the possibility that in some cases of B_{12} deficiency and elevated homocysteine, it is not possible to tell if the culprit is B_{12} deficiency as, in this study, low vitamin B_{12} concentrations were found to be linked to increased risk of coronary atherosclerosis, partly independently of plasma homocysteine.[50]

10.4.11 Treatment with Vitamin B_{12}

A report in the *European Journal of Hematology* examined sHcy in patients with reduced serum vitamin B_{12} and/or red cell folate (RCF) to determine its usefulness in diagnostic interpretation of reduced vitamin levels. Of 3,846 patients who had serum vitamin B_{12} and RCF assayed, 9% had reduced vitamin levels. There was a significant association between sHcy and serum creatinine, positive intrinsic factor (IF) antibody, or neutrophil hypersegmentation (NHS), increased MCV, and low RCF, but no relationship with the level of serum vitamin B_{12} or hemoglobin. After eliminating patients with renal impairment from the data pool, the distribution of the remaining patients with elevated sHcy was 25% with low serum vitamin B_{12} with or without low RCF, and 27% with low RCF alone.

sHcy correctly identified response to vitamin therapy in 94% of patients who had adequate parameters to assess response. The positive predictive values of IF antibody/NHS, macrocytosis and/or low RCF for elevated sHcy, were 100% and 34%, respectively. Twenty-four percent of patients with a low serum vitamin B_{12} and elevated sHcy had no abnormal hematologic parameters as determined by the routine laboratory tests. These results suggest to the authors that the usefulness of measuring sHcy in a routine diagnostic setting is limited, and a careful review of the peripheral blood for macrocytosis and NHS, plus determination of RCF, may be a more cost-effective process than sHcy assay to determine the presence of tissue deficiency in most instances.[51]

10.4.12 Treatment with Folic Acid/Folate

Homocysteine can impair endothelial function, but folate supplements can reduce tHcy levels by approximately 25%, as reported in *Cardiologia*.[52] Elevated homocysteine is causally linked to CVDs, and supplemental folic acid reduces the risk: A meta-analysis published in the *British Medical Journal* found that the association of sHcy concentration with CVD is causal. Furthermore, lowering homocysteine concentrations by 3 μmol/L from current levels, achievable by increasing folic acid intake, would reduce the risk of IHD by 16%, deep vein thrombosis by 25%, and stroke by 24%.[53]

The *American Journal of Epidemiology* reported a study concluding that folate is the most important determinant of plasma homocysteine, even in individuals with apparently adequate nutritional status of this vitamin. Dietary intakes of vitamins B_6, B_{12}, and folate were estimated from a food frequency questionnaire from patients under 76 years old, hospitalized with a first myocardial infarction. Plasma homocysteine level was 11% higher in patients than controls. Dietary and plasma levels of vitamin B_6 and folate were lower in patients, and these vitamins were inversely associated with the risk of myocardial infarction, independently of other potential risk factors. Vitamin B_{12} showed no clear association with myocardial infarction, although methylmalonic acid levels were significantly higher in many cases. Plasma folate and, to a lesser extent, plasma vitamin B_{12}, but not vitamin B_6, correlated inversely with plasma homocysteine, even for concentrations at the high end of normal values.[54]

These data provide further evidence that plasma homocysteine is an independent risk factor for myocardial infarction. Folate was the most important determinant of plasma homocysteine, even in subjects with apparently adequate nutritional status of this vitamin.

Serum folate, but not vitamin B_6 or B_{12}, strongly predicts plasma homocysteine: This case-reference observational study was reported in the journal *Nutrition*: The study included patients with at least 70% stenosis of one major coronary artery. Risk factors of CVD were evaluated, including age, sex, blood lipid profile, hypertension, smoking habits, and drinking habits. Plasma homocysteine, folate, P-5-P, and vitamin B_{12} were also recorded.

CAD patients were found to have significantly higher mean plasma homocysteine concentrations than did the control participants. There were no significant differences between groups with regard to the B vitamins; however, mean serum folate concentrations for those in the highest two quartiles of plasma homocysteine concentration were significantly lower than for those in the lowest two quartiles. Plasma homocysteine was strongly inversely associated with serum folate in both groups. Age, gender, other confounding factors, and B vitamin-adjusted odds ratios, were significantly increased in the highest quartile of homocysteine concentration. The elevation of 1 ng/mL in serum folate concentration was found to decrease plasma levels.

Serum folate, but not vitamin B_6 or B_{12}, was a strong predictor of plasma homocysteine—all patients and participant subjects had adequate B vitamin status.[55] Parenthetically, a study in the *British Journal of Urology International* reported that in an animal model (rabbit), folic acid restored relaxation of cavernosal tissue impaired by induced diabetes.[56]

Results from the Homocysteine and Atherosclerosis Reduction Trial (HART), appearing in the journal *Vascular Medicine,* were based on carotid atherosclerosis assessed by measurements of carotid intima-media thickness (IMT) and plaque calcification in patients with vascular disease or diabetes. An inverse relationship was found between plasma folate and mean max-carotid IMT. Plaque calcification score increased across quartiles of tHcy, and decreased across quartiles of plasma folate concentrations. In high-risk individuals, tHcy and low folate concentrations were only weakly associated with carotid IMT. In contrast, there was an independent association with plaque calcification score, a measure of more advanced atherosclerosis.[57]

10.4.13 Folic Acid Reverses Endothelial Nitric Oxide Synthase Uncoupling by Inadequate Tetrahydrobiopterin

This study in *Circulation Research* reported the direct effects of 5-methyltetrahydrofolate (MTHF), the active form of folic acid, on the enzymatic activity of NOS both in recombinant eNOS, as well as in cultured endothelial cells. Investigators assessed the effects of L- and D-5-MTHF, the active form of folic acid, on BH4-free and partially BH4-repleted endothelial eNOS. Superoxide production of eNOS, and the rate constants for trapping of superoxide by MTHF, were determined with electron paramagnetic resonance (EPR) using 5-diethoxyphosphoryl-5-methyl-L-pyrroline-N-oxide (DEPMPO) as spin trap for superoxide. NO production was measured with [(3)H]arginine-citrulline conversion or nitrite assay. The rate constants for scavenging of superoxide by L- and D-MTHF were similar, 1.4×10^4 ms^{-1}. In BH4-free eNOS, L- and D-MTHF have no effect on enzymatic activity.[58]

In contrast, in partially BH4-repleted eNOS, there was a 2-fold effect of MTHF on the enzymatic activity. First, superoxide production is reduced; second, NO production is enhanced. In cultured endothelial cells, a similar enhancement of NO production is induced by MTHF.

Folic acid reverses endothelial dysfunction induced by BH4 depletion independently of either the regeneration or stabilization of BH4 or an antioxidant effect. In this study of cultured aortic endothelial cells, appearing in the *European Journal of Pharmacology*, folic acid reversed both the endothelial dysfunction and increased production of superoxide following induced depletion of rabbit aortic ring BH4 levels with 2,4-diamino-6-hydroxy-pyrimidine (DAHP) and N-acetyl-5-hydroxy-tryptamine (NAS).[59]

10.4.14 Methylenetetrahydrofolate Reductase

Methylenetetrahydrofolate reductase (MTHFR) is the rate-limiting enzyme in the methyl cycle that catalyzes the conversion of 5,10-MTHF to 5-MTHF, a co-substrate for homocysteine re-methylation to methionine. Genetic variation in this gene may influence susceptibility to endothelial and occlusive vascular disease. The MTHFR A1298C mutation inhibits the utilization of 5-MTHF, or methylfolate, in forming BH4. Not only does BH4 play a major role in the production of NO, but it is also a cofactor in neurotransmitter production, including serotonin, dopamine, melatonin, epinephrine, and norepinephrine, and it plays a role in the production of NO. This deficiency may cause psychological or neurological problems, as well as CVD.

Mice with a gene aberration for forming MTHFR have a 1.5- to 2-fold elevation in plasma homocysteine, significantly reducing the number of circulating endothelial progenitor cells (EPCs) and their differentiation. MTHFR deficiency is also associated with increased ROS production and reduced NO generation.

The *American Journal of Physiology – Heart and Circulation Physiology* reported a study concerning the function and survival of EPCs (defined as sca1(+) c-kit(+) flk-1(+) bone marrow-derived cells). These cells, involved in neovascularization and endothelial regeneration, depend on suppression of ROS. Treatment of EPCs with sepiapterin, a precursor of BH4, significantly reduced ROS and improved NO production. mRNA and protein expression of eNOS, and the relative amount of eNOS dimer compared with monomer, were decreased by MTHFR deficiency. Impaired differentiation of EPCs induced by that deficiency saw increased senescence, decreased telomere length, and reduced expression of SIRT1.

Addition of sepiapterin maintained cell senescence and SIRT1 expression at normal levels. These results demonstrate that MTHFR deficiency impairs EPC formation and increases EPC senescence by eNOS uncoupling and downregulation of SIRT1.[60]

10.4.15 Methylenetetrahydrofolate Reductase Deficiency and Erectile Dysfunction

In a publication titled "Might erectile dysfunction be due to the thermolabile variant of methylenetetrahydrofolate reductase?" the *Journal of Endocrinological Investigation* reported a case study concerning a young man with homozygote genotype mutated with 5-MTHFR thermolabile variant (C677T). The patient developed ED refractory to phosphodiesterase type-5 (PDE5) inhibitor therapy, absent evidence of emotional/psychological cause. After one month of treatment with 5 mg/day folic acid and 1,000 μg/day vitamin B_{12}, he resumed 50 mg sildenafil, and attained satisfying erections during sexual intercourse. The authors aver that elevated homocysteine may lead to ED by reducing NO formation, thus impairing endothelial function.[61]

In a subsequent study, ED patients were interviewed using the IIEF. Blood samples were drawn to assess MTHFR gene C677T mutation, and homocysteine and folate levels. Penile color Doppler was used to assess penile vascular status. Patients were split into three groups (A, B, and C) on the basis of MTHFR genotype, and in a further group defined as "sildenafil nonresponders" (NR). Patients were given sildenafil citrate for 2 months, and the nonresponders were treated with combined sildenafil, vitamin B_6, and folic acid for 6 weeks.

Significant differences in baseline values for homocysteine and folic acid were found between groups A and B, and A and C. The NR group (patients from group A and B) presented high levels of homocysteine and low levels of folic acid. After combination treatment, 88.9% showed improvement on the IIEF questionnaire. Moreover, a significant difference for the NRs was found between the values of homocysteine and folic acid at the baseline, and at the end of the study.

The authors contend that in patients with this gene aberration, HHcy may interfere with erection, and where it is found together with low levels of folates, the administration of PDE5 inhibitors may fail if not preceded by the correction of the altered levels of homocysteine and folate.[62]

Analyses of the human genomes have shown that the most common genetic mutation worldwide is the inherited defect in the efficiency of the enzyme MTHFR. Given that it is the most common genetic defect in the human race, it has been suggested that patients with CVD and heart disease might be tested for the MTHFR mutations. The test, available from 23andME[63] is relatively inexpensive.

10.4.16 Consistently Elevated Plasma Level of Homocysteine *Is Not* a Health Hazard

American Heart Association (AHA) Recommendation was updated in January 2012 in "Homocysteine, Folic Acid and Cardiovascular Disease."[64] Synopsis:

- Elevated homocysteine is not a major risk factor for CVD.
- Supplementing folic acid and B vitamins is not recommended.
- Recommend a healthy, balanced diet that's rich in fruits and vegetables, whole grains, and fat-free or low-fat dairy products.
- Too much blood homocysteine is related to a higher risk of CHD, stroke, and peripheral vascular disease (PVD).
- Homocysteine may promote atherosclerosis by damaging the inner lining of arteries and promoting blood clots. However, a causal link hasn't been established.
- Folic acid and other B vitamins help break down homocysteine in the body. Dietary folic acid and vitamins B_6 and B_{12} have the greatest effects.
- Higher blood levels of B vitamins are related, at least in part, to lower concentrations of homocysteine.
- Low blood levels of folic acid are linked with a higher risk of fatal CHD and stroke.
- Screening for homocysteine levels in the blood may be useful in patients with a personal or family history of CVD but who don't have the well-established risk factors (smoking, high blood cholesterol, high blood pressure, physical inactivity, obesity, and diabetes).
- Evidence for the benefit of lowering homocysteine levels is lacking, but patients at high risk should be strongly advised to be sure to get enough folic acid and vitamins B_6 and B_{12} in their diet. They should eat fruits and green, leafy vegetables daily.

"Although evidence for the benefit of lowering homocysteine levels is lacking, patients at high risk should be strongly advised to be sure to get enough folic acid and vitamins B_6 and B_{12} in their diet. They should eat fruits and green, leafy vegetables daily."[64]

It is difficult to understand the 2012 AHA Recommendations. They seem to conclude that a healthy, balanced diet is all that's needed ... maybe, unless one is at high risk. This is not consistent with the many clinical studies, quite a few cited in this monograph, that either support or categorically dismiss the homocysteine−vitamin links.

10.4.17 Homocysteine Level Is Not a Useful Diagnostic Tool, and B Vitamins, and Folate, May Be Useless—Sample Reports

Evidence suggests that the homocysteine hypothesis is still relevant as a predictor of cardiovascular risk, but measuring the homocysteine level is not useful in guiding treatment. In the *Cleveland Clinic Journal of Medicine*: "Furthermore, studies of primary and secondary prevention are said to show no evidence that taking folic acid or other B vitamins lowers the risk of cardiovascular events."[65]

From a meta-analysis in *Archives of Internal Medicine*: Folic acid supplementation resulted in an average of 25% reduction in homocysteine levels, but no significant effects on vascular outcomes after 5 years.[66]

A clinical study reported in *JAMA* does not support B vitamins as secondary prevention in patients with CAD because no beneficial treatment effect of folic acid/vitamin B_{12} or vitamin B_6 on total mortality or cardiovascular events was found.[67]

A clinical study reported in *The New England Journal of Medicine* aimed to determine whether supplementation with 2.5 mg of folic acid, and 50 mg of B6, and 1 mg of B_{12}, for 5 years would reduce the risk of major cardiovascular events in patients with vascular disease or diabetes: Mean plasma homocysteine levels decreased by 2.4 μmol/L in the treatment group and increased by 0.8 μmol/L in the placebo group.

Primary outcome events occurred in 18.8% in the treatment group and 19.8% in the placebo group. Active treatment did not significantly decrease the risk of death from cardiovascular causes, myocardial infarction, or any of the secondary outcomes. Fewer patients assigned to active treatment than to placebo had a stroke. More patients in the active-treatment group were hospitalized for unstable angina.

It was concluded that supplementing folic acid and vitamins B_6 and B_{12} did not reduce the risk of major cardiovascular events in patients with vascular disease.[68,69]

10.4.18 Elevated Homocysteine Level Is Due to Kidney Dysfunction, and B Vitamins Won't Help

The *Mayo Clinic Proc*eedings reported that homocysteine is elevated in kidney disease, and renal dysfunction is a recognized risk factor for CVD.[70] The kidneys have an important role in homocysteine metabolism: as renal function declines, homocysteine concentrations rise.[71,72] In a recent *post-hoc* analysis of the Vitamins to Prevent Stroke (VITATOPS) trial, adjustment for renal function eliminated the relationship between tHcy and carotid IMT as well as flow-mediated dilatation (FMD) of the brachial artery.[73] Renal dysfunction may account for the epidemiologic association between mild HHcy and increased cardiovascular risk, and that lowering homocysteine levels with B vitamins would not eliminate the relationship between renal function and cardiovascular risk.

10.4.19 Folate Therapy May, in Fact, Be Harmful

According to a report in the *Journal of inherited Metabolic* Disease, no risk reduction is found in homocysteine-lowering trials that are mainly based on prescription of folic acid. A side effect of folic acid administration may be the promotion of proliferation and inflammation, both of which are crucial processes taking place in the atherosclerotic plaque. Intake of high amounts of folic acid may thus be beneficial via homocysteine lowering, but may also be harmful via destabilization of atherosclerotic plaque.[74]

10.4.20 Interactions with Medications

Folic acid supplements can interact with a number of different medications. Here are a few examples:

- Methotrexate (Rheumatrex®, Trexall®) for treatment of cancer, rheumatoid arthritis, or psoriasis.
- Antiepileptic medications, such as phenytoin (Dilantin®), carbamazepine (Carbatrol®, Tegretol®, Equetro®, Epitol®), and valproate (Depacon®), are used to treat epilepsy, psychiatric diseases, and other medical conditions. These medications can reduce serum folate levels. Folic acid supplements can reduce serum levels of these medications.
- Sulfasalazine (Azulfidine®), used primarily to treat ulcerative colitis, inhibits the intestinal absorption of folate, and can cause folate deficiency.

10.5 NIACIN (VITAMIN B₃)

Niacin is said to lower the risk of atherosclerosis and endothelium impairment, thus potentially enhancing erectile function. Not everyone agrees. For that reason, and because any successful treatment of atherosclerosis has wide ranging implications for treating or preventing ED, the controversy deserves some scrutiny.

Niacin was recognized early to be beneficial in combating atherosclerosis because it was shown to significantly lower harmful serum lipids. Niacin (nicotinic acid) is a water-soluble essential vitamin found in many foods, including liver, chicken, beef, fish, cereal, peanuts, and legumes. It is the oldest known lipid-lowering anti-atherosclerotic substance.[75] It was recently shown to reduce LDL-C, very low-density lipoprotein cholesterol (VLDL-C), and TGs, yet it raises HDL-C.[76] It reduces vascular inflammation and improves endothelial function and plaque stability.[77]

The Coronary Drug Project (CDP) reported in *JAMA* in 1975 was one of the first to study the long-term clinical lipid-lowering effect of niacin from the 1960s to early 1970s.[78] Recent meta-analyses support therapeutic effects of niacin alone, or in combination with other lipid modifying agents such as statins, to significantly reduce cardiovascular event and atherosclerosis progression: One such analysis on data preceding the use of statins as standard care found positive effects for secondary prevention with doses of 1–3 g/day,[79] and another meta-analysis likewise found support for niacin in secondary prevention of cardiovascular events.[80]

These findings are in agreement with current National Cholesterol Education Program (NCEP) recommendations on high cholesterol treatment. The NCEP recommends niacin alone for cardiovascular and atherogenic dyslipidemia in mild or normal LDL levels or in combination for higher LDL levels.[81] Dosages of 1,500 mg immediate release (IR) niacin daily were reported to result in 13% LDL, 20% LP, 10% TG reduction, and 19% HDL increase compared to placebo.[82]

In the latter study comparing niacin to Niaspan®, flushing events were more frequent, and flushing severity was slightly greater with Niaspan, but it was still well tolerated. The authors concluded that Niaspan may be an equivalent or better alternative to plain niacin at moderate doses in the management of hyperlipidemia. Extended-release niacin alone or with anti-flushing agent (laropiprant) shows similar effects.[83]

The ARBITER 6-HALTS study was reported at the 2009 annual meeting of the AHA and published in the *New England Journal of Medicine*.[84,85] It concluded that, when added to statins, 2,000 mg/day of extended-release niacin was more effective than ezetimibe (Zetia®) in reducing carotid IMT, a marker of atherosclerosis. Additionally, a recent meta-analysis found positive effects of niacin alone or in combination on all cardiovascular events and on development and progress of atherosclerosis.

On the other hand, a more recent study published in the *New England Journal of Medicine* concluded that in patients with atherosclerotic CVD and LDL-C levels of less than 70 mg per deciliter (1.81 mmol/L), there was no incremental clinical benefit from the addition of niacin to statin therapy during a 36-month follow-up period, despite significant improvements in HDL-C and TG levels.[86] The study was funded by the National Heart, Lung, and Blood Institute (NHLBI), AIM-HIGH ClinicalTrials.gov number, NCT00120289, and Abbott Laboratories.

It should be noted also that the 2011 "Atherothrombosis Intervention in Metabolic Syndrome With Low HDL/High Triglycerides: Impact on Global Health Outcomes" (AIM-HIGH) trial was halted early because patients showed no decrease in cardiovascular events, but also had an increase in the risk of stroke. These patients already had LDL levels well controlled by a statin drug, and the aim of the study was to evaluate extended-release niacin (2,000 mg per day) to see if raising HDL levels had an additional positive effect on risk. In this study, it did not have such an effect and, furthermore, it appeared to increase the risk of stroke.[87]

A previous chapter cited the effect of statins, both positive and negative, on erectile function. More recent studies aimed to show the beneficial effects of statins combined with niacin based on previous evidence that niacin promotes healthy blood lipid levels and thus, endothelial health and satisfactory erectile function, and the combination should potentiate.

Keeping in mind published skepticism, here follow some reports of clinical trials, again both the positive and the negative outcome studies that may help to form an opinion about the value of niacin as a supplement in improving endothelial and erectile function.

10.5.1 Niacin Promotes Endothelium Health

According to a clinical trial reported in *Arteriosclerosis, Thrombosis & Vascular Biology*, niacin inhibits CAD by reducing LDL and plasma TG levels and increasing HDL levels. The study presents evidence that the effects of niacin are independent of changes in plasma lipids. In this study, New Zealand white rabbits received normal chow or chow supplemented with 0.6% or 1.2% (wt/wt) niacin. This regimen had no effect on plasma cholesterol, TG, or HDL levels. Acute vascular inflammation and endothelial dysfunction were induced in the animals with a periarterial carotid collar.

At the 24-hour post collar-implantation, the endothelial expression of vascular cell adhesion molecule-1, intercellular adhesion molecule-1, and monocyte chemotactic protein-1 was markedly decreased in the niacin-supplemented animals compared with controls. Niacin also inhibited intima-media neutrophil recruitment, myeloperoxidase accumulation, enhanced endothelial-dependent vasorelaxation, enhanced cGMP production, increased vascular-reduced glutathione content, and it protected against hypochlorous acid-induced endothelial dysfunction and tumor necrosis factor alpha-induced vascular inflammation.[77]

Niacin is said to improve impaired endothelial vasoprotective effects of HDL in type-2 diabetes in a report in the journal *Circulation*,[88] and extended release (ER) niacin therapy (1,500 mg/day) raises HDL plasma levels and markedly improves endothelial-protective functions of HDL in type-2 diabetes patients.[89]

Treatment with niacin improves endothelial function in patients with CAD: Patients with CAD were treated with ER-niacin 1,000 mg/day for 12 weeks without effect on FMD or nitroglycerin-induced dilation (NMD), compared to placebo. However, *post-hoc* subgroup analysis revealed an improvement in FMD in patients with low HDL-C at baseline: ER-niacin treatment improves endothelial dysfunction in patients with CAD and low HDL-C, but not with normal HDL-C.[81]

Anti-atherogenic effects of ER niacin in the metabolic syndrome improves endothelial function and decreases vascular inflammation: Patients with metabolic syndrome (Adult Treatment Panel (ATP) III criteria) were given ER niacin (1,000 mg/day), or placebo for 52 weeks. Treatment resulted in a significant change in carotid IMT. Endothelial function improved by 22% in the niacin group; no significant changes were seen in the placebo group. High sensitivity C-reactive protein (CRP) decreased by 20% in the niacin group and significantly increased HDL-C and significantly decreased LDL-C and TGs. There were no adverse effects on fasting glucose levels after 52 weeks of treatment.[90]

10.5.2 Niacin Improves Erectile Function

The *Journal of Sexual Medicine* reports a study aimed to assess the effect of niacin on erectile function in patients suffering from both ED and dyslipidemia. Patients received 1,500 mg oral niacin daily or placebo for 12 weeks. The niacin treatment group showed a significant increase in both IIEF-Q3 scores and IIEF-Q4 scores, compared with baseline values. The placebo group also showed a significant increase in IIEF-Q3 scores, but not IIEF-Q4 scores. However, when patients were stratified according to the baseline severity of ED, the patients with moderate and severe ED who received niacin showed a significant improvement in IIEF-Q3 scores and IIEF-Q4 scores, respectively, compared with baseline values, but not so the placebo group. The improvement in IIEF-EF domain score for severe and moderate ED patients in the niacin group were 5.28 ± 5.94, and 3.31 ± 4.54, and in the placebo group were 2.65 ± 5.63 and 2.74 ± 5.59, respectively. The statistical tests are not indicated, and the statistics show extreme skew in these distributions. However, all differences were said to be statistically significant.

There was no improvement in erectile function for patients with mild and mild-to-moderate ED for both groups. For patients not receiving statin treatment, there was a significant improvement in IIEF-Q3 scores for the niacin group, but not for the placebo group. The authors concluded that niacin alone can improve the erectile function in patients suffering from moderate-to-severe ED and dyslipidemia.[91]

10.5.3 Nutraceutical Approach to Treatment of Erectile Dysfunction

The aim of this study titled "Propionyl-L-carnitine, L-arginine and niacin in sexual medicine: a nutraceutical approach to erectile dysfunction," published in *Andrologia*, was to evaluate the effects of a 3-month supplementation with propionyl-L-carnitine (PLC promotes blood circulation), L-arginine, and niacin on the sexual performance of men 35 to 75 years old in an ED clinic. All patients had the short IIEF questionnaire, global assessment questions (GAQs), and routine laboratory testing, at baseline and 3 months afterward. All patients received unidentified white packets containing granulated powder consisting of PLC (250 mg/day), niacin (20 mg/day) and L-arginine (2,500 mg/day) (provided by Sigma-Tau, Ezerex, Rome, Italy). Patients were instructed to take the powder orally, once daily, for 3 months, with one full glass of water.

After 3 months of treatment, a small, but statistically significant, improvement was found in total and single items of the IIEF: improved erections in 40% of cases, with a partial response occurring in up to 77% of participants.[92]

These preliminary findings indicate that the favorable cardiovascular effects of nutraceuticals might also reflect on male sexual function with possible implication in the treatment and prevention of ED. This study also reported a considerable interest in nutritional supplementation by patients as first-line or treatment adjunctive to PDE5 inhibitors that goes beyond the measurable improvement in penile rigidity.

This study is cited here also for what might be considered an unconventional "nutraceuticals" approach to treatment, underscoring the understanding that ED, to quote the well-known TV commercial for Cialis®, "... may be a blood flow problem." Treatment incorporates three constituents independently known to improve blood flow.

Furthermore, this study departs from the tradition of testing one treatment at a time, consistent with a strict research vs. treatment tradition that abhors *confounding* factors, including the interaction of treatment components, as well as the inability to extricate a quantitative measure of the partial contribution of each component.

In another study, patients aged between 50 and 60 years with ED and type-2 diabetes of 3 to 4 years duration were assigned for 12 weeks to one of four treatment groups. Group 1 was given the test formulation (PLC, -arginine, and nicotinic acid (Ezerex)) each day; group 2, 20 mg of vardenafil (Levitra®) twice a week; group 3, the test formulation plus vardenafil (20 mg) twice a week; and group 4, placebo. Endothelial function was evaluated by FMD, and erectile function was estimated with the IIEF5 questionnaire in all subjects.

At the end of treatment, group 1 improved by 2 points in the IIEF5; group 2 showed an increment of 4 points; group 3, given all the treatments, showed an increment of 5 points; and group 4, treated with placebo, showed no increment in the IIEF5. Despite the small number of participants, the data suggest that the test formulation may improve the endothelial situation in diabetes. The test formulation together with vardenafil was better than the PDE5 inhibitor alone.[93]

10.5.4 The Niacin Controversy

The National Heart, Lung and Blood Institute of the National Institutes of Health (NIH) issued a press release on November 15, 2011 titled "AIM-HIGH: Blinded Treatment Phase of Study Stopped." To wit: It has stopped the blinded treatment phase of the AIM-HIGH Clinical Trial 18 months earlier than planned. AIM-HIGH was designed to test whether raising HDL and lowering TGs in people with heart and vascular disease and well-controlled LDL levels would reduce the risk of repeat heart and vascular problems.

Half of the study participants took high-dose, extended-release niacin plus a statin drug, and the other half took a statin drug only. Interim results of the study concluded that there was no treatment-outcome difference between the two treatment groups.

The following reason was given for stopping the treatment phase AIM-HIGH: It was concluded that high dose, extended-release niacin offered no benefits beyond statin therapy alone in reducing cardiovascular-related complications in this trial because the rate of clinical events was the same in both treatment groups, and there was no evidence that this would change by continuing the trial. It was recommended that the blinded treatment phase of the study be terminated. In addition, "... a small and unexplained increase in ischemic stroke rates in the high-dose, extended-release niacin group was noted." For more information, see the NHLBI press release.[94,95]

On the other hand, the *Cleveland Clinic Journal of Medicine* recently published an article titled "Is niacin ineffective? Or did AIM-HIGH miss its target?" The author concluded that the study was terminated early because of concerns raised by the interim analysis: perhaps niacin is not a good preventive agent; perhaps raising levels of HDL-C is flawed as a preventive strategy; perhaps AIM-HIGH had methodologic flaws; or "perhaps statins are so good that once you prescribe one, anything else you do will not make much of a difference."

The AIM-HIGH trial found no cardiovascular benefit from taking extended-release niacin (Niaspan). In addition, a statistically non-significant increased risk of ischemic stroke was noted.[96]

10.5.5 Toxicity

Adverse reactions to pharmacological doses of niacin (1.5 to 6 g per day) include, but are not limited to:

- Skin flushing and itching, dry skin, and skin rashes including eczema exacerbation and skin thickening and hyperpigmentation (acanthosis nigricans).
- Flushing usually lasts for about 15 to 30 minutes, though it can sometimes last up to two hours. It is sometimes accompanied by a prickly or itching sensation (in particular) in areas covered by clothing.
- Gastrointestinal complaints, such as indigestion, nausea, and liver toxicity (fulminant hepatic failure has also been reported).
- High doses of niacin may elevate blood sugar, thereby worsening diabetes mellitus.
- Hyperuricemia is another side effect of taking high-dose niacin, and may exacerbate gout.
- Extremely high doses of niacin can also cause niacin maculopathy, a thickening of the macula and retina, which leads to blurred vision and blindness. This maculopathy is reversible after niacin intake ceases.

For additional information on side effects and adverse reactions to niacin, see References 97 and 98.

REFERENCES

1. Recommended Dietary Allowances: Tenth Edition. Subcommittee on the Tenth Edition of the Recommended Dietary Allowances, Food and Nutrition Board, Commission on Life Sciences, 1989. National Research Council.
2. Anderson RA. Chromium and polyphenols from cinnamon improve insulin sensitivity. *Proc Nutr Soc* 2008;**67**:48−53.
3. Iqbal N, Cardillo S, Volger S, Bloedon LT, Anderson RA, Boston R, et al. Chromium picolinate does not improve key features of metabolic syndrome in obese non-diabetic adults. *Metab Syndr Relat Disord* 2009;**7**:143−50.
4. Kleefstra N, Bilo HJ, Bakker SJ, Houweling ST. [Chromium and insulin resistance]. *Ned Tijdschr Geneeskd* 2004;**148**:217−20.
5. Shils ME, Olson JA, Shike M, editors. *Modern nutrition in health and disease.* 8th ed. Philadelphia: Lea & Febiger; 1994.
6. Mennen B. Dietary chromium: an overview. Chromium Information Bureau, 1998, <www.chromium.edu/intro.htm>; 2013 [accessed 16.11.13].
7. Reaven GM. Role of insulin resistance in human disease. *Diabetes* 1988;**37**:1595−607.
8. Hummel M, Standl E, Schnell O. Chromium in metabolic and cardiovascular disease. *Horm Metab Res* 2007;**39**:743−51.
9. Vladeva SV, Terzieva DD, Arabadjiiska DT. Effect of chromium on the insulin resistance in patients with type II diabetes mellitus. *Folia Med (Plovdiv)* 2005;**47**:59−62.
10. Cheng HH, Lai MH, Hou WC, Huang CL. Antioxidant effects of chromium supplementation with type 2 diabetes mellitus and euglycemic subjects. *J Agric Food Chem* 2004;**52**:1385−9.

11. Lai E, De Lepeleire I, Crumley TM, Liu F, Wenning LA, Michiels N, et al. Suppression of niacin-induced vasodilation with an antagonist to prostaglandin D2 receptor subtype 1. *Clin Pharm Ther* 2007;**81**:849−57.

12. Sharma S, Agrawal RP, Choudhary M, Jain S, Goyal S, Agarwal V. Beneficial effect of chromium supplementation on glucose, HbA1C and lipid variables in individuals with newly onset type-2 diabetes. *J Trace Elem Med Biol* 2011;**25**:149−53.

13. Anderson RA, Cheng N, Bryden NA, Polansky MM, Cheng N, Chi J, et al. Elevated intakes of supplemental chromium improve glucose and insulin variables in individuals with type 2 diabetes. *Diabetes* 1997;**46**:1786−91.

14. Kaats GR, Blum K, Fisher JA, Adelman JA. Effects of chromium picolinate supplementation on body composition: a randomized, double-masked, placebo-controlled study. *Curr Ther Res* 1996;**57**:747−65.

15. Gunton JE, Cheung NW, Hitchman R, Hams G, O'Sullivan C, Foster-Powell K, et al. Chromium supplementation does not improve glucose tolerance, insulin sensitivity, or lipid profile. A randomized, placebo-controlled, double-blind trial of supplementation in subjects with impaired glucose tolerance. *Diab Care* 2005;**28**:712−3.

16. Komorowski J, Juturu V. Chromium supplementation does not improve glucose tolerance, insulin sensitivity, or lipid profile: a randomized, placebo-controlled, double-blind trial of supplementation in subjects with impaired glucose tolerance. *Diab Care* 2005;**28**:1841−2 [Response to Gunton *et al.* (2005)].

17. Cefalu WT, Hu FB. Role of chromium in human health and in diabetes. *Diabetes Care* 2004;**27**:2741−51.

18. No authors listed (2003). Complementary Medicines Evaluation Committee (CMEC), meeting 41, 1 August 2003, public recommendation summary. <http://www.tga.gov.au/archive/committees-cmec-resolutions-41.htm>; 2013 [accessed 16.11.13].

19. Bansal TC, Guay AT, Jacobson J, Woods BO, Nesto RW. Incidence of metabolic syndrome and insulin resistance in a population with organic erectile dysfunction. *J Sex Med* 2005;**2**:96−103.

20. Francis SH, Corbin JD. PDE5 inhibitors: targeting erectile dysfunction in diabetics. *Curr Opin Pharmacol* 2011;**11**:683−8.

21. Trussell JC, Legro RS. Erectile dysfunction: does insulin resistance play a part? *Fertil Steril* 2007;**88**:771−8.

22. Masharani U, Gjerde C, McCoy S, Maddux BA, Hessler D, Goldfine ID, et al. Chromium supplementation in non-obese non-diabetic subjects is associated with a decline in insulin sensitivity. *BMC Endocrine Disorders* 2012;**12**:31. Available from: http://dx.doi.org/10.1186/1472−68.

23. Wang ZQ, Cefalu WT. Current concepts about chromium supplementation in type 2 diabetes and insulin resistance. *Curr Diab Rep* 2010;**10**:145−51.

24. Jarvill-Taylor KJ, Anderson RA, Graves DJ. A hydroxychalcone derived from cinnamon functions as a mimetic for insulin in 3T3-L1 adipocytes. *J Am Coll Nutr* 2001;**20**:327−36.

25. Anderson RA, Broadhurst CL, Polansky MM, Schmidt WF, Khan A, Flanagan VP, et al. Isolation and characterization of polyphenol type-A polymers from cinnamon with insulin-like biological activity. *J Agric Food Chem* 2004;**52**:65−70.

26. Subash Babu P, Prabuseenivasan S, Ignacimuthu S. Cinnamaldehyde—a potential antidiabetic agent. *Phytomed* 2007;**14**:15−22.

27. Verspohl EJ, Bauer K, Neddermann E. Antidiabetic effect of Cinnamomum cassia and Cinnamomum zeylanicum in vivo and in vitro. *Phytother Res* 2005;**19**:203−6.

28. Rafehi H, Ververis K, Karagiannis TC. Controversies surrounding the clinical potential of cinnamon for the management of diabetes. *Diabetes Obes Metab* 2012;**14**:493−9.

29. Ranasinghe P, Jayawardana R, Galappaththy P, Constantine GR, de Vas Gunawardana N, Katulanda P. Efficacy and safety of "true" cinnamon (Cinnamomum zeylanicum) as a pharmaceutical agent in diabetes: a systematic review and meta-analysis. *Diabet Med* 2012;**29**:1480−92.

30. Davis PA, Yokoyama W. Cinnamon intake lowers fasting blood glucose: meta-analysis. *J Med Food* 2011;**14**:884−9.

31. Khan A, Safdar M, Ali Khan MM, Khattak KN, Anderson RA. Cinnamon improves glucose and lipids of people with type 2 diabetes. *Diabetes Care* 2003;**26**:3215−8.

32. Mang B, Wolters M, Schmitt B, Kelb K, Lichtinghagen R, Stichtenoth DO, et al. Effects of a cinnamon extract on plasma glucose, HbA, and serum lipids in diabetes mellitus type 2. *Eur J Clin Invest* 2006;**36**:340−4.

33. <http://www.cinnulin.com/more_info.html>; 2013 [accessed 16.11.13].

34. Ziegenfuss TN, Hofheins JE, Mendel RW, Landis J, Anderson RA. Effects of a water-soluble cinnamon extract on body composition and features of the metabolic syndrome in pre-diabetic men and women. *J Int Soc Sports Nutr* 2006;**28**:45−53.

35. Magistrelli A, Chezem JC. Effect of ground cinnamon on postprandial blood glucose concentration in normal-weight and obese adults. *J Acad Nutr Diet* 2012;**112**:1806−9.

36. Akilen R, Tsiami A, Devendra D, Robinson N. Glycated haemoglobin and blood pressure-lowering effect of cinnamon in multi-ethnic Type 2 diabetic patients in the UK: a randomized, placebo-controlled, double-blind clinical trial. *Diabet Med* 2010;**27**:1159−67.

37. Kleefstra N, Logtenberg SJ, Houweling ST, Verhoeven S, Bilo HJ. [Cinnamon: not suitable for the treatment of diabetes mellitus—article in Dutch]. *Ned Tijdschr Geneeskd* 2007;**151**:2833−7.

38. Spence D. Mechanisms of thrombogenesis in atrial fibrillation. *Lancet* 2009;**373**:1006. Available from: http://dx.doi.org/10.1016/S0140−6736(09)60604-8.

39. Spence JD. Nutrition and stroke prevention. *Stroke* 2006;**37**:2430−5.

40. Jones RW, Jeremy JY, Koupparis A, Persad R, Shukla N. Cavernosal dysfunction in a rabbit model of hyperhomocysteinaemia. *BJU Int* 2005;**95**:125−30.

41. Demir T, Comlekçi A, Demir O, Gülcü A, Calýpkan S, Argun L, et al. Hyperhomocysteinemia: a novel risk factor for erectile dysfunction. *Metab* 2006;**55**:1564−8.

42. Alp NJ, Channon KM. Regulation of endothelial nitric oxide synthase by tetrahydrobiopterin in vascular disease. *Arterioscler Thromb Vasc Biol* 2004;**24**:413−20.

43. He L, Zeng H, Li F, Feng J, Liu S, Liu J, et al. Homocysteine impairs coronary artery endothelial function by inhibiting tetrahydrobiopterin in patients with hyperhomocysteinemia. *Am J Physiol Endocrinol Metab* 2010;**299**:E1061−5. [Epub 2010 Sep 21].

44. Alfthan G, Aro A, Gey KF. Plasma homocysteine and cardiovascular disease mortality. *Lancet* 1997;**349**:397.

45. Humphrey LL, Fu R, Rogers K, Freeman M, Helfand M. Homocysteine level and coronary heart disease incidence: a systematic review and meta-analysis. *Mayo Clin Proc* 2008;**83**:1203−12.

46. Wald NJ, Watt HC, Law MR, Weir DG, McPartlin J, Scott JM. Homocysteine and ischemic heart disease: results of a prospective study with implications regarding prevention. *Arch Intern Med* 1998;**158**:862−7.

47. Nygård O, Nordrehaug JE, Refsum H, Ueland PM, Farstad M, Vollset SE. Plasma homocysteine levels and mortality in patients with coronary artery disease. *N Engl J Med* 1997;**337**:230−6.

48. Selhub J, Jacques PF, Wilson PWF, Rush D, Rosenberg IH. Vitamin status and intake as primary determinants of homocysteinaemia in an elderly population. *J Am Med Assoc* 1993;**270**:2693−8.

49. Robinson K, Mayer EL, Miller DP, Green R, van Lente F, Gupta A, et al. Hyperhomocysteinemia and low pyridoxal phosphate. Common and independent reversible risk factors for coronary artery disease. *Circulation* 1995;**92**:2825−30.

50. Siri PW, Verhoef P, Kok FJ. Vitamins B6, B12, and folate: association with plasma total homocysteine and risk of coronary atherosclerosis. *J Am Coll Nutr* 1998;**17**:435−41.

51. Curtis D, Sparrow R, Brennan L, Van der Weyden MB. Elevated serum homocysteine as a predictor for vitamin B12 or folate deficiency. *Eur J Haematol* 1994;**52**:227−32.

52. Andreotti F, Burzotta F, Mazza A, Manzoli A, Robinson K, Maseri A. Homocysteine and arterial occlusive disease: a concise review. *Cardiologia* 1996;**44**:341−5.

53. Wald DS, Law M, Morris JK. Homocysteine and cardiovascular disease: evidence on causality from a meta-analysis. *Brit Med J, 23* 2002;**325**(7374):1202.

54. Verhoef P, Stampfer MJ, Buring JE, Gaziano JM, Allen RH, Stabler SP, et al. Homocysteine metabolism and risk of myocardial infarction: relation with vitamins B6, B12, and folate. *Am J Epidemiol* 1996;**143**:845−59.

55. Lee BJ, Lin PT, Liaw YP, Chang SJ, Cheng CH, Huang YC. Homocysteine and risk of coronary artery disease: folate is the important determinant of plasma homocysteine concentration. *Nutrition* 2003;**19**:577−83.

56. Shukla N, Hotston M, Persad R, Angelini GD, Jeremy JY. The administration of folic acid improves erectile function and reduces intracavernosal oxidative stress in the diabetic rabbit. *BJU Int* 2009;**103**:98−103.

57. Held C, Sumner G, Sheridan P, McQueen M, Smith S, Dagenais G, et al. Correlations between plasma homocysteine and folate concentrations and carotid atherosclerosis in high-risk individuals: baseline data from the Homocysteine and Atherosclerosis Reduction Trial (HART). *Vasc Med* 2008;**13**:245−53.

58. Stroes ES, van Faassen EE, Yo M, Martasek P, Boer P, Govers R, et al. Folic acid reverts dysfunction of endothelial nitric oxide synthase. *Circ Res* 2000;**86**:1129−34.

59. Moat SJ, Clarke ZL, Madhavan AK, Lewis MJ, Lang D. Folic acid reverses endothelial dysfunction induced by inhibition of tetrahydrobiopterin. *Eur J Pharmacol* 2006;**530**:250−8.

60. Lemarié CA, Shbat L, Marchesi C, Angulo OJ, Deschênes ME, Blostein MD, et al. Mthfr deficiency induces endothelial progenitor cell senescence via uncoupling of eNOS and downregulation of SIRT1. *Am J Physiol Heart Circ Physiol* 2011;**300**:H745−53.

61. Lombardo F, Sgrò P, Gandini L, Dondero F, Jannini EA, Lenzi A. Might erectile dysfunction be due to the thermolabile variant of methylenetetrahydrofolate reductase? *J Endocrinol Invest* 2004;**27**:883−5.

62. Lombardo F, Tsamatropoulos P, Piroli E, Culasso F, Jannini EA, Dondero F, et al. Treatment of erectile dysfunction due to C677T mutation of the MTHFR gene with vitamin B6 and folic acid in patients non responders to PDE5i. *J Sex Med* 2010;**7**(1 Pt 1): 216−23.

63. <http://www.23andme.com> [accessed 16.11.13].

64. <http://www.heart.org/HEARTORG/GettingHealthy/NutritionCenter/Homocysteine-Folic-Acid-and-Cardiovascular-Disease_UCM_305997_Article.jsp>; [accessed 16.11.13].

65. Abraham JM, Cho L. The homocysteine hypothesis: still relevant to the prevention and treatment of cardiovascular disease? *Clev Clin J Med* 2010;**77**:12911−8.

66. Clarke R, Halsey J, Lewington S, Lonn E, Armitage J, Manson JE, et al. Effects of lowering homocysteine levels with B vitamins on cardiovascular disease, cancer, and cause-specific mortality: Meta-analysis of 8 randomized trials involving 37,485 individuals. *Arch Intern Med* 2010;**170**:1622−31.

67. Ebbing M, Bleie Ø, Ueland PM, Nordrehaug JE, Nilsen DW, Vollset SE, et al. Mortality and cardiovascular events in patients treated with homocysteine-lowering B vitamins after coronary angiography: a randomized controlled trial. *J Am Med Assoc (JAMA)* 2008;**300**:795−804.

68. Lonn E, Yusuf S, Arnold MJ, Sheridan P, Pogue J, Micks M, et al. Heart Outcomes Prevention Evaluation (HOPE) 2. Homocysteine lowering with folic acid and B vitamins in vascular disease. *N Engl J Med* 2006;**354**:1567−77.

69. <http://www.ClinicalTrials.gov>; 2013; number, NCT00106886; Current Controlled Trials number, ISRCTN14017017 [accessed 16.11.13].

70. Milani RV, Lavie CJ. Homocysteine: the Rubik's cube of cardiovascular risk factors. *Mayo Clin Proc* 2008;**83**:1200−12.

71. Friedman AN, Bostom AG, Selhub J, Levey AS, Rosenberg IH. The kidney and homocysteine metabolism. *J Am Soc Nephrol* 2001;**12**:2181−9.

72. Rodionov RN, Lentz SR. The homocysteine paradox [editorial]. *Arterioscler Thromb Vasc Biol* 2008;**28**:1031−3.

73. Potter K, Hankey GJ, Green DJ, Eikelboom JW, Arnolda LF. Homocysteine or renal impairment: which is the real cardiovascular risk factor? *Arterioscler Thromb Vasc Biol* 2008;**28**:1158−64.

74. Blom HJ, Smulders Y. Overview of homocysteine and folate metabolism. With special references to cardiovascular disease and neural tube defects. *J Inherit Metab Dis* 2010;**34**:75−81.

75. Altschul R, Hoffer A, Stephen JD. Influence of nicotinic acid on serum cholesterol in man. *Arch biochem biophys* 1955;**54**:558−9.

76. Villines TC, Kim AS, Gore RS, Taylor AJ. Niacin: the evidence, clinical use, and future directions. *Curr Atheroscler Rep* 2012;**14**:49−59.

77. Wu BJ, Yan L, Charlton F, Witting P, Barter PJ, Rye KA. Evidence that niacin inhibits acute vascular inflammation and improves endothelial dysfunction independent of changes in plasma lipids. *Arterioscler Thromb Vasc Biol* 2010;**30**:968−75.

78. No authors. Clofibrate and niacin in coronary heart disease. *JAMA* 1975;**231**(4):360−81.

79. Bruckert E, Labreuche J, Amarenco P. Meta-analysis of the effect of nicotinic acid alone or in combination on cardiovascular events and atherosclerosis. *Ather* 2010;**210**:353−61.

80. Duggal JK, Singh M, Attri N, Singh PP, Ahmed N, Pahwa S, et al. Effect of niacin therapy on cardiovascular outcomes in patients with coronary artery disease. *J Cardiovasc Pharmacol Ther* 2010;**15**:158−66.

81. Warnholtz A, Wild P, Ostad MA, Elsner V, Stieber F, Schinzel R, et al. Effects of oral niacin on endothelial dysfunction in patients with coronary artery disease: results of the randomized, double-blind, placebo-controlled INEF study. *Atherosclerosis* 2009;**204**:216−21.

82. No authors listed. NCEP. Third report of the National Cholesterol Education Program (NCEP) expert panel on detection, evaluation, and treatment of high blood cholesterol in adults (adult treatment panel III) final report. *Circulation* 2002;**106**:3143−421.

83. Knopp RH, Alagona P, Davidson M, Goldberg A, Kafonek SD, Kashyap M, et al. Equivalent efficacy of a time-release form of niacin (Niaspan) given once-a-night versus

　　　　plain niacin in the management of hyperlipidemia. *Metabolism: clinical experimental*
　　　　1998;**47**:1097−104.

84.　Bays H, Shah A, Dong Q, McCrary Sisk C, MacCubbin D. Extended-release niacin/laro-
　　　　piprant lipid-altering consistency across patient subgroups. *Int J Clin Pract* 2011;**65**:
　　　　436−45.

85.　<http://www.ClinicalTrials.gov>; 2013. Identifier: NCT00397657 [accessed 16.11.13].

86.　Boden WE, Probstfield JL, Anderson T, Chaitman BR, Desvignes-Nickens P, Koprowicz
　　　　K, et al. Niacin in patients with low HDL cholesterol levels receiving intensive statin ther-
　　　　apy. *N Engl J Med* 2011;**365**:2255−67.

87.　Lai MH. Antioxidant effects and insulin resistance improvement of chromium combined
　　　　with vitamin C and e supplementation for type 2 diabetes mellitus. *J Clin Biochem Nutr*
　　　　2008;**43**:191−8.

88.　Sorrentino SA, Besler C, Rohrer L, Meyer M, Heinrich K, Bahlmann FH, et al.
　　　　Endothelial-vasoprotective effects of high-density lipoprotein are impaired in patients with
　　　　type 2 diabetes mellitus but are improved after extended-release niacin therapy. *Circulation*
　　　　2010;**121**:110−22.

89.　<http://www.Clinicaltrials.gov>; 2013. Identifier: NCT00346970 [accessed 16.11.13].

90.　Thoenes M, Oguchi A, Nagamia S, Vaccari CS, Hammoud R, Umpierrez GE, et al. The
　　　　effects of extended-release niacin on carotid intimal media thickness, endothelial function
　　　　and inflammatory markers in patients with the metabolic syndrome. *Int J Clin Pract*
　　　　2007;**61**:1942−8.

91.　Ng CF, Lee CP, Ho AL, Lee VW. Effect of niacin on erectile function in men suffering
　　　　erectile dysfunction and dyslipidemia. *J Sex Med* 2011;**8**:2883−93.

92.　Gianfrilli D, Lauretta R, Di Dato C, Graziadio C, Pozza C, De Larichaudy J, et al.
　　　　Propionyl-L-carnitine, L-arginine and niacin in sexual medicine: a nutraceutical approach
　　　　to erectile dysfunction. *Andrologia* 2012;**44**(Suppl, 1):600−4.

93.　Gentile V, Antonini G, Antonella Bertozzi M, Dinelli N, Rizzo C, Ashraf Virmani M, et
　　　　al. Effect of propionyl-L-carnitine, L-arginine and nicotinic acid on the efficacy of vardena-
　　　　fil in the treatment of erectile dysfunction in diabetes. *Curr Med Res Opin* 2009;**25**:
　　　　2223−8.

94.　<http://www.aimhigh-heart.com/>; 2013 [accessed 16.11.13].

95.　<http://www.nih.gov/news/health/may2011/nhlbi-26.htm>; 2013 [accessed 16.11.13].

96.　Nicholls SJ. Is niacin ineffective? Or did AIM-HIGH miss its target? *Clev Clin J Med*
　　　　2012;**79**:38−43.

97.　University of Maryland Medical Center: <http://www.umm.edu/altmed/articles/vitamin-b3-
　　　　000335.htm> [accessed 16.11.13].

98.　<http://www.drugs.com/sfx/niacin-side-effects.html>; 2013 [accessed 16.11.13].

Index

Note: Page numbers followed by "*f*" and "*t*" refer to figures and tables, respectively.

Printed in the United States
By Bookmasters